Periglacial and Paraglacial Processes and Environments

Geological Society books refereeing procedures

The Society makes every effort to ensure that the scientific and production quality of its books matches that of its journals. Since 1997, all book proposals have been refereed by specialist reviewers as well as by the Society's Books Editorial Committee. If the referees identify weaknesses in the proposal, these must be addressed before the proposal is accepted.

Once the book is accepted, the Society Book Editors ensure that the volume editors follow strict guidelines on refereeing and quality control. We insist that individual papers can only be accepted after satisfactory review by two independent referees. The questions on the review forms are similar to those for *Journal of the Geological Society*. The referees' forms and comments must be available to the Society's Book Editors on request.

Although many of the books result from meetings, the editors are expected to commission papers that were not presented at the meeting to ensure that the book provides a balanced coverage of the subject. Being accepted for presentation at the meeting does not guarantee inclusion in the book.

More information about submitting a proposal and producing a book for the Society can be found on its web site: www.geolsoc.org.uk.

It is recommended that reference to all or part of this book should be made in one of the following ways:

Knight, J. & Harrison, S. (eds) 2009. *Periglacial and Paraglacial Processes and Environments.* Geological Society, London, Special Publications, **320**.

Curry, A. M., Sands, T. B. & Porter, P. R. 2009. Geotechnical controls on a steep lateral moraine undergoing paraglacial slope adjustment. *In*: Knight, J. & Harrison, S. (eds) *Periglacial and Paraglacial Processes and Environments.* Geological Society, London, Special Publications, **320**, 181–197.

GEOLOGICAL SOCIETY SPECIAL PUBLICATION NO. 320

Periglacial and Paraglacial Processes and Environments

EDITED BY

J. KNIGHT
University of Exeter, UK

and

S. HARRISON
University of Exeter, UK

2009
Published by
The Geological Society
London

THE GEOLOGICAL SOCIETY

The Geological Society of London (GSL) was founded in 1807. It is the oldest national geological society in the world and the largest in Europe. It was incorporated under Royal Charter in 1825 and is Registered Charity 210161.

The Society is the UK national learned and professional society for geology with a worldwide Fellowship (FGS) of over 9000. The Society has the power to confer Chartered status on suitably qualified Fellows, and about 2000 of the Fellowship carry the title (CGeol). Chartered Geologists may also obtain the equivalent European title, European Geologist (EurGeol). One fifth of the Society's fellowship resides outside the UK. To find out more about the Society, log on to www.geolsoc.org.uk.

The Geological Society Publishing House (Bath, UK) produces the Society's international journals and books, and acts as European distributor for selected publications of the American Association of Petroleum Geologists (AAPG), the Indonesian Petroleum Association (IPA), the Geological Society of America (GSA), the Society for Sedimentary Geology (SEPM) and the Geologists' Association (GA). Joint marketing agreements ensure that GSL Fellows may purchase these societies' publications at a discount. The Society's online bookshop (accessible from www.geolsoc.org.uk) offers secure book purchasing with your credit or debit card.

To find out about joining the Society and benefiting from substantial discounts on publications of GSL and other societies worldwide, consult www.geolsoc.org.uk, or contact the Fellowship Department at: The Geological Society, Burlington House, Piccadilly, London W1J 0BG: Tel. +44 (0)20 7434 9944; Fax +44 (0)20 7439 8975; E-mail: enquiries@geolsoc.org.uk.

For information about the Society's meetings, consult *Events* on www.geolsoc.org.uk. To find out more about the Society's Corporate Affiliates Scheme, write to enquiries@geolsoc.org.uk.

Published by The Geological Society from:
The Geological Society Publishing House, Unit 7, Brassmill Enterprise Centre, Brassmill Lane, Bath BA1 3JN, UK

(*Orders*: Tel. +44 (0)1225 445046, Fax +44 (0)1225 442836)
Online bookshop: www.geolsoc.org.uk/bookshop

British Library Cataloguing in Publication Data

A catalogue record for this book is available from the British Library.
ISBN 978-1-86239-281-6

Typeset by Techset Composition Ltd., Salisbury, UK
Printed by MPG Books Ltd, Bodmin, UK

Distributors

North America
For trade and institutional orders:
The Geological Society, c/o AIDC, 82 Winter Sport Lane, Williston, VT 05495, USA
Orders: Tel. +1 800-972-9892
Fax +1 802-864-7626
E-mail: gsl.orders@aidcvt.com

For individual and corporate orders:
AAPG Bookstore, PO Box 979, Tulsa, OK 74101-0979, USA
Orders: Tel. +1 918-584-2555
Fax +1 918-560-2652
E-mail: bookstore@aapg.org
Website: http://bookstore.aapg.org

India
Affiliated East-West Press Private Ltd, Marketing Division, G-1/16 Ansari Road, Darya Ganj, New Delhi 110 002, India
Orders: Tel. +91 11 2327-9113/2326-4180
Fax +91 11 2326-0538
E-mail: affiliat@vsnl.com

Contents

Periglacial and paraglacial environments: a view from the past into the future

JASPER KNIGHT* & STEPHAN HARRISON

Department of Geography, University of Exeter, Cornwall Campus, Penryn, Cornwall TR10 9EZ, UK

**Corresponding author (e-mail: j.knight@exeter.ac.uk)*

Periglacial and paraglacial (cold-climate) environments, located outside the margins of past and present ice sheets but responding to similar climate forcings, are key to identifying climate change effects upon the Earth system (Warburton 2007). These environments are relicts of cold Earth processes and thus are most sensitive to climate change that took place during the last glacial–interglacial transition, and at the present time under enhanced global climate warming. These effects include changes in humidity/aridity and radiation balance, which are most significant in the higher latitudes and at high elevations where periglacial and paraglacial environments are most common and where these environments occur near their climatic limits (Harris 1994; Matsuoka 2001). Variations in humidity and radiation balance have implications for heat budgets, water balance, land surface stability, downslope sediment supply, biodiversity and biogeochemical cycling (e.g. Schneider *et al.* 1999; Scott *et al.* 2008). The dynamics of cold-climate environments are, therefore, strongly controlled by external climatic forcing; and hence periglacial and paraglacial processes (and the landforms and sediments that result from them) can be considered as a transient response to the landscape disturbance and land surface instability that accompanies climatic change (Hewitt *et al.* 2002).

This view of a transient landscape responding to environmental disturbance is significant because it underpins influential deterministic and steady-state models in cold-climate science (Church & Slaymaker 1989; André 2003; Warburton 2007). These models predict a rapid increase in sediment yield (which results from land surface disturbance) associated with initial climate forcing, followed by exponential decay of sediment yield towards background rates which are achieved as land surfaces are stabilized (Church & Ryder 1972; Ballantyne 2002). Such a view of climatic causality is useful because it can be used to consider the magnitude and longevity of landscape impacts of past and future climate changes, respectively.

These views of land surface response to deglaciation are based on the premise that the processes and climates associated with glaciation are related to an increase in sediment generation (by glacial processes themselves, and by enhanced weathering) (Kirkby 1995). In reality, landscape responses are more subtle and strongly conditioned by local-scale geological and topographical factors that lie outside of these models.

The limitations of uniformitarianism

Our view of the processes and products of present-day periglaciation and paraglaciation is set within the context of evidence preserved in the geological record, in particular glacial–interglacial cycles duing the late Pleistocene (Raymo 1997; Tziperman *et al.* 2006). In turn, these have given rise to repeated cycles of sediment generation and delivery downslope into lowland basins and coastal margins (Bridgland 2002; Van der Zwan 2002; Warburton 2007). These are manifested stratigraphically as stacked sequences of periglacial and paraglacial sediments and structures, which are observed in many locations worldwide (e.g. Blikra & Nemec 1998; van Vliet-Lanöe *et al.* 2000; Matsuoka 2001). These climate-driven sediment cycles can be used to help interpret temporal patterns and processes of sediment accumulation in local-scale depocentres, and can, therefore, help distinguish between climatic and non-climatic (such as local geological, topographical, etc.) controls on sediment fluxes (van Vliet-Lanöe *et al.* 2000). This uniformitarian approach can be used effectively in order to evaluate climate-driven sediment patterns over centennial–millennial timescales.

Much of our understanding of past periglacial and paraglacial processes and environments comes from a synthesis of observations drawn from contemporary environments and from preserved geological evidence from the last glacial–interglacial transition (and into the early Holocene). Very little

From: Knight, J. & Harrison, S. (eds) *Periglacial and Paraglacial Processes and Environments*.
The Geological Society, London, Special Publications, **320**, 1–4.
DOI: 10.1144/SP320.1 0305-8719/09/$15.00

is known about the extent, dynamics and evolution of periglacial and paraglacial environments associated with older glacial cycles. This is probably owing to low preservation potential in areas that were overridden by ice in later glaciations. In addition, as interglacials progress, pre-existing periglacial and paraglacial sediments and structures are probably destroyed by plant growth and soil development. These limitations suggest that little is known about past macroscale dynamics of periglacial and paraglacial environments, and that the principle of uniformitarianism is not always appropriate to apply.

Interglaciations, including the present Holocene, are characterized by low continental ice volume and land surfaces dominated by plants. As interglacials develop, therefore, the geographical zones in which periglacial and paraglacial processes operate retreat towards core high-latitude and high-altitude areas. This means that these processes are time transgressive across the landscape as climate belts shift, and that their environments become smaller and more geographically isolated over time. As interglacials develop, geological records from these environments reflect local-scale controls rather than regional-scale climate, and there are associated problems of correlation. Interglacial records are therefore sparser and their interpretations limited. The present situation of anthropogenically enhanced climate changes (global warming), superimposed upon the already warm Holocene, is unprecedented. The net future climatic effects (in both precipitation and in air and ground temperature) are uncertain (Rosenzweig *et al.* 2008; Scott *et al.* 2008). This poses many questions as to how periglacial and paraglacial processes and environments will respond, and how quickly (Warburton 2007), under climatic contexts for which there is no preserved analogue. This clearly illustrates the limitations of uniformitarianism as a tool to understand future changes in periglacial and paraglacial environments.

Discussion and outlook to the future

Human impacts on late Holocene climate and Earth systems have dramatically affected land surface stability and associated sediment fluxes, and led to the late Holocene period being informally termed the Anthropocene (Ruddiman 2003). The impact of human activity on landscape dynamics has been discussed in a number of studies (e.g. Hooke 2000; Ehlen *et al.* 2005; Wilkinson 2005). Other studies have focused more specifically on human-related changes in sediment budgets in different physical settings. For example, Hooke (1999) and Wilkinson & McElroy (2007) argued that human activity has helped shift the focus of highest sediment fluxes from upland (river headwater) to lowland parts of catchments, which has implications for the capacity of river systems to respond to climatic v. anthropogenic forcings (Meybeck 2003; Juen *et al.* 2007). Further, Wilkinson & McElroy (2007) argued that current rates of continental denudation are far higher than background rates over past glacial–interglacial cycles, hence that human activity is more significant than other processes in shaping Holocene landscapes. This is significant because it suggests that paraglacial sediment systems are being (or have been since the late Holocene) overwhelmed by a direct anthropogenic overprint controlled by deforestation, ecosystem changes, etc. In addition, future enhanced global warming (and changing temperature and precipitation regimes) is going to impact most strongly on the climatically determined environments where periglacial and paraglacial processes take place, in particular in upland and glaciated catchments.

A probable effect of anthropogenic climate warming is that the present interglacial is extended beyond the timescale determined solely by Milankovitch forcing (Mitchell 1972; Mörner 1972), which has been largely responsible for controlled interglacial length in the past (e.g. Tziperman *et al.* 2006). As periglacial and paraglacial processes are, on the macroscale, determined by climate, it is to be anticipated that sediment generation and supply will decrease over time as the land area under these favourable climates decreases also. This follows the paraglacial sediment exhaustion model of Ballantyne (2002). Under an extended (and warmer) interglacial, it is probable that sediment fluxes from the headwaters of mid-latitude glaciated basins will decrease dramatically, leading to sediment starvation and, eventually, cannibalization of river lowlands and coastal fringes. In high-latitude areas, permafrost melt and reduced sea ice protection is already leading to enhanced coastal erosion and sediment supply (Lawrence *et al.* 2008). Global warming, therefore, is already leading to a decrease in the continuity and interconnectedness of permafrost and associated periglacial processes (Lemke *et al.* 2007). A sediment budget approach (e.g. Syvitski *et al.* 2003; Phillips & Slattery 2006) can help monitor the progression of this breakup.

Imperatives in the understanding of periglacial and paraglacial environments

The foregoing discussion identifies the subtle interrelationships between periglacial and paraglacial environments and climates of the past and future.

Understanding these interrelationships is important because present decreases in the distribution and thickness of permafrost, particularly in continental interiors (Camill 2005), have implications for ecosystem and landscape stability, human activities and engineering solutions, and CO_2 degassing from thawing permafrost (Lawrence *et al.* 2008). This is mirrored in sensitive and marginal periglacial Alpine environments that are presently experiencing increased rockfall and mass movement, including solifluction, rock glacier instability and changes in sediment release to downstream rivers (Juen *et al.* 2007; Warburton 2007). Likewise, a major initiative in sensitive glaciated mountain environments is to understand the processes of geomorphic change, the rate of landscape modification and the nature of resulting paraglacial landsystems.

In considering how periglacial and paraglacial environments are going to respond to future climate changes, two key questions present themselves. First, given that renewed paraglaciation will accompany future glacier retreat and decreased extent of periglacial environments under global warming, how will we accommodate such geomorphological instability into our models of economic and social use of both mountain and lowland cold-climate regions? Second, how far can models of paraglaciation, and periglacial slope processes, be used to interpret the geomorphic evolution of these landscapes under future climate scenarios? These questions, and related issues, are explored in this volume in an inter- and multidisciplinary framework, through case studies from both contemporary and Quaternary periglacial and paraglacial settings.

This volume is organized into three sections. The first section focuses on periglacial processes and environments. The paper by **André** sets periglacial studies into a wider and historical context. Periglacial weathering and palsa processes are examined in the papers by **Nicholson**, and **Seppälä & Kujala**, respectively. The papers by **Waller *et al.*** and **Slaymaker** discuss the interrelationships between periglacial and glacial processes and environments.

The second section of the book focuses on paraglacial environments and processes in the British Isles. The paper by **Whalley** discusses how periglacial and paraglacial sediments and structures can be used to reconstruct past climate changes. The succeeding papers by **Jarman**, **Wilson**, **Passmore & Waddington** and **Knight** provide evidence for paraglacial processes and environmental change during the late Quaternary and early Holocene, using examples from upland areas of Britain and Ireland.

The final section of the book examines paraglacial processes, climate change and related issues of sediment supply using examples from Europe, North America and Asia. The paper by **Curry *et al.*** considers the geotechnical and geomorphic implications on ongoing paraglaciation. Specific examples of paraglacial landscape responses from British Columbia are shown in the papers by **Wilkie & Clague** and **Friele & Clague**. The paper by **Hewitt** considers paraglaciation in Pakistan as a transient landscape response to climatic disturbance. The final paper in the volume, by **Harrison**, addresses the sensitivity of periglacial and paraglacial geomorphic systems to climatic forcing, which is particularly important when one considers that these environments are most at threat from future climate change.

The papers in this volume are largely the outcome of a meeting held at the Geological Society, London (UK) in January 2007, on the theme of Periglacial and Paraglacial Processes and Environments, Past, Present and Future. The meeting was held jointly between the Geological Society of London and the Quaternary Research Association who are thanked for their financial support.

The editors wish to thank the authors for their contributions, and acknowledge the following reviewers: N. Betts, J. Boelhouwers, J. Catt, J. Carrivick, P. Christoffersen, M. Clark, J. Desloges, J. Dixon, B. Etzelmuller, A. Findlayson, D. Giles, S. Gurney, K. Hall, P. Hughes, K. Huntington, O. Humlum, J. Kemp, M. Konen, O. Korup, W. Mitchell, A. Nesje, S. Payette, R. Pine, B. Rea, A. Strom, D. Swift, R. Tipping, F. Tweed, C. Whiteman and C. Zangerl.

References

André, M.-F. 2003. Do periglacial landscapes evolve under periglacial conditions? *Geomorphology*, **52**, 149–164.

Ballantyne, C. K. 2002. Paraglacial geomorphology. *Quaternary Science Reviews*, **21**, 1935–2017.

Blikra, L. H. & Nemec, W. 1998. Postglacial colluvium in western Norway: depositional processes, facies and palaeoclimatic record. *Sedimentology*, **45**, 909–959.

Bridgland, D. R. 2002. Fluvial deposition on periodically emergent shelves in the Quaternary: examples records from the shelf around Britain. *Quaternary International*, **92**, 25–34.

Camill, P. 2005. Permafrost thaw accelerates in boreal peatlands during late-20th century climate warming. *Climatic Change*, **68**, 135–152.

Church, M. & Ryder, J. M. 1972. Paraglacial sedimentation: a consideration of fluvial processes conditioned by glaciation. *Bulletin of the Geological Society of America*, **83**, 3059–3071.

Church, M. & Slaymaker, O. 1989. Disequilibrium of Holocene sediment yield in glaciated British Columbia. *Nature*, **337**, 452–454.

Ehlen, J., Haneberg, W. C. & Larson, R. A. (eds). 2005. *Humans as Geological Agents*. Reviews in Engineering Geology, **16**. Geological Society of America, Boulder, CO.

HARRIS, S. A. 1994. Climatic zonality of periglacial landforms in mountain areas. *Arctic*, **47**, 184–192.
HEWITT, K., BRYNE, M.-L., ENGLISH, M. & YOUNG, G. (eds). 2002. *Landscapes in Transition: Landform Assemblages and Transformations in Cold Regions*. Kluwer Academic, London.
HOOKE, R. LEB. 1999. Spatial distribution of human geomorphic activity in the United States: Comparison with rivers. *Earth Surface Processes and Landforms*, **24**(8), 687–692.
HOOKE, R. LEB. 2000. On the history of humans as geomorphic agents. *Geology*, **28**(9), 843–846.
JUEN, I., KASER, G. & GEORGES, C. 2007. Modelling observed and future runoff from a glacierized tropical catchment (Cordillera Blanca, Perú). *Global and Planetary Change*, **59**, 37–48.
KIRKBY, M. J. 1995. A model for variations in gelifluction rates with temperature and topography: implications for global change. *Geografiska Annaler*, **77A**, 269–278.
LAWRENCE, D. M., SLATER, A. G., TOMAS, R. A., HOLLAND, M. M. & DESER, C. 2008. Accelerated Arctic land warming and permafrost degradation during rapid sea ice loss. *Geophysical Research Letters*, **35**, L11506, doi: 10.1029/2008GL033985.
LEMKE, P., REN, J. *ET AL*. 2007. Observations: changes in snow, ice and frozen ground. *In*: SOLOMON, S., QIN, D. *ET AL*. (eds) *Climate Change 2007: The Physical Science Basis. Contribution of Working Group I to the Fourth Assessment Report of the Intergovernmental Panel on Climate Change*. Cambridge University Press, Cambridge, 337–383.
MATSUOKA, N. 2001. Solifluction rates, processes and landforms: a global review. *Earth-Science Reviews*, **55**, 107–134.
MEYBECK, M. 2003. Global analysis of river systems: from Earth system controls to Anthropocene syndromes. *Philosophical Transactions of the Royal Society of London*, Series B, **358**, 1935–1955.
MITCHELL, J. M., JR. 1972. The natural breakdown of the present interglacial and its possible intervention by human activities. *Quaternary Research*, **2**, 436–445.
MÖRNER, N.-A. 1972. When will the present interglacial end? *Quaternary Research*, **2**, 341–349.
PHILLIPS, J. D. & SLATTERY, M. C. 2006. Sediment storage, sea level, and sediment delivery to the ocean by coastal plain rivers. *Progress in Physical Geography*, **30**, 513–530.
RAYMO, M. E. 1997. The timing of major climate terminations. *Paleoceanography*, **12**, 577–585.
ROSENZWEIG, C., KAROLY, D. *ET AL*. 2008. Attributing physical and biological impacts to anthropogenic climate change. *Nature*, **453**, 353–357.
RUDDIMAN, W. F. 2003. The Anthropogenic greenhouse era began thousands of years ago. *Climatic Change*, **61**, 261–293.
SCHNEIDER, E. K., KIRTMAN, B. P. & LINDZEN, R. S. 1999. Tropospheric water vapour and climate sensitivity. *Journal of the Atmospheric Sciences*, **56**, 1649–1658.
SCOTT, J. R., SOKOLOV, A. P., STONE, P. H. & WEBSTER, M. D. 2008. Relative roles of climate sensitivity and forcing in defining the ocean circulation response to climate change. *Climate Dynamics*, **30**, 441–454.
SYVITSKI, J. P. M., PECKHAM, S. D., HILBERMAN, R. & MULDER, T. 2003. Predicting the terrestrial flux of sediment to the global ocean: a planetary perspective. *Sedimentary Geology*, **162**, 5–24.
TZIPERMAN, E., RAYMO, M. E., HUYBERS, P. & WUNSCH, C. 2006. Consequences of pacing the Pleistocene 100 kyr ice ages by nonlinear phase locking to Milankovitch forcing. *Paleoceanography*, **21**, PA4206, doi: 10.1029/2005PA001241.
VAN DER ZWAN, C. J. 2002. The impact of Milankovitch-scale climatic forcing on sediment supply. *Sedimentary Geology*, **147**, 271–294.
VAN VLIET-LANOË, B., LAURENT, M. *ET AL*. 2000. Middle Pleistocene raised beach anomalies in the English Channel: regional and global stratigraphic implications. *Journal of Geodynamics*, **29**, 15–41.
WARBURTON, J. 2007. Sediment budgets and rates of sediment transfer across cold environments in Europe: a commentary. *Geografiska Annaler*, **89A**, 95–100.
WILKINSON, B. H. 2005. Humans as geological agents: A deep-time perspective. *Geology*, **33**, 161–169.
WILKINSON, B. H. & MCELROY, B. J. 2007. The impact of humans on continental erosion and sedimentation. *Geological Society of America Bulletin*, **119**, 140–156.

From climatic to global change geomorphology: contemporary shifts in periglacial geomorphology

MARIE-FRANÇOISE ANDRÉ

Laboratory of Physical and Environmental Geography, GEOLAB-UMR 6042 CNRS, Blaise Pascal University, MSH – 4 rue Ledru, 63057 Clermont-Ferrand, Cedex 1, France (e-mail: m-francoise.andre@univ-bpclermont.fr)

Abstract: Periglacial geomorphology developed in the 1940s–1960s as a branch of climatic geomorphology, focusing first on Quaternary studies and palaeoenvironmental reconstructions, then on current geomorphic activity in cold regions. The 'periglacial fever' of the 1960s–1970s was dominated by the 'freeze–thaw dogma': periglacial areas were regarded as necessarily submitted to efficient frost-driven processes ruling over the geomorphic activity. Such a view was severely criticized in the 1980s–1990s based both on monitoring studies and on time–space multiscale approaches that pointed to the need to cross the 'smokescreen of the periglacial scenery' to search for the real past and present processes responsible for the landform geometry. The role of non-cold-related processes in the making of 'periglacial' landcapes was re-evaluated, and the necessity to better take into account the rock properties and the pre-Quaternary history of slope systems was emphasized. Whereas the part of the cold-related processes was being minimized, the interest of genuine periglacial landforms as geoindicators of climate change was growing, providing a new legitimacy to periglacial geomorphology. Polar and Alpine regions are nowadays considered as key observatories of ongoing climate change, and periglacial geomorphologists are involved in the detection, monitoring and prediction of environmental changes. Finally, the evolution of 'periglacial geomorphology' over the past six decades is in accordance with the development of the whole geomorphology. Based on the quantitative and technological revolution, it tends to find a balance between the functional and historical approaches.

The main objective of this paper is to provide some insights into the development of periglacial geomorphology since World War II. Excellent overviews covering the period have been produced by key actors of the periglacial community (e.g. Barsch 1993; French 2003; French & Thorn 2006). The idea is to deliver an additional and somewhat continental view, characterized by the long-lasting influence of historical geomorphology and the pervasiveness of geological factors as the main control on landform evolution. The perspective will be illustrated by selected examples, mainly from polar regions. Particular attention will be paid to changing ideas concerning weathering processes responsible for rock breakdown in cold regions, where weathering can be seen as 'a fundamental geomorphic input that embraces both geomorphic work and landform initiation' (Thorn 2004, p. 10).

Five temporal trends will be distinguished:

- 1940s–1960s: periglacial geomorphology as a branch of climatic geomorphology;
- 1960s–1980s: the 'periglacial fever';
- 1980s–1990s: the freeze–thaw dogma under pressure;
- 1990s–2000s: from periglacial geomorphology to cold-region geomorphology;
- 2000s–: periglacial geomorphology and the 'Global Change fever'.

Periglacial geomorphology as a branch of climatic geomorphology (1940s–1960s)

Three decades after Lozinski's 1909 seminal paper, periglacial geomorphology played a prominent role in the development of climatic geomorphology as illustrated by two benchmark papers published by Büdel and Troll in the same issue of the *Geologische Rundschau* (Büdel 1944; Troll 1944). These papers clearly illustrate the burst of Pleistocene palaeogeographical reconstructions in Europe and prefigure the launching of emblematic pioneer process studies in periglacial environments.

Pleistocene studies in Europe

In the 1940s–1950s the main trend in periglacial studies was based on the advances of Quaternary geology. Inventories of periglacial deposits and geomorphic features such as sand wedges and grèzes litées (Fig. 1a, b) were conducted throughout Europe. Among the most outstanding palaeoclimatic reconstructions deriving from these inventories stands the one by Poser (1948), who produced

From: KNIGHT, J. & HARRISON, S. (eds) *Periglacial and Paraglacial Processes and Environments.* The Geological Society, London, Special Publications, **320**, 5–28.
DOI: 10.1144/SP320.2 0305-8719/09/$15.00

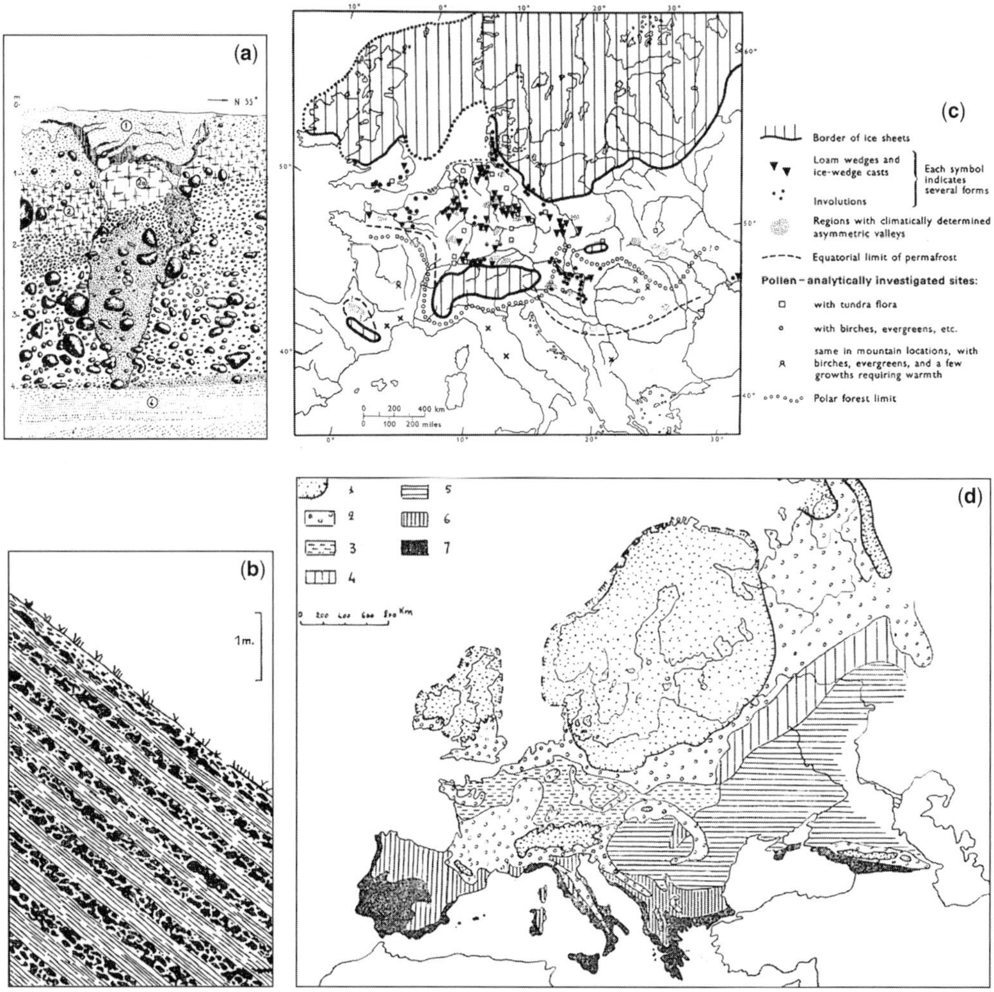

Fig. 1. Pleistocene periglacial features in Europe: their distribution and palaeoenvironmental significance. (**a**) Sand wedge, Slawecin, Poland (Dylikowa 1956). (**b**) 'Grèzes litées', Crussol, Rhône Valley, France (Tricart 1967, fig. 91, p. 238). (**c**) Distribution of Würm periglacial features in Europe (Poser 1948). (**d**) Würmian bioclimatic zones in Europe after Büdel, Troll and Tricart (Tricart 1967, fig. 5, p. 35), where: 1, glaciers; 2, frostschuttundra (polar desert); 3–4, loess-covered areas; 5–6, steppic areas; 7, forested areas.

a map of climatic regions in Europe during the Würm glaciation, based on the mapping of periglacial features (Fig. 1c). In the following years audacious periglacial interpretations were proposed, such as the one of remnants of Pleistocene pingos for rampart-enclosed depressions found throughout the Belgian uplands (Pissart 1956).

As stressed by French (2003), the growth of Pleistocene periglacial studies in Europe owes much to Cailleux, the first secretary of the International Geographical Union Periglacial Commission established in 1949, who employed sedimentological techniques and microscopy to study past and present niveo-aeolian deposits. His articles were later expanded in a book entitled *Cryopédologie*, the first formal periglacial text published in French (Cailleux & Taylor 1954). During the same period Tricart was completing his thesis about the Pleistocene periglacial legacy in the Eastern Paris Basin. His lectures at the Sorbonne were published under the title *Le modelé périglaciaire*, modified later on into *Le modelé des regions périglaciaires* (Tricart 1950, 1967). It includes a reconstruction of Pleistocene 'morphoclimatic provinces' west of the Urals (Fig. 1d).

In the foreword to this book, Tricart emphasized the drastic change that had occurred since World War II in the perception of periglacial phenomena among the international community of geomorphologists: 'Les manifestations périglaciaires, considérées il y a 20 ans comme des curiosités pittoresques, des bizarreries de la nature, sont devenues maintenant des phénomènes admis comme fondamentaux sur près de 40% du globe' (in Tricart 1967, p. 5). Indeed, it is not by chance that the *Biuletyn Peryglacjalny*, the first international journal devoted entirely to periglacial geomorphology, was launched in 1954. Two years later, when Dylik, its founder, took over the presidency of the IGU Periglacial Commission, 'an international periglacial community has been established possessing a strong European and Pleistocene orientation' (French 2003, p. 38).

Pioneer process studies in present-day periglacial environments

In parallel with Pleistocene reconstructions, process studies in present-day periglacial environments were initiated in the 1950s by a number of scientists such as Mackay in Canada, Péwé in Alaska and Antarctica, and Jahn on Spitsbergen. Of special significance are the quantitative pioneer studies carried out by Rapp in Scandinavia and by Washburn in Greenland.

Based on a 9-year monitoring study in the Swedish Kärkevagge site, and additional investigations and measurements at the Tempelfjord site on Spitsbergen, Rapp provided a huge amount of quantitative data on the current geomorphic activity operating on Arctic slopes (Rapp 1960*a*, *b*) (see Fig. 2). What is particularly striking in his contribution is the novelty and the variety of methods used to meet his objectives: including the use of painted lines, natural and artificial debris traps (vegetation and snow covers; sack carpets), photogrammetry, a rockfall calendar correlated with climate data and detailed geomorphological mapping. His approach was also innovative in so far as he addressed the relative significance of 'extreme' events and more continuous processes in the making of landforms, proposed a hierarchy of geomorphological processes based on quantitative data (Table 1) and drew attention to the importance of chemical activity in cold areas (a quite iconoclastic view in the 1950s!). In Tempelfjord, he also pointed out the discrepancy between the negligible contemporary rates of rockwall retreat based on photogrammetry and the significant Holocene rates inferred from the volumes of scree cones and rock glaciers. Rapp's theses have become classics and his approach is still an inexhaustible source of inspiration for Arctic and Alpine researchers (Luckman 2000).

Of high interest are also the instrumental observations of mass-wasting carried out by Wahburn in the Mesters Vig district of NE Greenland from 1956 to 1961 (Washburn 1967). Experimental sites ranging in mean gradient from 2.5° to 25° were established in various conditions regarding grain size, moisture and vegetational characteristics. Theodolite readings of cone targets on wood pegs inserted in the ground provided valuable information concerning the possibility of distinguishing quantitatively frost creep from gelifluction, and demonstrated the prominent role played by moisture conditions in the rates of movement.

The prevalence of climate control

Both monitoring studies and palaeogeographical reconstructions in periglacial environments point to cold climates as the key drivers of rapid changes in landscape geometry. Derived from Davis's cyclic concept, Peltier's 'cycle of periglacial erosion' leading to ultimate 'cryoplanation' of the landscapes (Peltier 1950) has been widely adopted in the international periglacial community. Despite Rapp's insights into the importance of solute loads, most researchers take for granted the prevalence of mechanical frost-driven processes of rock breakdown. Even the extensive Scandinavian upland blockfields are interpreted as post-glacial landforms due to very active frost shattering (e.g. Dahl 1966). Such an interpretation fits with Boyé's hypothesis of the 'défonçage périglaciaire', which suggests a possible renewal of coarse debris stocks during the brief interglacial phases (Boyé 1949). The 'periglacial fever' has started.

The 'periglacial fever' (1960s–1980s)

The 1960s–1980s were marked by an unprecedented development of periglacial geomorphology based on laboratory experiments and field investigations carried out in various Arctic and Alpine areas. Undoubtedly, this 'periglacial fever', which marks the acme of climatic geomorphology, benefited from the quantitative revolution initiated by fluvial geomorphologists (e.g. Horton 1945; Leopold *et al.* 1964).

A series of periglacial textbooks published by geomorphologists arose from this development, such as Tricart (1963, 1967, 1970), Washburn (1973, 1979), Embleton & King (1975), French (1976) and Pissart (1987). These books offer an extensive overview of periglacial studies carried out during this period, of which demonstrative examples will be mentioned later to illustrate some major research trends.

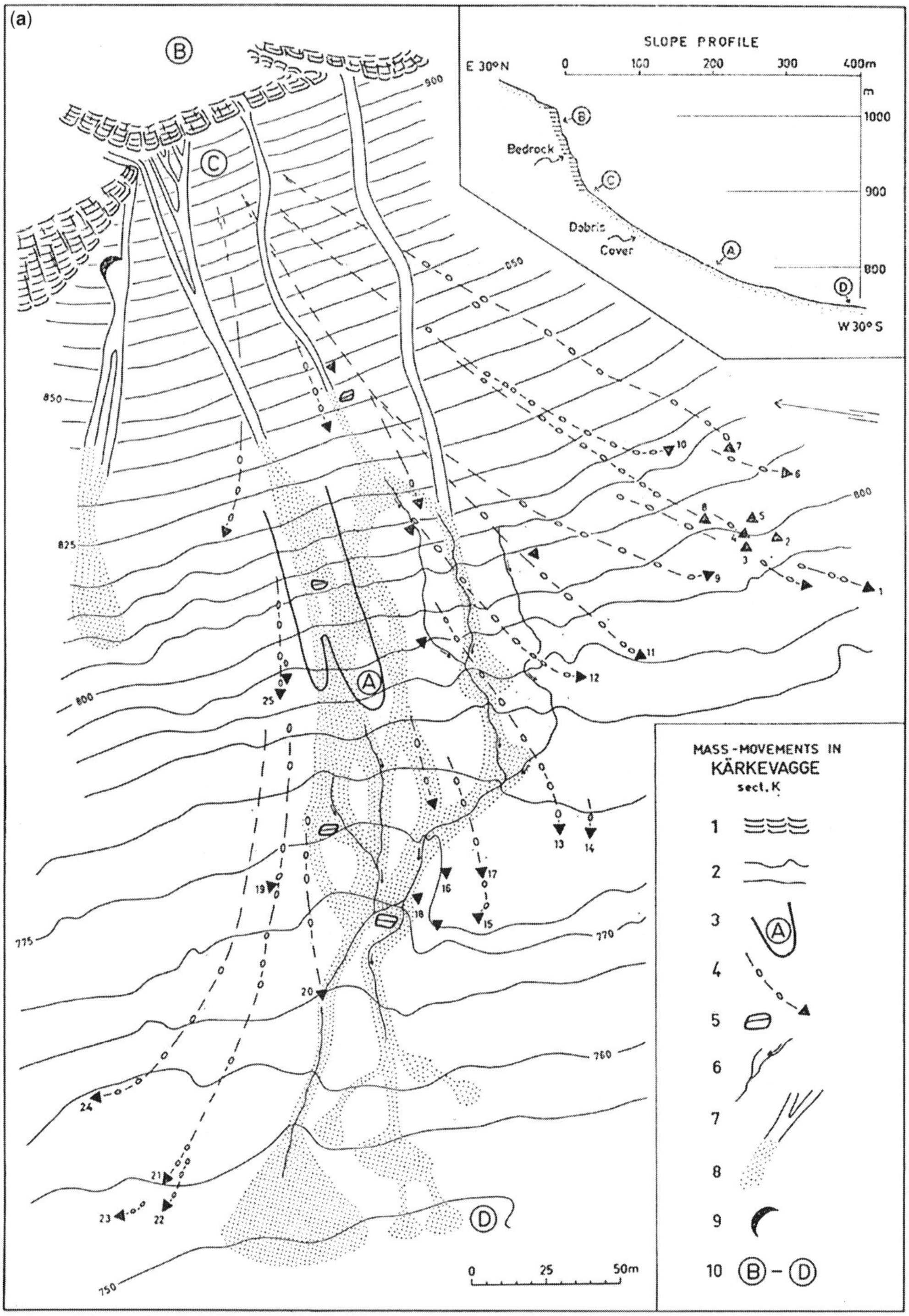

Fig. 2. Anders Rapp in Kärkevagge: a pioneer process and climate monitoring study in a periglacial environment. (**a**) Record of slope dynamics over the 1952–1960 observation period (Rapp 1960*a*, fig. 11, p. 98), where: 1, rock wall; 2, contours; 3, dirty snow avalanche K36 of 4 May 1953; 4, boulder-fall; 5, old, big boulder; 6, runnels; 7, gullies,

(b)

(c)

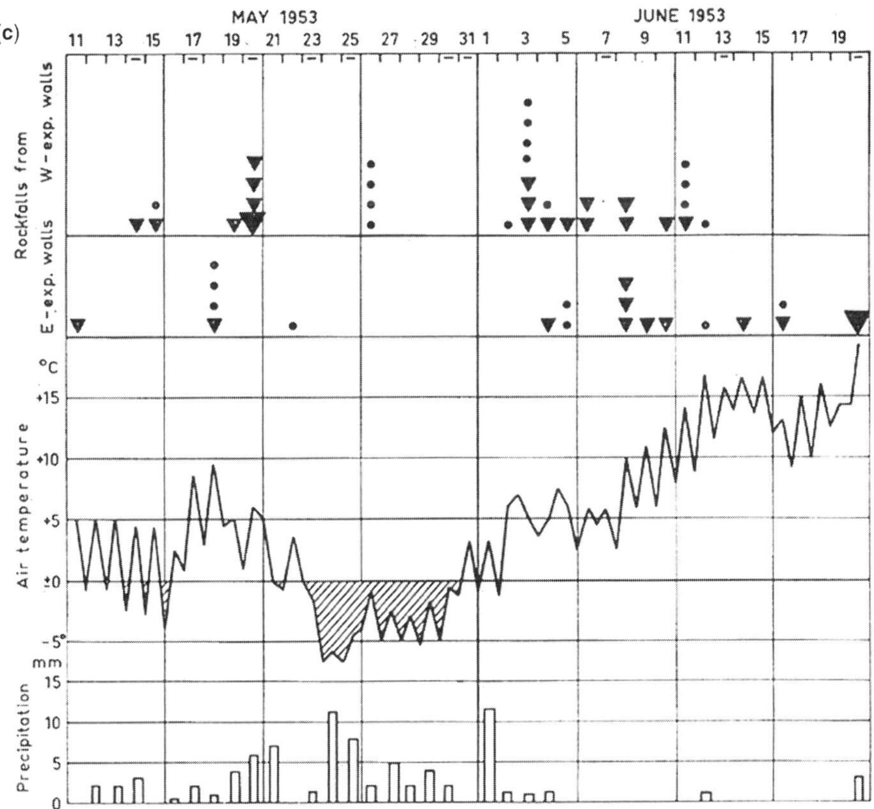

Fig. 2. (*Continued*) enlarged on 8 October 1959; 8, mudflow; 9, earth-slide scar (ncs 8 and 9 of 10 October 1959); 10, points on the profile, Point B. Release scar of boulder-fall K37. (**b**) Anders Rapp leading a field excursion during the 1960 IGU Meeting, Swedish Lapland (collection: Birgit Rapp). (**c**) Correlations between rockfalls, air temperature and precipitation in May and June 1953 (Rapp 1960*a*, fig. 18, p. 105).

Table 1. *Denudation of slopes in Kärkevagge, Swedish Lapland, 1952–1960 (Rapp 1960*a*, table 32, p. 185)*

Process	Volume (m^3)	Density	Tons (t)	Tons per km^2	Average movement (m)	Average gradient (°)	Ton-metres (vertical)
Rockfalls							
Pebble-falls	5	2.6	13	1	90	45	845
Small boulder-falls	10	2.6	26	1.7	225	45	4160
Big boulder-falls	35	2.6	91	6	225	45	14 560
Avalanches							
Small avalanches	8	2.6	21	1.4	100	30	1050
Big avalanches (slushers)	80	2.6	208	14	200	30	20 800
Earth-slides etc.							
Bowl-slides	170	1.8	300	20	0.5	30	75
Sheet-slides	190	1.8	340	23	12–420	30	20 000
Sheet-slides + mudflows	150	1.8	270	18	70–600	30	70 000
Other mudflows	70	1.8	126	8.4	100	30	6300
Creep							
Talus-creep	300 000	1.8	–	–	0.01	30	2700
Solifluction	550 000	1.8	–	–	0.02	15	5300
Running water							
Dissolved salts	150	2.6	390	26	700	30	136 500
Slope wash	?			?			?

Fascinating periglacial features

The fascination exerted by frost mounds and patterned ground on geomorphologists arise from their intriguing geometry, particularly striking from the air. The highest concentration of pingos (more than 1300) is found in the Tuktoyaktuk Peninsula area (Mackenzie Delta, Canada), resulting from the favourable conditions found there (thick permafrost, coarse-grained sediments, frequent draining of thermokarst lakes). Detailed long-term field studies of pingo growth and decay were undertaken by Mackay in the 1970s (e.g. Mackay 1973, 1977, 1979). Mackay's monitoring studies revealed a pattern of rapid early growth (1.5 m $year^{-1}$), followed by decreasing growth rates of pingos (2–3 cm $year^{-1}$). This work also showed that such hydrostatic (closed) system pingos (Fig. 3a, b) often exhibited pulsating patterns of heave owing to the build-up of water lenses forming under pressure beneath the growing pingos.

During the same period Mackay carried out comprehensive investigations of ice-wedge polygons in the western Canadian Arctic (e.g. Mackay 1974, 1986). Timing and direction of cracking were established thanks to electronic crack detectors. Interestingly, Mackay pointed out the occurrence of discrepancies between field data and theoretical considerations, such as those developed by Lachenbruch (1962). In Mackay's opinion, the frequency of polygonal cracking appears to be site-specific owing to changing conditions in microrelief, vegetation and snow cover induced by crack formation. Based on results from the 1967–1987 monitoring period, it appears that polygonal cracking does not obey universal laws governed by air temperature changes, but, instead, should be regarded as a random process as defined by the 'chaos theory' (Mackay 1992).

In the 1970s–1980s the pure geometry of sorted circles on Spitsbergen fascinated geomorphologists and soil scientists, and provoked stimulating discussions about their origin. Whereas Hallet *et al.* (1988) inferred circulatory patterns from measurements of soil movement within sorted circles, Van Vliet-Lanoë (1985, 1988) supported Sharp's hypothesis of differential frost heave as the cause of cryoturbation. She proposed two scenarios of sorted circle formation depending on the granulometrical pattern of the affected deposits (Fig. 3c, d). Her approach to periglacial phenomena, initiated in the 1970s (e.g. Van Vliet-Lanoë 1976), is particularly innovative and integrative in so far as it combines field and micromorphological observations in active and inherited contexts. She pointed out segregation ice as the driving force of a variety of periglacial phenomena and demonstrated that a continuum leads from frost creep to gelifluction, based on the study of microfabrics created by ice lensing.

Burst of laboratory freeze–thaw experiments

In parallel with field studies, frost experiments developed tremendously in the 1960s–1980s. A prominent role was played by the Centre de Géomorphologie de Caen in Normandy, where researchers from various European and North American

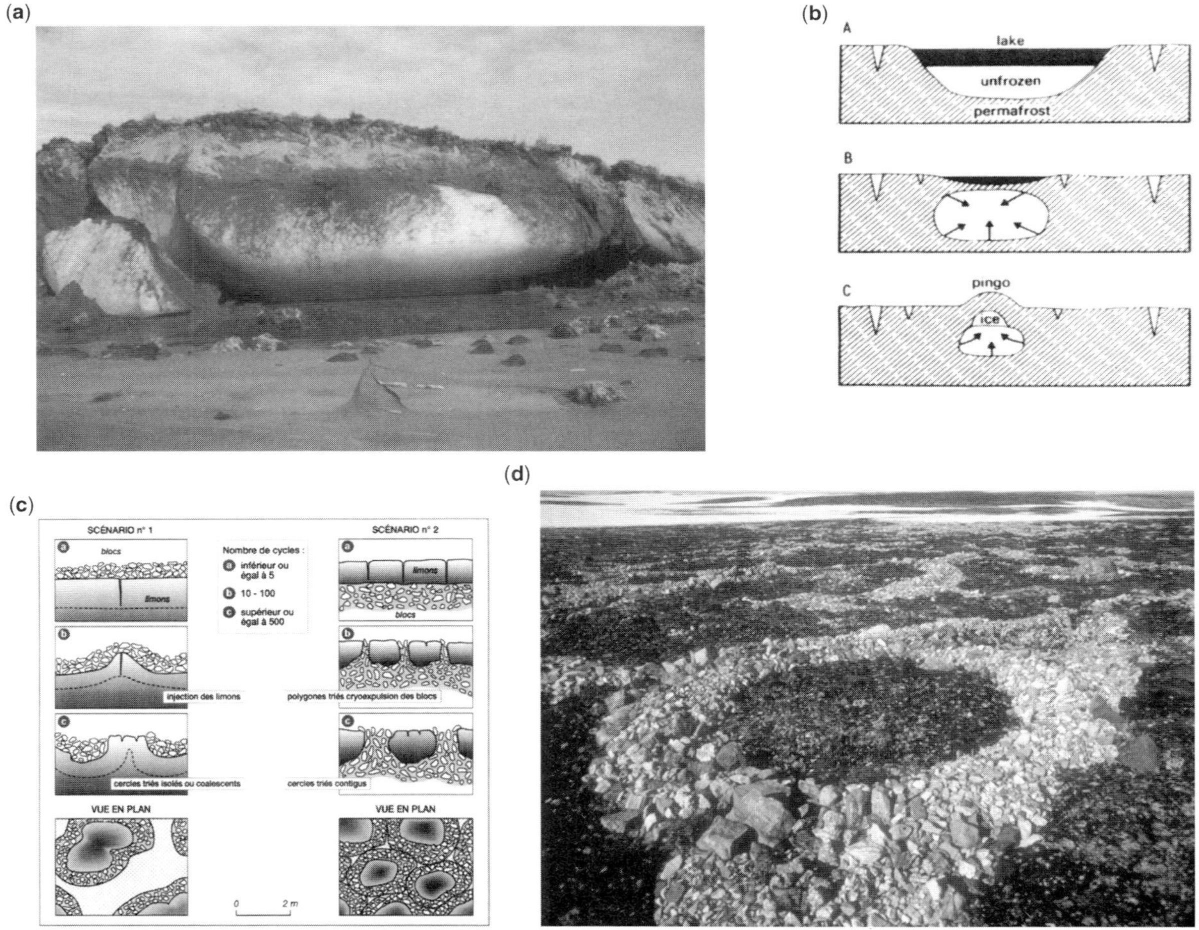

Fig. 3. Fascinating pingos and patterned ground. (**a**) Injection ice in a pingo of western Arctic Canada. © J. R. Mackay. (**b**) Formation of a closed-system pingo (Mackay 1982). (**c**) Differential frost heave as the cause of stone circle formation (after Van Vliet-Lanoë 1988, p. 526). (**d**) Stone circles in beach deposits, Kvadehuk, Spitsbergen, © M. F. André.

countries set up experiments and exchanged ideas, and still do (e.g. Murton *et al.* 2000) (see Fig. 4d). Thousands of rock samples belonging to various lithologies were submitted to freeze–thaw cycles in Caen's cold rooms: sandstones (e.g. from Fontainebleau, Brive and Vosges), limestones (e.g. from Normandy and Spitsbergen), granites (e.g. from Massif Central, Portugal, Finland and Norway), schists (e.g. from Corsica, Wales and Labrador) and basalts (e.g. from Iceland).

A summary of the most significant findings is found in Lautridou & Ozouf (1982) (see Fig. 4). A number of parameters were considered in the derivation of scales of frost susceptibility, and the ability of frost shattering to produce fine-grained material was demonstrated. Even crystalline rocks proved to be frost-susceptible, provided they had been either microfissured or chemically weathered during their geological and/or palaeoclimatic history. Experiments were also conducted to study the action of freeze–thaw on soils, and to understand the mechanisms of frost sorting and differential frost heaving (e.g. Coutard *et al.* 1988). On the whole, experiments in Caen were particularly innovative in so far as they were conducted in the framework of programmes combining basic and applied research. They also aimed to transfer experimental results for a better understanding of inherited periglacial deposits. For instance, the size and morphology of the experimental gelifracts were fruitfully compared with those coming from Quaternary deposits (e.g. Ozouf 1983).

Freeze–thaw seen as the key driver of morphodynamics in cold areas

During the 1960s–1980s 'periglacial fever', based on field and laboratory studies, cold-climate regions were considered as parts of the world undergoing rapid geomorphological evolution, governed by freeze–thaw cycles. A variety of frost-related processes – frost shattering, frost wedging, frost bursting, frost heaving, gelifluction, nivation and cryoplanation – was systematically emphasized, whereas the role of non-cold-related processes was either ignored or minimized. Etienne's analysis of the contents of major periglacial textbooks clearly illustrates this trend (Etienne 2004). The presumption of frost having been responsible for the formation of blockfields and Richter slopes was taken for granted both in Mediterranean and in subtropical areas, where Pleistocene periglacial conditions were thought to explain the abundance of angular debris.

Some isolated scientists had openly criticized the overrating of the role of freeze–thaw mechanisms. Czeppe (1964) had stated that frost processes failed to account for rock flaking on Spitsbergen. Later, Malaurie (1966) had refuted the défonçage périglaciaire theory – closely related to the cryoplanation concept – developed by Boyé (1949), who replied to these criticisms in the *Biuletyn Peryglacjalny* (Boyé 1968). Both Malaurie (1968) and Meckelein (1974) pointed out the similarity of flaking phenomena, tafonis and duricrusts found in both cold and hot deserts, and suggested salt and thermal weathering as key processes in rock breakdown in both these types of arid environments. Last, but not least, at the zenith of the 'periglacial fever', Gray downplayed the efficiency of periglacial processes by publishing very slow rates of rock wall retreat in Yukon (Gray 1972).

The overall community of geomorphologists did not pay much attention to these dissonant voices, and continued to designate frost action as *the* driver of rock breakdown in periglacial environments.

The 'freeze–thaw dogma' under pressure (1980s–1990s)

The 'periglacial fever' had been a very fruitful period, especially regarding the development of new methods of monitoring and experimentation, but had resulted in a biased vision of the geomorphological activity occurring in periglacial regions. Some kind of 'cultural revolution' was on the move, which came to light in the 1980s.

Sources of changing ideas and benchmark papers

The vigorous reaction against the 'freeze–thaw dogma', which originated from various sources, was triggered by a combination of the following factors:

- new findings from monitoring studies and associated rock temperature data;
- renewed interest in rock control (the golden age of climatic geomorphology was left far behind);
- survival of some historical geomorphology, looking back far into the past;
- cross-breeding between the periglacial and the weathering geomorphological communities, and between the cold and hot desert communities.

Providing an exhaustive overview of the resulting literature is beyond the scope of this article. Some benchmark papers will be mentioned, which followed insightful reviews of ongoing weathering studies published in the 1980s in *Progress in Physical Geography* (e.g. McGreevy 1981; Whalley & McGreevy 1983, 1985).

Thorn (1988, 1992) was among the first to strike hard. He questioned 'the status of a sacred cow' occupied by freeze–thaw weathering within periglacial geomorphology, and called nivation a 'geomorphic

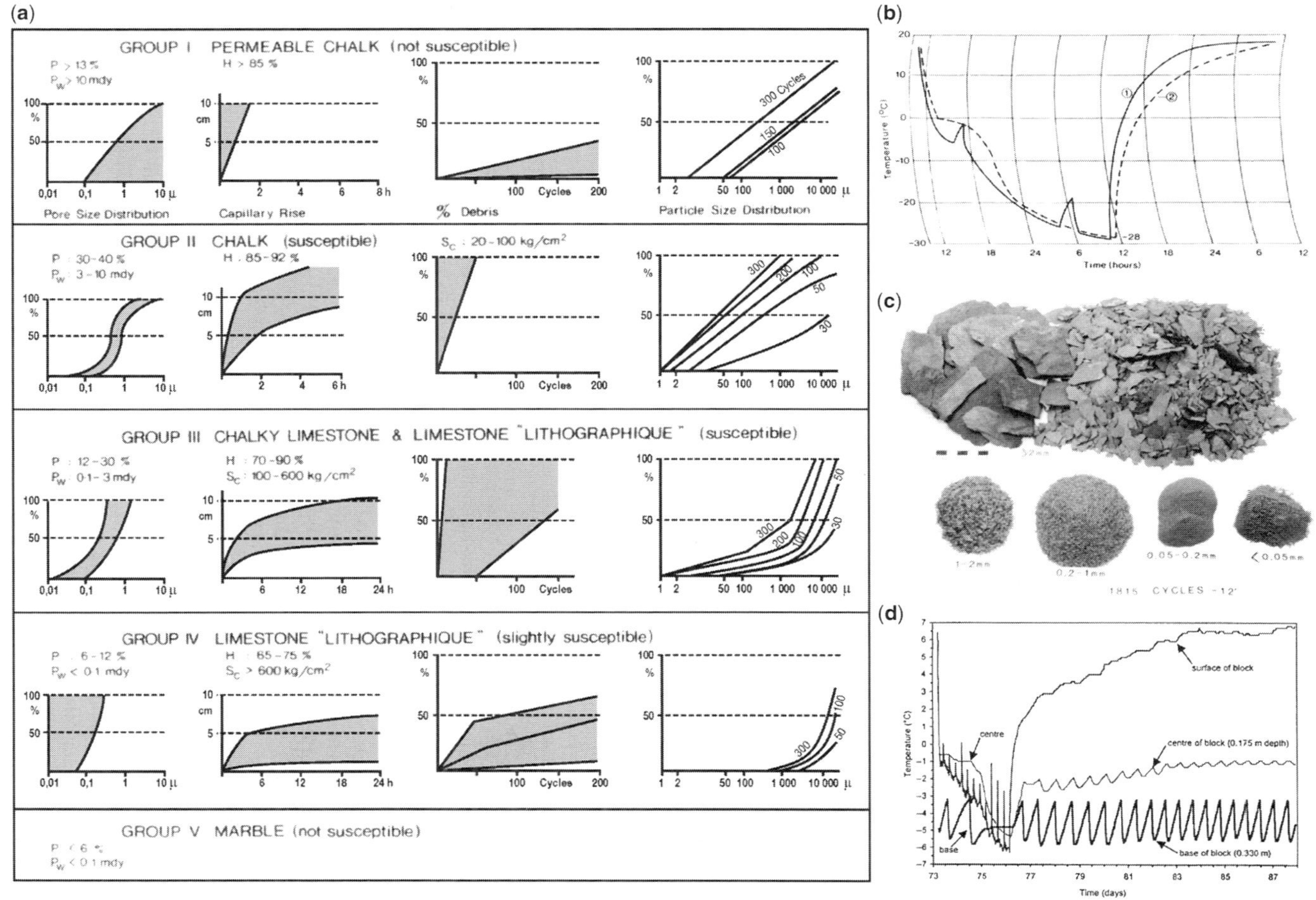

Fig. 4. Frost experiments at the Centre de Géomorphologie of Caen, Normandy. (**a**) Scale of frost susceptibility (Lautridou & Ozouf 1982, fig. 2, p. 219), where P is total porosity, H is the saturation coefficient, P_w is permeability and S_c is compressive strength. (**b**) Example of the temperature regime used in cold rooms in the 1980s, where curve 1 is air temperature and curve 2 is rock temperature (Lautridou & Ozouf 1982, fig. 1, p. 216). (**c**) Frost-shattered shales from Labrador after 1815 cycles at − 12 °C in Caen (Y. Delehaye, collection: M.-F. André). (**d**) Rock temperatures recorded during a pilot experimental study carried out in Caen in 1999–2000 (Murton *et al.* 2000).

chimera', based on a critical review of the literature and on findings from his monitoring studies in the Colorado Front Range. His demonstration was successful in so far as it resulted in the following statement that appears in the second edition of French's *Periglacial Environment*: 'Modern use of the term nivation is not recommended' (French 1996, p. 159). Hall (1991, 1995, 1997, 1999) severely criticized the cold-region 'panacea' (alias freeze–thaw weathering), the circular reasoning linking the angularity of debris to cold-climate conditions, and he investigated the possibility of thermal weathering as an alternative process of rock breakdown in cold-climate regions. André (1999) questioned the ability of Quaternary cold conditions to modify significantly the geometry of slope systems in basement regions, and called for 'the smokescreen of the periglacial scenery' to be crossed in order to search for the *real* processes responsible for the formation of landforms and deposits found in cold regions.

Evidence for the non-universality of freeze–thaw mechanisms

In the 1980s–1990s a growing body of data was published, which supported the iconoclastic idea of downplaying the part played by frost action in cold-region geomorphology. First, it was suggested that an unrealistic approach of rock moisture conditions in simulations had led to the exaggeration of frost weathering effects, and that the lack of moisture limited the efficiency of freeze–thaw cycles at many sites (e.g. McGreevy & Whalley 1985; Hall 1986; Coutard & Francou 1989). Second, rock properties were increasingly emphasized as major controls on free-face debris-fall activity (e.g. Godard 1983; André 1991; Douglas *et al.* 1991), whereas climatic control was minimized. It was pointed out that most outcrops in polar regions are made of massive and non-porous crystalline rocks – granite, syenite, gneiss, amphibolite, etc. – that do not show any particular susceptibility to frost shattering (André 1995, 1996). After 10 000 years, such deglaciated outcrops still exhibit nicely polished and striated surfaces (Fig. 5). Based on the inefficiency of freeze–thaw weathering on basement rocks of Scandinavia over the Holocene period, André refuted Boyé's idea of a possible renewal of debris stocks under periglacial conditions occurring between glacial phases.

Chemical and biogenic weathering re-evaluated

Whereas the efficiency of freeze–thaw was increasingly questioned, non-cold-related processes were

Fig. 5. Preservation of glacial polish on Precambrian syenite outcrops over the Holocene period, Björnfjell–Riksgränsen area, Lapland, 68°N. Photograph © M. F. André.

shown to operate in periglacial environments. For instance, chemical and biological weathering processes, usually invoked for warmer environments, proved to be ubiquitous processes that actively contributed to rock breakdown in periglacial environments.

As early as the 1960s–1970s, influential soil scientists had criticized the widespread presentation of Arctic soils as 'lithosols' and 'cryosols', and drawn attention to the diversity of chemical processes operating in high latitudes (see reviews in Tedrow 1977; Ugolini 1986). It appeared quite clear that brunification, oxidation, decarbonization, podzolization and salinization were not restricted to temperate regions, but are currently found operating in periglacial environments. Weathering rinds, rock flaking and salt efflorescences (Fig. 6a, b), clay minerals and grus material, solution figures (both macroscopic and microscopic), rock coatings and solute loads are among the expressions of this chemical activity operating in cold regions. A number of field studies demonstrated that chemical processes contributed very efficiently to rock breakdown in various periglacial contexts, from Alaska to the Himalayas (e.g. Dixon *et al.* 1984; Whalley *et al.* 1984).

Of particular interest is the multidisciplinary weathering experiment and associated microclimate recordings, and soil and water analyses (Fig. 6c, d), that started in the mid-1990s in Kärkevagge, Swedish Lapland (Dixon *et al.* 2001; Thorn *et al.* 2006; Darmody *et al.* 2007). This 10-year study of 'potential' weathering of 103 dolomite, granite and limestone discs buried beneath various vegetation cover types resulted in highly interesting findings concerning: (1) the inter-rock type comparisons of weathering rates (Table 2); (2) the impact of vegetation cover and microenvironmental conditions (drainage and soil pH) on weathering rates; (3) the influence of climatic fluctuations over the observation period; and (4) the prominent role of pyrite-rich rocks and secondary sulphate minerals in creating, in Kärkevagge, anomalously aggressive conditions for rock weathering. The 10-year Kärkevagge experiment is of special interest not only because it confirms the significant influence of chemical processes on landscape evolution in cold-climate regions, but also because it provides insights into the nature of control categories across various spatial scales (Dixon *et al.* 2008).

Although biologists had drawn early attention to the influence of micro-organisms on rock weathering from the Russian high mountains to the Antarctic Dry Valleys (e.g. Glazovskaya 1952; Polynov 1953; Friedmann 1977, 1982), most geomorphologists used to consider biological weathering as a 'trifling' process in cold environments (Etienne 2002). This attitude changed in the 1990s. Hall & Otte (1990) demonstrated that the alternating drying and wetting of chasmolithic algae mucilage was responsible for granite flaking on the nunataks of the Juneau Icefield in Alaska. In central Spitsbergen, André (1991) suggested that amphibolite flaking was induced by epilithic lichens; and, in northern Scandinavia, granular disintegration of glacially scoured outcrops and morainic boulders in various lithologies (granite, quartzite, amphibolite, gabbro) was found to be partly caused by saxicolous lichen communities (André 1995, 1996; McCarroll & Viles 1995). In particular, crustose and foliose lichens like *Lecidea auriculata* and *Parmelia centrifuga* were found to penetrate the rock surface, incorporate it within their thallus and probably expel flakes of rocks (mainly ferro-magnesian minerals such as biotite).

Of particular interest is the in-depth study carried out by Etienne on the influence exerted by fungal communities on basalt weathering in south Iceland since the Little Ice Age (Etienne 2002, 2004; Etienne & Dupont 2002). Weathering rind formation was shown by Etienne to be a consequence of biological activity, and comparisons between *in vitro* experiments and *in vivo* observations proved useful in assessing the role of fungi in mineral etching, flaking and precipitation (Fig. 7a, b). Etienne's conceptual model for the formation and erosion of weathering rinds (Fig. 7c) was extended to hot desert environments and longer timescales by Gordon & Dorn (2005).

From periglacial geomorphology to cold-region geomorphology (1990s–2000s)

Downplaying the importance of frost action led many geomorphologists to promote cold-region geomorphology instead of (or besides) periglacial geomorphology, although this trend is more effective amongst polar, rather than Alpine, geomorphologists. Cold-region geomorphology can be seen as an areal component of geomorphology, which includes periglacial geomorphology and 'embraces a mix of glacial, periglacial, and azonal processes' (French & Thorn 2006) (see Fig. 8). However, cold-region geomorphology does not only pay attention to current processes, but also to inherited weathering products and landforms by looking as far back in the geological past as the Tertiary period. After and besides the 'monitoring fever', it is time to come back to the big picture, although bridging the gap between process and historical geomorphology remains a challenging, if not impossible, task.

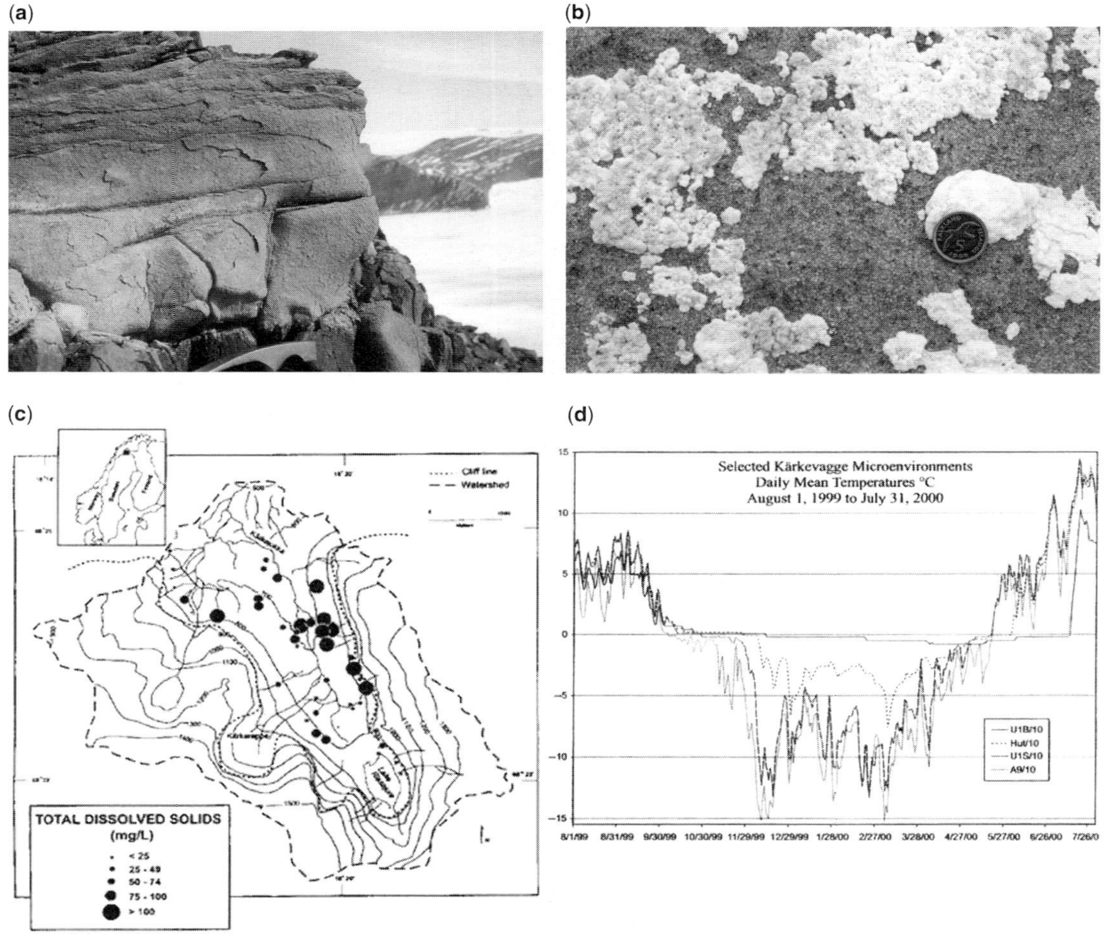

Fig. 6. Evidence for chemical weathering activity in polar regions. (**a**) and (**b**) Sandstone flaking and aragonite efflorescences in nunataks of Alexander Island, Antarctic Peninsula. Photographs © M. F. André. (**c**) and (**d**) The Kärkevagge pilot weathering experiment, Swedish Lapland (1994–2004). (c) Distribution of total dissolved solids showing the influence of pyrite-rich rocks on the east side of the valley (Campbell *et al.* 2001, fig. 6, p. 175). (d) Example of daily ground temperature recordings at four sites (Thorn *et al.* 2002, fig. 4, p. 295).

Table 2. *Decadal record of* in situ *experimental rock weathering in Kärkevagge, Swedish Lapland, 1994–2004. Results are expressed in percentage disk mass loss by rock type (Thorn* et al. *2006, table II, p. 995)*

Lithology	Period 1 (1994–1999)				Period 2 (1999–2004)			
	Minimum	Mean	Maximum	*n*	Minimum	Mean	Maximum	*n*
Dolomite	0.06	2.36	10.12	48	0.03	3.90	15.97	48
Granite	0.07	0.16	0.51	37	0.01	0.08	0.25	36
Limestone	0.06	5.52	14.27	19	0.10	7.23	20.06	16

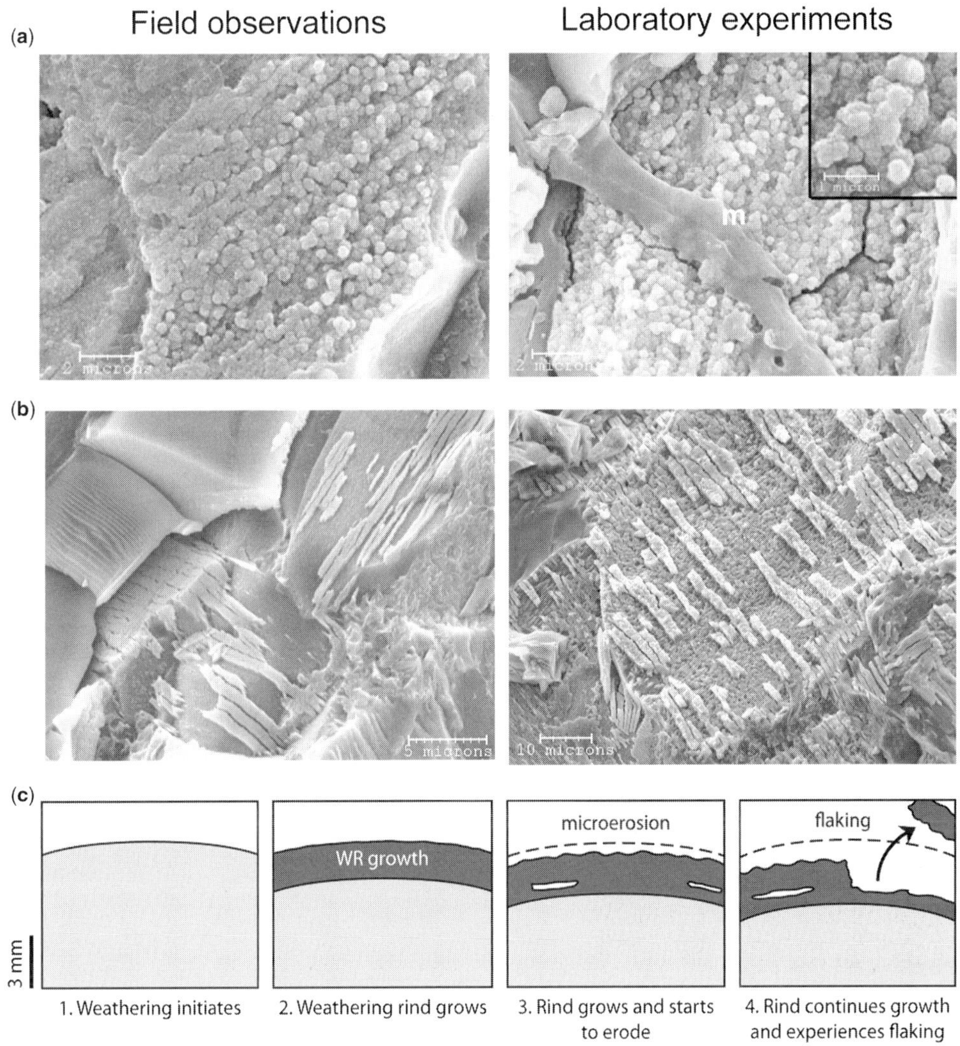

Fig. 7. Basalt bioweathering in Iceland (Etienne 2002; Etienne & Dupont 2002). (**a**) Iron-rich aggregates observed in natural weathering rinds and produced during experiments using *Apiospora montagnei* (Etienne & Dupont 2002, figs 9 and 10, p. 744). (**b**) Plagioclase flaking observed in natural weathering rindsand obtained in experimental weathering using *Aspergillus niger* (Etienne & Dupont 2002, figs 8 and 11, pp. 743 and 745). (**c**) Etienne's conceptual model for the formation and erosion of weathering rinds (Gordon & Dorn 2005, fig. 1, p. 99, based on Etienne 2002, fig. 9, p. 83).

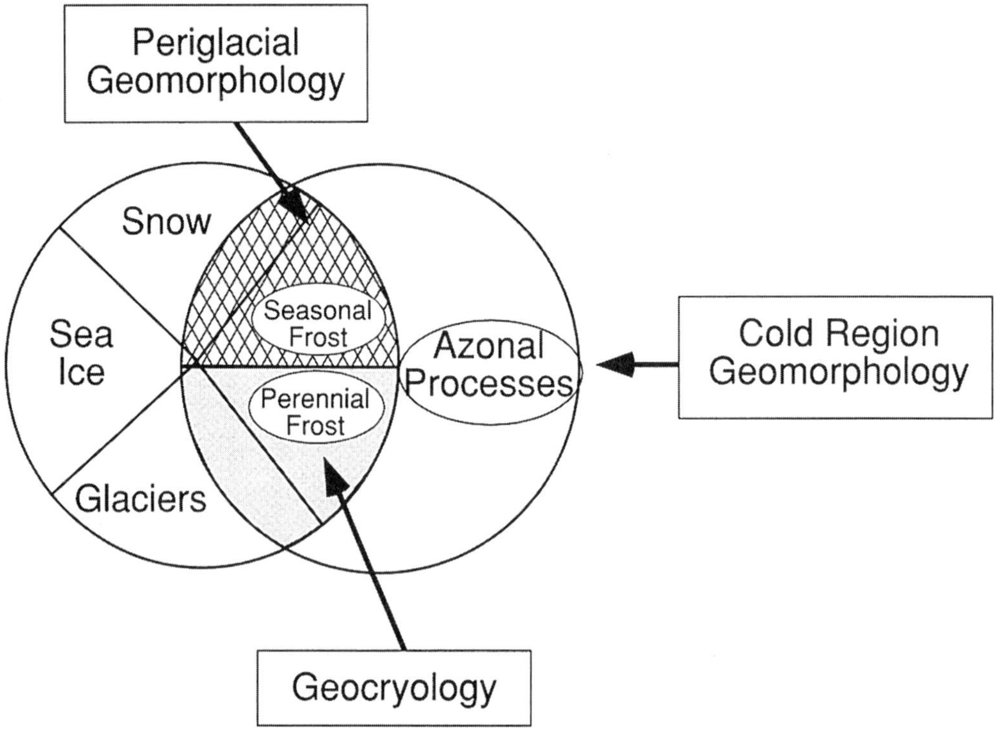

Fig. 8. Periglacial geomorphology and its overlap with certain cryospheric sciences (French & Thorn 2006, fig. 1, p. 170).

The present: the variability in the hierarchy of weathering processes in cold-climate regions

Regarding ongoing weathering, the aim of cold-region geomorphology is to stop focusing on a presupposed key process (freeze–thaw), to search for a variety of combinations and suites of processes (see Hall *et al.* 2002), which depend primarily on rock control and on (macro–micro–nano) environmental conditions. In this perspective, the dominant weathering process – if any – should be different from one site to another.

Such an attempt was made by Etienne & André (2003), based on their field experience in various North Atlantic periglacial environments (Iceland, Spitsbergen, Lapland and Labrador). From this survey, it appears that rock control greatly influences weathering signatures. Densely jointed and porous rocks are frost-sensitive and produce the so-called 'periglacial' landscapes (rhyolite in central Iceland, slates in Labrador, limestones on Spitsbergen). In contrast, massive and non-porous rocks are refractory to frost-driven processes, thus providing opportunities for alternative processes to operate. Biological weathering, for instance, dominates on the massive granitic roches moutonnées of Swedish Lapland and the young basaltic plains of south Iceland (André 1995, Etienne 2004). In coastal areas, marine salts can also blur the zonal input, as shown by the similar honeycomb patterns developed on basaltic lavas flows of Iceland peninsulas and on the gneissic outcrops along Labrador fjords (Etienne & André 2003). Such characteristic landscapes must not overshadow the fact that local morphogenic agents can modify the weathering landscape. For instance, katabatic winds in Iceland inhibit biogenic rind production on exposed rock surfaces, whereas this process remains active in nearby sheltered areas.

These different examples 'urge for caution when considering the hierarchy of weathering processes in a zonal perspective' (Etienne & André 2003). As stressed by Twidale & Lageat (1994), 'the climatic factor in landform development is by no means as clear cut and simple as was once thought', and both rock control and microenvironmental conditions occur to totally neutralize the potential effects of regional climate (see Pope *et al.* 1995 concerning the microenvironments). In this perspective, working at various scales is a must. In the Antarctic Dry Valleys, for instance, the regional aridity accounts for the importance of thermal weathering and wind erosion. However, at a local scale, salt distribution explains the importance of alveolar weathering. Lastly, at a nanoscale, life can be responsible for rock flaking where cryptoendoliths meet suitable conditions for colonization (i.e. translucent and porous rocks). This example pleads also for the need for an in-depth study of rock control, by adding to the conventional rock attributes (joint density, porosity, mineralogical/chemical composition, etc.) the optical properties that operate either synergistically or antagonistically with other rock properties (André *et al.* 2004).

The past: back to the big picture

The contemporary decline of historical geomorphology, which accompanied the tremendous development of process geomorphology, led most of the 'periglacial' community to show a propensity for explaining the observed landscapes by Present and/or post-glacial cold conditions. Fortunately, in the 1990s, *time* made its comeback on the geomorphological stage through the reassessment of the concepts of landform 'persistence' and 'lifetime' (Brunsden 1993), and the renewed vision of landforms as palimpsests, with particular reference to glacial landscapes (Kleman 1992). In this context, field and laboratory investigations carried out in the 1990s–2000s in Scandinavia, Canada and the Falklands resulted in a reassessment of the age of blockfields, which to date, had been mainly interpreted as the fresh products of Holocene frost shattering.

In Sweden, Kleman & Borgström (1990) found blockfields that had been preserved for several glacial phases below cold-based ice. In northern Norway, Rea *et al.* (1996) and Whalley *et al.* (1997) provided evidence for the pre-Quaternary origin of blockfields, mainly based on the clay content and the thickness of the blockfield soils. They interpreted blockfields as residual deposits derived from a weathering mantle, left after most of the finer material had been eroded away. In north Québec–Labrador, Marquette *et al.* (2004) and Gray *et al.* (2005) obtained ^{10}Be and ^{26}Al cosmogenic dates up to 340 ka for blockfields of the Torngat and Kaumajet mountains. Moreover, in blockfield soils they found evidence of intense chemical weathering, such as etch-marks on quartz grains, oxide concentrations and secondary minerals, like gibbsite and kaolinite, inherited from warmer interglacial and, possibly, Tertiary climates.

In the Falkland Islands, where blockstreams (known as 'stone runs') were previously attributed to frost wedging and other periglacial processes, André *et al.* (2008) interpreted them as complex polygenetic landforms derived from the stripping and accumulation downslope of a regolith, possibly Tertiary in age. Access to the internal structure of blockstreams provided evidence for the existence of a threefold profile, with clear vertical size gradation presenting striking similarities with an inverted weathering profile. Based on micromorphological analyses – scanning electron microscopy (SEM), X-ray diffraction (XRD), thin sections and grain-size analyses – André *et al.* (2008) proposed a six-stage scenario of stone run formation (Fig. 9), with periglacial reworking of the stone run material occurring only at a final stage.

Cosmogenic dating is undoubtedly a promising tool for a reappraisal of the age of cold-region landforms and deposits, previously interpreted as Late Quaternary. Interestingly, recent datings indicate that the 'lifetime' of small-scale geomorphic features such as roches moutonnées can exceed 2 Ma (Matsuoka *et al.* 2006). It challenges the common viewpoint on the youth of minor landforms, and illustrates the prominent role of rock control and moisture conditions in landform evolution, in polar areas as well as elsewhere.

Periglacial geomorphology and the 'Global Change fever' (2000s–)

It is both ironic and comforting to realize that in parallel to the comeback of historical and structural geomorphology, a new 'fever' arose from the contemporary environmental concerns, the 'Global Change fever', which contributed to somewhat restoring the legitimity of climatic geomorphology. As the Arctic is known as a climate-change hotspot, there is no surprise that periglacial geomorphology benefited from this trend. Indeed, the global warming concern provided an appropriate framework for an unprecedented international effort among the multidisciplinary community involved in permafrost science, including the so-called 'periglacial geomorphologists'. Another expression of the influence of this concern is the burst of 'paraglacial geomorphology'.

Periglacial features as geoindicators of global warming

Frozen ground was included by the International Union of Geological Sciences (IUGS) in its list of 'geoindicators' to be used to detect and assess environmental changes over relatively short periods (Berger & Iams 1996). In this context, landscape changes due to permafrost thaw are of special interest and have been extensively studied since the early 1990s (see the review in André & Anisimov 2009). Of the multiple landscape changes induced

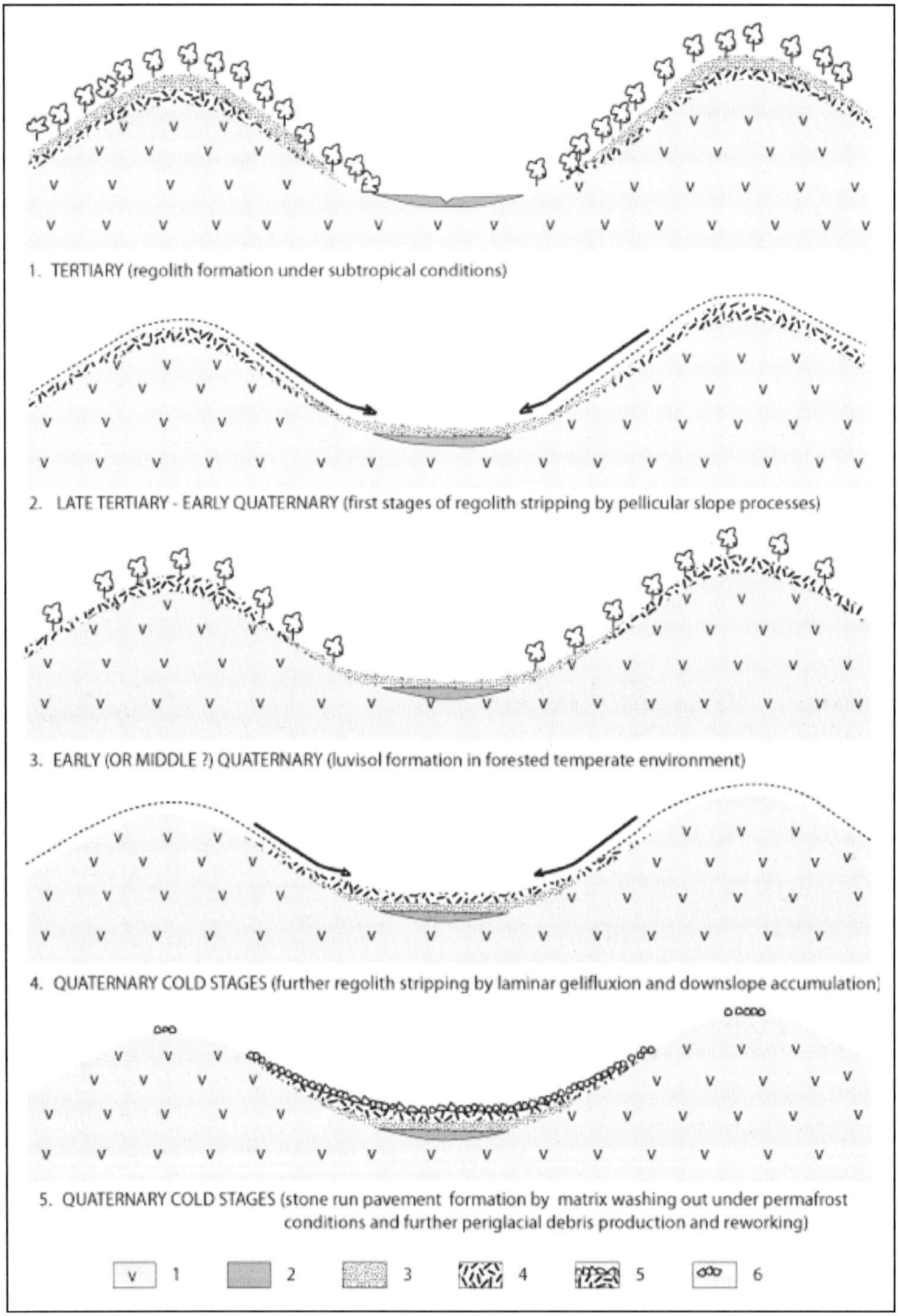

Fig. 9. Tentative scenario of stone run formation in the Falkland Islands: back to the Tertiary? (André *et al.* 2008, fig. 14, p. 537). 1, bedrock (mainly quartzite); 2, alluvial deposits, 3, grus (initially top of regolith); 4, block-rich material with fine-grained matrix (initially bottom of regolith); 5, valley-bottom blockstream with stone run pavement derived from matrix washing-out of the upper part of unit 4; 6, hillslope blockfield (André *et al.* 2008, fig. 14, p. 357).

by permafrost degradation, two are of particular interest in so far as they cover extensive areas and can be traced through remote sensing. The first one is the drastic change from networks of ice-wedge polygons into groups of hills called 'baydzherakh', separated by gullies following the collapsing ice wedges, which are actively forming in the Canadian Arctic owing to permafrost degradation (Fortier *et al.* 2007). The second landscape change, from frost mounds to thaw ponds, is widespread in more southern subarctic regions such as the lowlands of Canada (Fig. 10) and Scandinavia.

In northern Sweden and Finnish Lapland, warm and humid summers combined with increased snowfall favoured rapid decay of palsa complexes, with almost complete collapse of individual palsas occurring within 5–10 years (Zuidhoff 2002; Luoto & Seppälä 2003). In the southernmost palsa mire of Sweden, palsa extension decreased by about 50% between 1960 and 1997 (Zuidhoff & Kolstrup 2000). In northern Québec, the key driver of palsa decay over the last 50 years has been the reduction of frost penetration caused by increased snow precipitation, and since the mid-1990s accelerated thawing has been facilitated by the additional temperature rise (Payette *et al.* 2004). On the whole, the recent increase in thermokarst development from palsas/lithalsas indicates the high sensitivity of these frost mounds to changes in temperature and precipitation, and predictive models of palsa distribution in subarctic Fennoscandia are currently being developed (Fronzck *et al.* 2006).

Undoubtedly, 'periglacial geomorphologists' are surfing on the Global Change wave. Such an attitude is stimulating, although risky, because it is tempting to attribute to climate warming landscape changes that have different causes.

Global Change issues and the permafrost scientific community

Permafrost science involves a multidisciplinary and international network of scientists, including the so-called 'periglacial geomorphologists' who have been increasingly involved in international research programmes and networks on landscape responses to climate change (IPA WG, IPA/IASC-ACD, ESF-PACE, ESF-SEDIFLUX, etc.). To obtain a more comprehensive picture of the spatial and

Fig. 10. Lithalsa collapse as geoindicator of climate warming in Québec. Photograph © F. Calmels.

temporal changes in permafrost, the Global Terrestrial Network for Permafrost (GTN-P) was developed in the 1990s under the auspices of the International Permafrost Association (Burgess *et al.* 2000). The Circumpolar Active Layer Monitoring (CALM) Program (Brown *et al.* 2000) and the European permafrost observatory developed within the Permafrost and Climate in Europe (PACE) Project (Harris *et al.* 2001) mainly focused their activity on the ground thermal regimes, i.e. the active layer depths and the permafrost temperatures. Active layer monitoring (Fig. 11a, b) is a promising tool to detect, characterize and quantify ongoing changes, and is of interest to basic and applied research, for certain changes may have a direct impact on northern ecosystems, communities and/or infrastructure.

Of special interest to the geomorphologists is the recent attempt of a new IPA working group on 'Periglacial Landforms, Processes and Climate' (co-chairs: Ole Humlum and Norikazu Matsuoka) to construct a monitoring network highlighting geomorphic processes associated with the ground thermal regimes. Such a global monitoring campaign requires standardization of monitoring parameters and techniques, and priority should be given to compact, cold-resistant and maintenance-free instruments to expand the network to remote periglacial sites (Matsuoka 2006). A model experimental site is under construction in Svalbard that

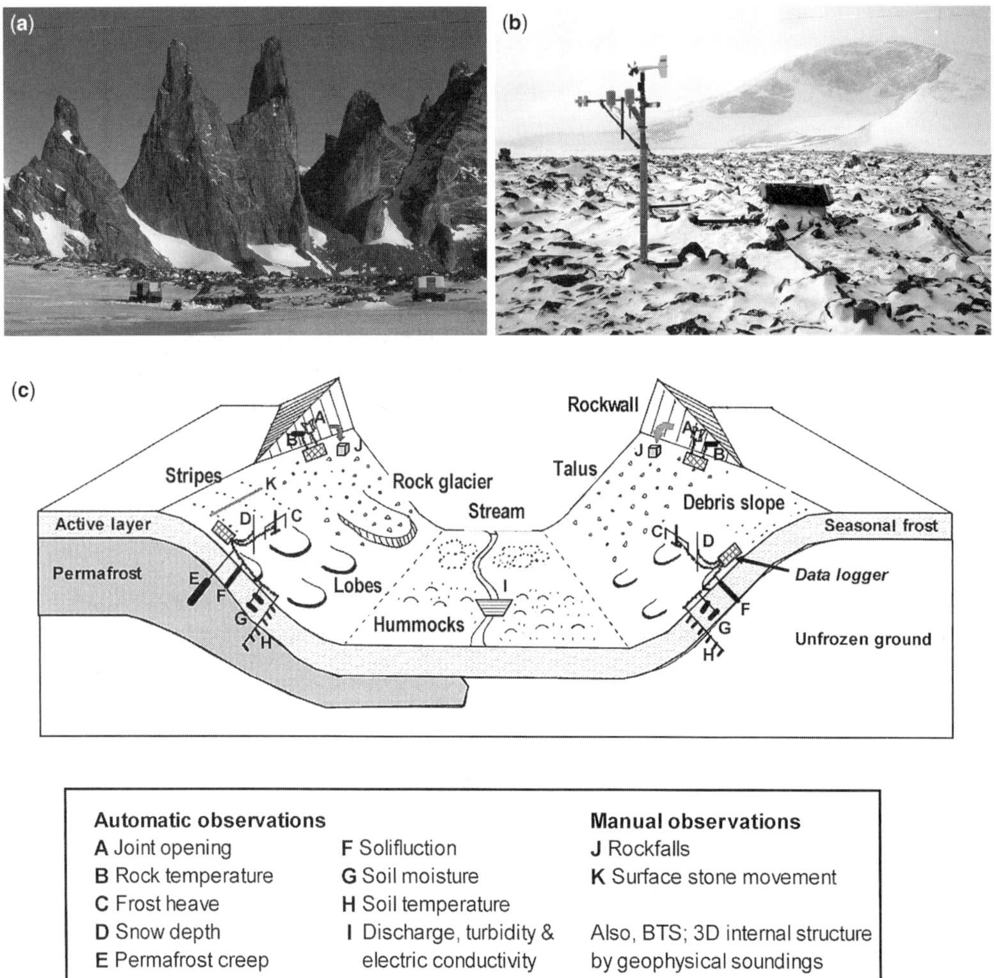

Fig. 11. Monitoring periglacial processes in a Global Change context: towards standardized designs? Monitoring sites: (**a**) in Antarctica (photograph © N. Matsuoka) and (**b**) in Scandinavia (photograph © B. H. Juvvasshoe). (**c**) Scheme of processes and landforms in a periglacial catchment and parameters for monitoring (Matsuoka 2006, fig. 2, p. 22).

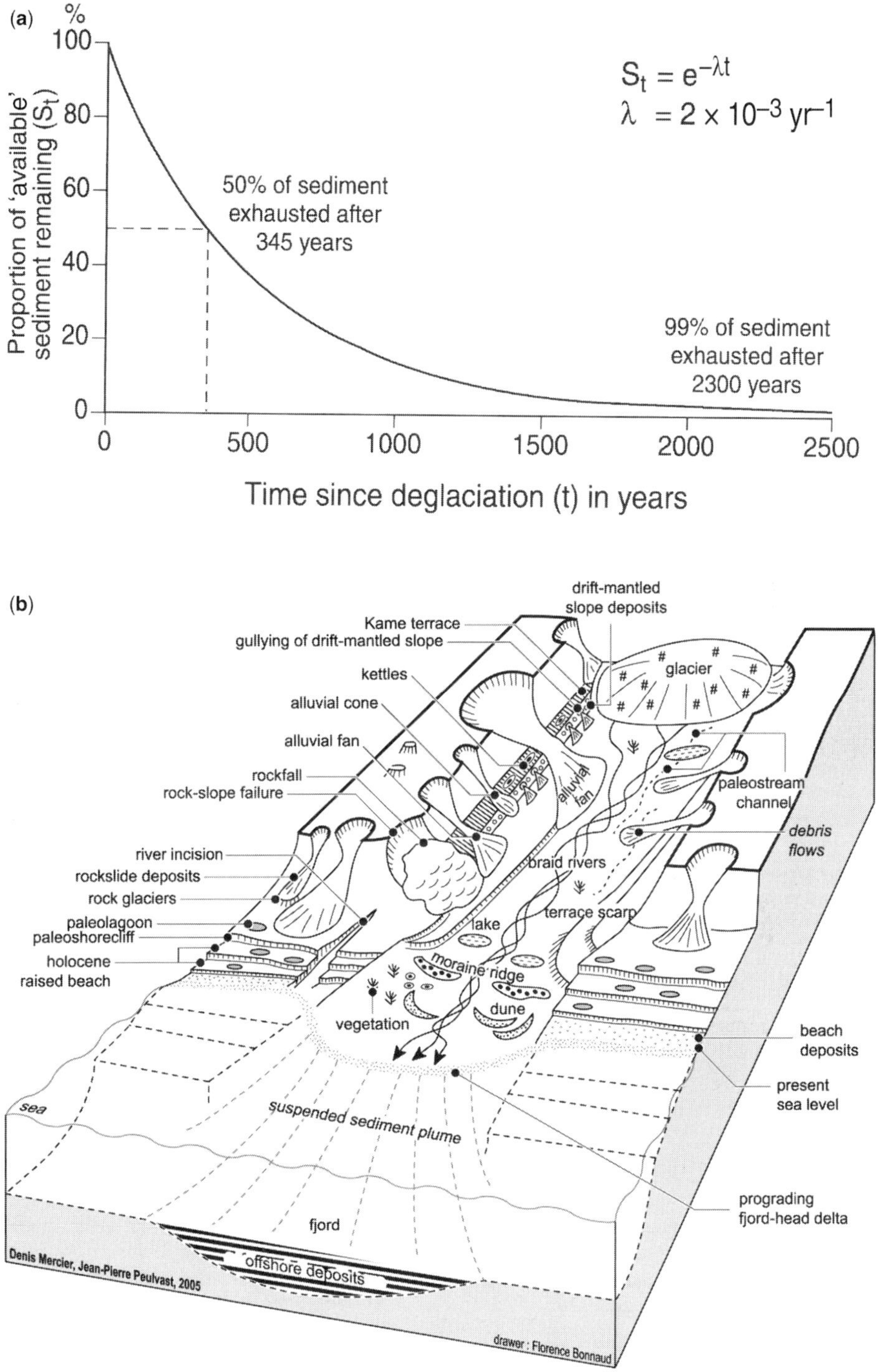

Fig. 12. The paraglacial landsystem: a new panacea? (**a**) Exhaustion model of paraglacial sediment release (Ballantyne 2002*b*, fig. 2, p. 372); (**b**) paraglacial landforms and processes (Mercier 2007, fig. 3, p. 347).

meets the 'standards' of a favourable monitoring site (Fig. 11c), as it exhibits a variety of processes coexisting within a small catchment.

The 'paraglacial fever'

Part of the success achieved among the international community by 'paraglacial geomorphology' is probably due to the 'Global Change fever'. Following the Little Ice Age, the twentieth century has known two thermal peaks in the late 1930s and the 1990s, well marked in the Scandinavian Arctic where active runoff generated high erosion rates owing to the abundance of loose debris and meltwater (Ballantyne & Benn 1994; Curry 1999; Mercier 2001). Rapid shore progradation over the last 30 years is one more signature of this 'paraglacial' context, as shown on Spitsbergen by Mercier & Laffly (2005) based on field and remote sensing data.

First considered as a particular episode in the evolution of previously glaciated regions, the 'paraglacial' developed into a unifying concept when Ballantyne (2002*a*, *b*) proposed his exhaustion model of paraglacial sediment release (Fig. 12a). The expressions of 'paraglacial landsystems' and 'paraglacial landform assemblages' (i.e. Iturrizaga 2008; Mercier & Etienne 2008) are now in common use, and, despite the fact that Ballantyne (2003, p. 432) wrote that 'the paraglacial concept cannot be defined by process', the expression of 'paraglacial process' is referred to frequently (e.g. Mercier 2007) (see Fig. 12b). The hegemonic trend of 'paraglacial geomorphology', which leaves less and less space to periglacial and even glacial geomorphology, is subject to hot debates between those who surf on the 'paraglacial wave' and those who are tempted to reject the whole concept. Coming back to the notion of 'paraglacial context' or 'paraglacial crisis' seems preferable to part of the community of geomorphologists who outline the fugacity of the 'paraglacial dynamics' in a number of cases. For instance, on Spitsbergen, the activity of gullies created by meltwater from ice-cored lateral moraines lasts no more than 20 years (Mercier 1997). The question of timescales is crucial, although complex, for genuine paraglacial processes such as rock slope failures due to glacial debuttressing can operate over millennia, i.e. contemporaneously with a wide range of non-paraglacial geomorphic processes.

Conclusion

This overview of the past half-century indicates that periglacial v. cold-region geomorphology might be considered as a 'fashion science' (see the insightful analysis by Etienne (2004, pp. 57–69) of the application to geomorphology of Sperber's fashion model by Sherman 1996). Undoubtedly, the periglacial, paraglacial and Global Change 'fevers' have influenced the development of cold-region geomorphology. If resulting in somewhat biased visions of the complex reality of polar and Alpine environments, they proved to be fruitful from a methodological point of view and/or as an intellectual stimulus for the whole community of cold-region geomorphologists. Moreover, the successive 'fevers' and their consecutive reactions had cumulative effects. They generated within the community a series of research trends that are more and more intricate. Matsuoka's approach to periglacial environments is emblematic of this promising trend in so far as it combines the development of the most up-to-date monitoring designs and the use of cosmogenic dating to improve our knowledge of the long-term landscape evolution.

The author expresses her warmest thanks to the members of the international community of geomorphologists, geocryologists and soil scientists with whom fruitful exchanges have been had over the last 30 years. She is grateful to E. Roussel, J.-P. Magnier and A. Decaulne for their graphical assistance, and to the editors and referees for their useful comments upon an earlier draft of this article.

References

André, M. F. 1991. *Dynamique actuelle et évolution holocène des versants du Spitsberg (Kongsfjord–Wijdefjord, 79°Nord).* Thèse de Doctorat d'Etat, Université de Paris I – Panthéon-Sorbonne.

André, M. F. 1995. Postglacial microweathering of granite *roches moutonnées* in northern Scandinavia (Riksgränsen area, 68°N). *In*: Slaymaker, O. (ed.) *Steeplands Geomorphology*. Wiley, Chichester, 103–127.

André, M. F. 1996. Rock weathering rates in arctic and subarctic environments (Abisko Mts, Swedish Lapland). *Zeitschrift für Geomorphologie*, **40**, 499–517.

André, M. F. 1999. La livrée périglaciaire des paysages polaires: l'arbre qui cache la forêt? *Géomorphologie*, **3**, 231–252.

André, M. F. & Anisimov, O. 2009. Tundra and permafrost dominated taiga. *In*: Slaymaker, O., Spencer, T. & Embleton-Hamann, C. (eds) *Geomorphology and Global Environmental Change*. Cambridge University Press, **13**, 344–367.

André, M. F., Hall, K., Bertran, P. & Arocena, J. 2008. Stone runs in the Falkland Islands: periglacial or tropical? *Geomorphology*, **95**, 524–543.

André, M. F., Hall, K. & Comte, V. 2004. Optical rock properties and weathering processes in polar environments (with special reference to Antarctica). *Polar Geography*, **28**, 43–62.

BALLANTYNE, C. K. 2002*a*. A general model of paraglacial landscape response. *The Holocene*, **12**, 371–376.

BALLANTYNE, C. K. 2002*b*. Paraglacial geomorphology. *Quaternary Science Reviews*, **21**, 1935–2017.

BALLANTYNE, C. K. 2003. Paraglacial landsystems. *In*: EVANS, D. J. A. (ed.) *Glacial Landsystems*. Arnold, London, 432–461.

BALLANTYNE, C. K. & BENN, D. I. 1994. Paraglacial slope adjustment and resedimentation following recent glacier retreat, Fåbergstölsdalen, Norway. *Arctic and Alpine Research*, **26**, 255–269.

BARSCH, D. 1993. Periglacial geomorphology in the 21st century. *In*: VITEK, J. D. & GIARDINO, J. R. (eds) *Geomorphology: The Research Frontier and Beyond*. Elsevier, Amsterdam, 141–163.

BERGER, A. R. & IAMS, W. J. (eds). 1996. *Geoindicators: Assessing Rapid Environmental Changes in the Earth System*. Balkema, Rotterdam.

BOYÉ, M. 1949. Importance du défonçage périglaciaire dans l'élaboration des formes glaciaires. *Comptes Rendus à l'Académie des Sciences de Paris*, **229**, 723–724.

BOYÉ, M. 1968. Défense et illustration de l'hypothèse du 'défonçage périglaciaire'. *Biuletyn Peryglacjalny*, **17**, 5–56.

BROWN, J., HINKEL, K. M. & NELSON, F. E. 2000. The Circumpolar Active Layer Monitoring (CALM) program: research designs and initial results. *Polar Geography*, **24**, 165–258.

BRUNSDEN, D. 1993. The persistance of landforms. *Zeitschrift für Geomorphologie*, Suppl.-Bd, **93**, 13–28.

BÜDEL, J. 1944. *Die morphologische Wirkungen des Eiszeitklimas im gletscherfreien Gebiet. Geologische Rundschau*, **34**, 482–519.

BURGESS, M. M., SMITH, S. L., BROWN, J., ROMANOVSKY, V. & HINKEL, K. 2000. The Global Terrestrial Network for Permafrost (GTN-P): permafrost monitoring contributing to global climate observations. *Geological Survey of Canada Current Research*, **2000-E14**. Available online at http://www.nrcan.gc.ca/gsc/bookstore.

CAILLEUX, A. & TAYLOR, G. 1954. *Cryopédologie, étude des sols gelés*. Expéditions Polaires Françaises, **IV**, Hermann, Paris.

CAMPBELL, S. W., DIXON, J. C., DARMODY, R. G. & THORN, C. E. 2001. Spatial variation of early season surface water chemistry in Kärkevagge, Swedish Lapland. *Geografiska Annaler*, **83A**, 169–178.

COUTARD, J. P. & FRANCOU, B. 1989. Rock temperature measurements in two alpine environments: implications for frost shattering. *Arctic and Alpine Research*, **21**, 399–416.

COUTARD, J. P., VAN VLIET-LANOË, B. & AUZET, A. V. 1988. Frost heaving and frost creep on an experimental slope. Results for soil structures and sorted stripes. *Zeitschrift für Geomorphologie*, Suppl.-Bd, **71**, 13–23.

CURRY, A. M. 1999. Paraglacial modification of slope form. *Earth Surface Processes and Landforms*, **24**, 1213–1228.

CZEPPE, Z. 1964. Exfoliation in a periglacial climate. *Geographia Polonica*, **2**, 5–10.

DAHL, R. 1966. Block fields, weathering pits and tor-like forms in the Narvik Mountains, Nordland, Norway. *Geografiska Annaler*, **48A**, 55–85.

DARMODY, R. G., THORN, C. E. & DIXON, J. C. 2007. Pyrite-enhanced chemical weathering in Kärkevagge, Swedish Lapland. *Geological Society of America Bulletin*, **119**, 1477–1485.

DIXON, J. C., THORN, C. E. & DARMODY, R. G. 1984. Chemical weathering processes on the Vantage peak nunatak, Juneau icefield, southern Alaska. *Physical Geography*, **5**, 111–131.

DIXON, J. C., THORN, C. E. & DARMODY, R. G. 2008. Spatial scale and chemical weathering in Kärkevagge: Influences on landscape evolution. *Zeitschrift für Geomorphologie*, **52**, Suppl. 1, 27–49.

DIXON, J. C., THORN, C. E., DARMODY, R. G. & SCHLYTER, P. 2001. Weathering rates of fine pebbles at the soil surface in Kärkevagge, Swedish Lapland. *Catena*, **45**, 273–286.

DOUGLAS, G. R., WHALLEY, W. B. & MCGREEVY, J. P. 1991. Rock properties as controls on free-face debris fall activity. *Permafrost and Periglacial Processes*, **2**, 311–319.

DYLIKOWA, A. 1956. The ice-wedges at Stawecin. *Biuletyn Peryglacjalny*, **3**, 129–133.

EMBLETON, C. & KING, C. A. M. 1975. *Periglacial Geomorphology*. Edward Arnold, London.

ETIENNE, S. 2002. The role of biological weathering in periglacial areas: a study of weathering rinds in South Iceland. *Geomorphology*, **47**, 75–86.

ETIENNE, S. 2004. *Islande – Biogéomorphologie d'un milieu périglaciaire humide*. Thèse, Université de Paris I – Panthéon-Sorbonne (version condensée).

ETIENNE, S. & ANDRÉ, M. F. 2003. La variabilité de la hiérarchie des processus de météorisation à travers les bilans météoriques de divers milieux périglaciaires nord-atlantiques (Islande, Labrador, Laponie, Spitsberg). *Géomorphologie*, **3**, 177–190.

ETIENNE, S. & DUPONT, J. 2002. Fungal weathering of basaltic rocks in a cold oceanic environment (Iceland): comparison between experimental and field observations. *Earth Surface Processes and Landforms*, **27**, 737–748.

FORTIER, D., ALLARD, M. & SHUR, Y. 2007. Observation of rapid drainage system development by thermal erosion of ice wedge on Bylot Island, Canadian Arctic Archipelago. *Permafrost and Periglacial Processes*, **18**, 229–243.

FRENCH, H. M. 1976. *The Periglacial Environment*. Longman, London.

FRENCH, H. M. 1996. *The Periglacial Environment*, 2nd edn. Longman, Harlow.

FRENCH, H. M. 2003. The development of periglacial geomorphology: 1 – up to 1965. *Permafrost and Periglacial Processes*, **14**, 29–60.

FRENCH, H. M. & THORN, C. E. 2006. The changing nature of periglacial geomorphology. *Géomorphologie*, **3**, 165–174.

FRIEDMANN, E. I. 1977. Microorganisms in Antarctic desert rocks from dry valleys and Dufek Massif. *Antarctic Journal of the United States*, **12**, 26–30.

FRIEDMANN, E. I. 1982. Endolithic microorganisms in the Antarctic cold desert. *Science*, **215**, 1045–1053.

FRONZEK, S., LUOTO, M. & CARTER, T. R. 2006. Potential effect of climate change on the distribution of palsa mires in subarctic Fennoscandia. *Climate Research*, **32**, 1–12.

GLAZOVSKAYA, M. A. 1952. Biological factors of weathering in high mountain regions. *Priroda*, **12**, 106–110 (in Russian).

GODARD, A. 1983. Le rôle de la structure dans les éboulis des hautes latitudes. *In*: SOUTADÉ, G. & BARTHÉLÉMY, L. (eds) *Eboulis et environnement géographique passé et actuel*. Actes du Colloque A. G. F. du 8 janvier 1983, Université de Paris X. Association les Amis du Centre de Géographie Physique Henri Elhaï, Nanterre.

GORDON, S. J. & DORN, R. I. 2005. *In situ* weathering rind erosion. *Geomorphology*, **67**, 97–113.

GRAY, J., MARQUETTE, G. C., GOSSE, J., STAIGER, J. & HENAULT-TESSIER, M. 2005. The extent, origins and age of felsenmeer surfaces of the Torngat, Kaumajet and Kiglapait Mountains of Northern Québec-Labrador, Eastern Canada. *In*: *Sixth International Conference on Geomorphology, Abstracts Volume, Zaragoza*, **8**, Planetearth.

GRAY, J. T. 1972. Debris accretion on talus slopes in the central Yukon Territory. *In*: SLAYMAKER, O. & MCPHERSON, H. J. (eds) *Mountain Geomorphology: Geomorphological Processes in the Canadian Cordillera*. Tantalus Press, Vancouver, **14**, 75–84.

HALL, K. 1986. Rock moisture content in the field and the laboratory and its relationship to mechanical weathering studies. *Earth Surface Processes and Landforms*, **11**, 131–142.

HALL, K. 1991. The allocation of the freeze–thaw weathering mechanism in geocryological studies: a critical comment. *South African Geographical Journal*, **73**, 10–13.

HALL, K. 1995. Freeze–thaw weathering: the cold region 'panacea'. *Polar Geography and Geology*, **19**, 79–87.

HALL, K. 1997. Rock temperatures and implications for cold region weathering: I. New data from Viking Valley, Alexander Island, Antarctica. *Permafrost and Periglacial Processes*, **8**, 69–90.

HALL, K. 1999. The role of thermal stress fatigue in the breakdown of rock in cold regions. *Geomorphology*, **31**, 47–63.

HALL, K. & OTTE, W. 1990. A note on biological weathering on nunataks of the Juneau icefield, Alaska. *Permafrost and Periglacial Processes*, **1**, 189–196.

HALL, K., THORN, C. E., MATSUOKA, N. & PRICK, A. 2002. Weathering in cold regions: some thoughts and perspectives. *Progress in Physical Geography*, **26**, 577–603.

HALLET, B., ANDERSON, S. P., STUBBS, C. W. & GREGORY, E. C. 1988. Surface soil displacements in sorted circles, Western Spitsbergen. *In*: SENNESET, K. (ed.) *Permafrost, Proceedings of the Fifth International Conference on Permafrost, 2–5 August*, Volume 1. Tapir, Trondheim, 770–775.

HARRIS, C., HAEBERLI, W., VONDER MÜHLL, D. & KING, L. 2001. Permafrost monitoring in the high mountains of Europe: the PACE Project in its global context. *Permafrost and Periglacial Processes*, **12**, 3–11.

HORTON, R. E. 1945. Erosional development of streams and their drainage basins: hydrophysical approach to quantitative morphology. *Geological Society of America Bulletin*, **56**, 275–370.

ITURRIZAGA, L. 2008. Paraglacial landform assemblages in the Hindukush and Karakoram Mountains. *Geomorphology*, **95**, 27–47.

KLEMAN, J. 1992. The palimpsest glacial landscape in northwestern Sweden. *Geografiska Annaler*, **74A**, 305–325.

KLEMAN, J. & BORGSTRÖM, I. 1990. The boulder fields of Mt. Fulufjället, West-Central Sweden. *Geografiska Annaler*, **72A**, 63–78.

LACHENBRUCH, A. 1962. *Mechanics of Thermal-contraction Cracks and Ice-wedge Polygons in Permafrost*. Geological Society of America, Special Paper, **70**.

LAUTRIDOU, J. P. & OZOUF, J. C. 1982. Experimental frost shattering: 15 years of research at the Centre de Géomorphologie du CNRS. *Progress in Physical Geography*, **6**, 215–232.

LEOPOLD, L. B., WOLMAN, M. G. & MILLER, J. P. 1964. *Fluvial Processes in Geomorphology*. W. H. Freeman, San Francisco, CA.

LOZINSKI, W. VON. 1909. Über die mechanische Verwitterung der Sandsteine im gemässigten Klima. *Académie des Sciences de Cracovie, Classe des Sciences Mathématiques et Naturelles*, **1**, 1–25.

LUCKMAN, B. H. 2000. Classics in physical geography revisited – Rapp, A. 1960: Recent development of mountain slopes in Kärkevagge and surroundings, northern Scandinavia. Geografiska Annaler XLII, 71–200. *Progress in Physical Geography*, **24**(1), 97–101.

LUOTO, M. & SEPPÄLÄ, M. 2003. Thermokarst ponds as indicators of the former distribution of palsas in Finnish Lapland. *Permafrost and Periglacial Processes*, **14**, 19–27.

MACKAY, J. R. 1973. The growth of pingos, Western Arctic Coast, Canada. *Canadian Journal of Earth Sciences*, **10**, 979–1004.

MACKAY, J. R. 1974. Ice-wedge cracks, Garry Island, Northwest Territories. *Canadian Journal of Earth Sciences*, **11**, 1366–1383.

MACKAY, J. R. 1977. Pulsating pingos, Tuktoyaktuk Peninsula, N.W.T. *Canadian Journal of Earth Sciences*, **14**, 209–222.

MACKAY, J. R. 1979. Pingos of the Tuktoyaktuk Peninsula area, Northwest Territories. *Géographie physique et Quaternaire*, **33**, 3–61.

MACKAY, J. R. 1982. Active layer growth, Illisarvik experimentaldrained lake site, Richards Island, Northwest Territories. *Geological Survey of Canada, Paper*, **82-1A**, 123–126.

MACKAY, J. R. 1986. The first 7 years (1978–1985) of ice-wedge growth, Illisarvik experimental drained lake site, western Arctic coast. *Canadian Journal of Earth Sciences*, **23**, 1782–1795.

MACKAY, J. R. 1992. The frequency of ice-wedge cracking (1967–1987) at Garry Island, western Arctic coast. *Canadian Journal of Earth Sciences*, **29**, 236–248.

MALAURIE, J. 1966. Problèmes de géographie dans l'Arctique américain et groenlandais. *Bulletin de l'Association de Géographes Français*, **340–341**, 53–68.

MALAURIE, J. 1968. *Thèmes de recherche géomorphologique dans le nord-ouest du Groenland*. Thèse de Doctorat d'Etat, Mémoires et Documents du CNRS, Paris.

MARQUETTE, G. C., GRAY, J. T., GOSSE, J. C., COURCHESNE, F., STOCKLI, L., MACPHERSON, G. & FINKEL, R. 2004. Felsenmeer persistence under non-erosive ice in the Torngat and Kaumajet mountains, Quebec and Labrador, as determined by soil weathering and cosmogenic nuclide exposure. *Canadian Journal of Earth Sciences*, **41**, 19–38.

MATSUOKA, N. 2006. Monitoring periglacial processes: towards construction of a global network. *Geomorphology*, **80**, 20–31.

MATSUOKA, N., THOMACHOT, C. E., OGUCHI, C. T., HATTA, T., ABE, M. & MATSUZAKI, H. 2006. Quaternary bedrock erosion and landscape evolution in the Sør Rondane Mountains, East Antarctica: reevaluating rates and processes. *Geomorphology*, **81**, 408–420.

MCCARROLL, D. & VILES, H. 1995. Rock-weathering by the lichen *Lecidea auriculata* in an arctic alpine environment. *Earth Surface Processes and Landforms*, **20**, 199–206.

MCGREEVY, J. P. 1981. Some perspectives on frost shattering. *Progress in Physical Geography*, **5**, 56–75.

MCGREEVY, J. P. & WHALLEY, W. B. 1985. Rock moisture content and frost weathering under natural and experimental conditions: a comparative discussion. *Arctic and Alpine Research*, **17**, 337–346.

MECKELEIN, W. 1974. Aride Verwitterung in Polargebieten im Vergleich zum subtropischen Wüstengürtel. *Zeitschrift für Geomorphologie*, Suppl.-Bd, **20**, 178–188.

MERCIER, D. 1997. L'impact du ruissellement sur les moraines latérales du glacier du Roi (Colletthögda, Spitsberg, 79°N). *Norois*, **44**, 549–566.

MERCIER, D. 2001. *Le ruissellement au Spitsberg – le monde polaire face aux changements climatiques*. Thèse, Université de Clermont-Ferrand (version condensée). Presses Universitaires Blaise Pascal, Clermont-Ferrand.

MERCIER, D. 2007. Le paraglaciaire, évolution d'un concept. *In*: ANDRÉ, M.-F., ETIENNE, S., LAGEAT, Y., LE CŒUR, C. & MERCIER, D. (eds) *Du continent au bassin versant – théories et pratiques en géographie physique (Hommage au Professeur Alain Godard)*. Presses Universitaires Blaise Pascal, Clermont-Ferrand, Collection 'Nature & Sociétés', **4**, 341–353.

MERCIER, D. & ETIENNE, S. (eds). 2008. Paraglacial geomorphology: processes and paraglacial context. *Geomorphology*, **95**, 1–102.

MERCIER, D. & LAFFLY, D. 2005. Actual paraglacial progradation of the coastal zone in the Kongsfjorden area, western Spitsbergen (Svalbard). *In*: HARRIS, C. & MURTON, J. B. (eds) *Cryospheric Systems: Glaciers and Permafrost*. Geological Society, London, Special Publications, **242**, 111–117.

MURTON, J. B., COUTARD, J. P. *ET AL*. 2000. Experimental design for a pilot study on bedrock weathering near the permafrost table. *Earth Surface Processes and Landforms*, **25**, 1281–1294.

OZOUF, J. C. 1983. *Comparaison de gélifracts naturels de grèzes litées charentaises et de gélifracts fabriqués*. Thèse, Université de Caen.

PAYETTE, S., DELWAIDE, A., CACCIANIGA, M. & BEAUCHEMIN, M. 2004. Accelerated thawing of subarctic peatland permafrost over the last 50 years. *Geophysical Research Letters*, **31**, 1–4.

PELTIER, L. C. 1950. The geographical cycle in periglacial regions as it is related to climatic geomorphology. *Annals of the Association of American Geographers*, **40**, 214–236.

PISSART, A. 1956. L'origine périglaciaire des viviers des Hautes Fagnes. *Annales de la Société Géologique de Belgique*, **LXXIX**, 119–131.

PISSART, A. 1987. *Géomorphologie périglaciaire*. Laboratoire de Géomorphologie et de Géologie du Quaternaire de l'Université de Liège.

POLYNOV, B. B. 1953. The geological role of organisms. *Voprosy Geografii*, **33**, 45–64 (in Russian).

POPE, G. A., DORN, R. I. & DIXON, J. C. 1995. A new conceptual model for understanding geographical variations in weathering. *Annals of the Association of American Geographers*, **85**, 38–64.

POSER, H. 1948. Boden und Klimaverhältnisse im Mittel- und WestEuropa während der Würmeiszeit. *Erdkunde*, **2**, 53–68.

RAPP, A. 1960*a*. Recent development of mountain slopes in Kärkevagge and surroundings, northern Scandinavia. *Geografiska Annaler*, **XLII**, 71–200.

RAPP, A. 1960*b*. *Talus Slopes and Mountain Walls at Tempelfjorden, Spitsbergen*. Meddelanden från Uppsala Universitets Geografiska Institution, ser. A, **155**.

REA, B. R., WHALLEY, W. B., RAINEY, M. M. & GORDON, J. E. 1996. Blockfields, old or new? Evidence and implications from some plateaus in northern Norway. *Geomorphology*, **15**, 109–121.

SHERMAN, D. J. 1996. Fashion in geomorphology. *In*: RHOADS, B. L. & THORN, C. E. (eds) *The Scientific Nature of Geomorphology*, Wiley, Chichester, 87–114.

TEDROW, J. C. F. 1977. *Soils of the Polar Landscapes*. Rutgers University Press, New Brunswick, NJ.

THORN, C. E. 1988. Nivation: a geomorphic chimera. *In*: CLARK, M. J. (ed.) *Advances in Periglacial Geomorphology*. Wiley, Chichester, 3–31.

THORN, C. E. 1992. Periglacial geomorphology: what, where, when? *In*: DIXON, J. C. & ABRAHAMS, A. D. (eds) *Periglacial Geomorphology*. Wiley, Chichester, 1–30.

THORN, C. E. 2004. Whither, or wither, periglacial studies? *Polar Geography*, **28**, 4–12.

THORN, C. E., DARMODY, R. G., ALLEN, C. E. & DIXON, J. C. 2002. Near-surface ground temperature regime variability in selected microenvironments, Kärkevagge, Swedish Lapland. *Geografiska Annaler*, **84A**, 289–300.

THORN, C. E., DIXON, J. C., DARMODY, R. G. & ALLEN, C. E. 2006. Ten years (1994–2004) of 'potential' weathering in Kärkevagge, Swedish Lapland. *Earth Surface Processes and Landforms*, **31**, 992–1002.

TRICART, J. 1950. *Le modelé périglaciaire*. Cours de Géomorphologie, **I**(1), CDU, Paris.

TRICART, J. 1963. *Géomorphologie des régions froides*. PUF, Paris.

TRICART, J. 1967. *Le modelé des régions périglaciaires*. Traité de Géomorphologie (J. TRICART & A. CAILLEUX), Volume 2. SEDES, Paris.

TRICART, J. 1970. *Geomorphology of Cold Environments* (trans. E. WATSON). Macmillan, St Martin's Press, New York.

TWIDALE, C. R. & LAGEAT, Y. 1994. Climatic geomorphology: a critique. *Progress in Physical Geography*, **18**, 319–334.

TROLL, C. 1944. Strukturböden, Solifluktion und Frostklimate der Erde. *Geologische Rundschau*, **34**, 545–694.

UGOLINI, F. C. 1986. Pedogenic zonation in the well-drained soils of the Arctic regions. *Quaternary Research*, **26**, 100–120.

VAN VLIET-LANOË, B. 1976. Traces de ségrégation de glace en lentilles associées aux sols et phénomènes périglaciaires fossiles. *Biuletyn Peryglacjalny*, **26**, 41–54.

VAN VLIET-LANOË, B. 1985. Frost effects in soils. *In*: BOARDMAN, J. (ed.) *Soils and Quaternary Landscape Evolution*. Wiley, Chichester, 117–158.

VAN VLIET-LANOË, B. 1988. *Le rôle de la glace de ségrégation dans les formations superficielles de l'Europe de l'ouest – processus et héritages*. Thèse de doctorat d'Etat, Université de Paris I (2 vols). Editec, Caen.

WASHBURN, A. L. 1967. *Instrumental Observations of Mass-wasting in the Mesters Vig District, Northeast Greenland*. Meddelelser om Grønland, **166**.

WASHBURN, A. L. 1973. *Periglacial Processes and Environments*. Edward Arnold, London.

WASHBURN, A. L. 1979. *Geocryology: A Survey of Periglacial Processes and Environments*. Edward Arnold, London.

WHALLEY, W. B., REA, B. R., RAINEY, M. M. & MCALISTER, J. J. 1997. Rock weathering in blockfields: Some preliminary data from mountain plateaus in North Norway. *In*: WIDDOWSON, M. (ed.) *Palaeosurfaces: Recognition, Reconstruction and Interpretation*. Geological Society, London, Special Publications, **129**, 133–145.

WHALLEY, W. B. & MCGREEVY, J. P. 1983. Weathering. *Progress in Physical Geography*, **7**, 559–586.

WHALLEY, W. B. & MCGREEVY, J. P. 1985. Weathering. *Progress in Physical Geography*, **9**, 559–581.

WHALLEY, W. B., MCGREEVY, J. P. & FERGUSON, R. I. 1984. Rock temperature observations and chemical weathering in the Hunza region, Karakoram; preliminary data. *In*: MILLER, K. J. (ed.) *Proceedings of the International Karakoram Project*, Volume 2, Cambridge University Press, Cambridge, 616–633.

ZUIDHOFF, F. S. 2002. Recent decay of a single palsa in relation to weather conditions between 1996 and 2000 in Laivadalen, northern Sweden. *Geografiska Annaler*, **84A**, 103–111.

ZUIDHOFF, F. S. & KOLSTRUP, E. 2000. Changes in palsa distribution in relation to climate change in Laivadalen, northern Sweden, especially 1960–1997. *Permafrost and Periglacial Processes*, **11**, 55–69.

Holocene microweathering rates and processes on ice-eroded bedrock, Røldal area, Hardangervidda, southern Norway

DAWN T. NICHOLSON

Department of Environmental and Geographical Sciences, Manchester Metropolitan University, Chester Street, Manchester M1 5GD, UK (e-mail: d.nicholson@mmu.ac.uk)

Abstract: Post-glacial weathering of ice-eroded metamorphic bedrock was investigated in the Røldal area (60°N) of the Hardangervidda Plateau in southern Norway. Quartz veins were used as reference surfaces to determine a mean post-glacial surface lowering rate of 0.55 mm ka^{-1}. Chemical characteristics of late-season runoff were determined for one catchment (Snøskar) and a chemical erosion rate of 4.9 t km^{-2} a^{-1} was obtained. A mean *in situ* fracture enlargement due to microweathering of 0.12 mm ka^{-1} was also determined. These rates are low, although comparable with similar environments in cold regions, and suggest that microweathering has had relatively little impact on Holocene landscape evolution. Weathering rind thickness was found to be less on fracture walls than on exposed bedrock surfaces, suggesting fractures have not played a significant role in microweathering. Observations of weathering morphology reveal a range of forms including shallow spalling, tafoni and pseudokarren, indicating locally intense weathering activity. Analysis of interrelationships between multiple weathering indices points to the importance of bedrock microweathering as a precursor to macro-breakdown and landform evolution. The research reasserts the importance of chemical activity in cold environments and the importance of moisture supply for effective microweathering.

Weathering of ice-eroded outcrops has the potential to significantly influence landscape evolution in periglacial environments during the post-glacial period. However, there has been relatively little work to determine rates of post-glacial bedrock weathering in cold environments. Traditionally, weathering in cold environments was seen to be the work of mechanical processes, specifically freeze–thaw (Tricart 1969), and low temperatures were thought to inhibit chemical weathering activity (e.g. Peltier 1950). However, in recent years there has been a major change of emphasis in studies of weathering in cold environments away from traditional concepts to greater emphasis on the role of chemical (Anderson *et al.* 1997; Darmody *et al.* 2000; Thorn *et al.* 2001; Campbell *et al.* 2002*a*, *b*; Dixon & Thorn 2005; Owen *et al.* 2006) and biological processes (André 1995, 2002; Etienne 2002; Etienne & Dupont 2002). In addition, there is an increasing range of studies looking at the role of lithological and structural controls on cold-environment weathering (Glasser *et al.* 1998; Olvmo & Johansson 2002; Whalley *et al.* 2004; Nicholson 2008). These changes are reflected in the call from Hall & André (2001) and Hall *et al.* (2002) to reconsider our assumptions about the nature of weathering in cold environments.

There have been relatively few attempts to determine rates of post-glacial weathering in cold environments, with greater emphasis being placed on temperate and tropical regions. In the seminal study by Rapp (1960), transport rates determined from mass movements were compared against those for solute fluxes in Kärkevagge, and he concluded that solution was of greater importance than mass movement. This pointed, implicitly, to the significance of chemical weathering in periglacial environments. More recent work in Kärkevagge (e.g. Darmody *et al.* 2000) and the neighbouring valley of Latnjavagge (Beylich *et al.* 2004) has determined weathering rates from the analysis of solutes in runoff together with studies of pedogenesis and weathering rinds and coatings. Following on from Dahl's work in 1967, André (2002) has also determined weathering rates in the Abisko-Riksgränsen area of northern Sweden from measurements of surface lowering with reference to upstanding quartz veins. Owen *et al.* (2006) also used this technique to determine rates of weathering on calcitic rocks along a lake shoreline. However, it has been difficult to determine precisely what the impact of weathering has been on landscape evolution.

It is helpful when there is some temporal constraint on landscape processes to enable an assessment of the impact of weathering and, to this end, studies of post-glacial bedrock weathering are particularly useful. Notable examples include the attempts by André (1995, 1996*a*, *b*) and Dahl (1967) to investigate post-glacial bedrock microweathering in Swedish Lapland. Sumner *et al.* (2002) and Shakesby *et al.* (2006) have also used

From: KNIGHT, J. & HARRISON, S. (eds) *Periglacial and Paraglacial Processes and Environments*.
The Geological Society, London, Special Publications, **320**, 29–49.
DOI: 10.1144/SP320.3 0305-8719/09/$15.00

indices of post-glacial weathering to provide relative ageing of glaciated bedrock surfaces in southern Norway and the sub-Antarctic.

The overall aim of the research presented here is to examine microweathering rates and processes on ice-smoothed bedrock forms in an active periglacial environment and to examine their influence on landscape denudation. A further aim is to contribute to the growing body of research on the weathering of crystalline rocks in cold environments. The objectives are threefold: (i) to synthesize weathering rates determined for the post-glacial period from measurement of surface lowering, fracture enlargement and solute fluxes, and to compare these rates with those obtained for other, comparable locations; (ii) to infer weathering processes from measurement of surface weakening and weathering rind development, and to make comparisons between processes at the surface and those occurring on fracture walls; and (iii) to describe microweathering morphology and make further inferences about weathering processes.

Study area

Geomorphic environment

The study area is located in the Røldal region at 6°58′E and 59°53′N in the SW corner of Hardangervidda (the largest high mountain plateau in northern Europe) in southern Norway (Fig. 1). The plateau generally has subdued relief, but the SW corner is more deeply dissected, with deep valleys and steep-sided ridges. Generally, the plateau has tundra

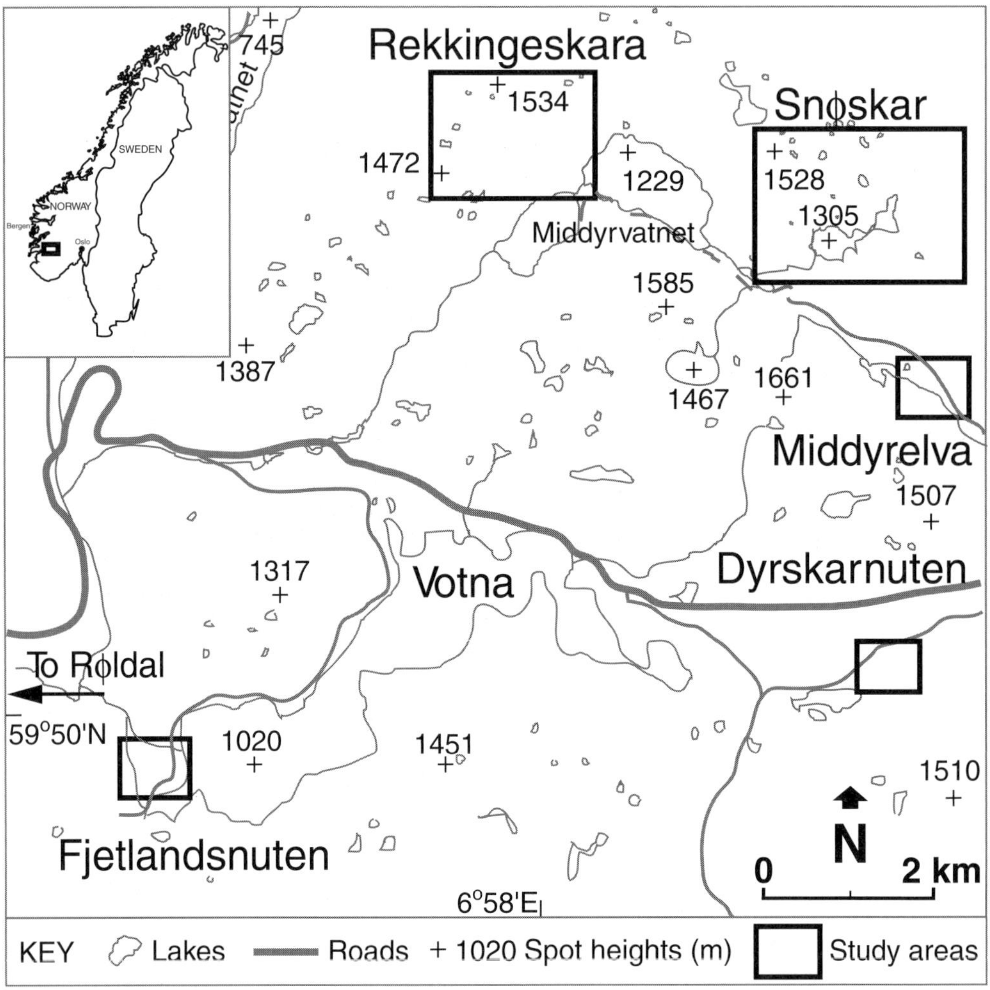

Fig. 1. Map of general area and five study locations.

vegetation, but the study area is dominated by exposed, ice-scoured bedrock. The study area comprises five sites totalling 4.5 km^2, which are characterized by roches moutonnées, whalebacks, scattered erratics and landforms of periglacial ground activity (Fig. 2). Many surfaces display glacial erosional features such as striations, grooves and P-forms. The area is believed to have been deglaciated approximately 10 000 years ago (Anderson 1980).

The five study sites are located at Rekkingeskara, Snøskar, Middyrelva, Dyrskarnuten and Fjetlandsnuten (Fig. 1). Altitude ranges from 950 to 1520 m, and all of the sites lie above the upper limit of boreal forest typified by high alpine tundra with very modest vegetation and many bare bedrock surfaces. Bedrock is frequently colonized by crustose lichen including *Rhizocarpon* species, and mosses and bryophytes are common. There is some ericaceous heath with occasional coarse tussock grasses, *Vaccinium* species and *Empetrum*. Soil development is generally incipient and immature, lacking organic matter, and being dominated by silt and accumulations of grus, plus limited patches of glacially derived sediment. Mean annual temperature for the area is $-2\,^{\circ}C$ and annual precipitation is about 1525 mm. Permanent and semi-permanent snowpatches are common and the Nupsfonn plateau glacier (*c*. 3 km^2), only 5–10 km to the east and at an altitude of 1600 m, is a shrunken remnant of a more extensive Little Ice Age glaciation that did not reach any of the study area.

The bedrock lithology consists entirely of metamorphic rocks, either Precambrian basement from the Stavsnuten and Dyrskarnuten Nappe complexes of Proterozoic age or, at Middyrelva, from the Mannevatn Nappe of Ordovician age. Bedrock at Rekkingeskara is mainly granitic gneiss and schist, with limited outcrops of amphibolite. The greatest variation in bedrock occurs at Snøskar, with hornblende amphibolite and amphibolitic schist, pockets of feldspathic quartzite and quartzo-feldspathic gneiss, and various forms of mica schist (including muscovite, chlorite and biotite–epidote varieties). Bedrock at Middyrelva generally consists of chlorite-rich mica schist with numerous quartz veins, while at Dyrskarnuten the bedrock is chlorite-rich amphibolite schist. Bedrock at Fjetlandsnuten consists of actinolite–amphibolite gneiss, probably derived from metamorphosed basalt.

Fig. 2. Typical landscape of the Røldal study area.

The Rekkingeskara study site (1.5 km^2) is a broad valley located at an altitude of 1250–1450 m between the Valldalsvatnet and Middyrvatnet basins. The valley is dominated by blockstreams that appear to 'flow' around upstanding roches moutonnées, several of which are breaking up and contribute to the supply of blockstream material. Occasionally there are flatter areas with plentiful moisture supply that support solifluction lobes and patterned ground. The Snøskar study site (3 km^2) is an enclosed, circular basin with two central lakes and a single outlet to the SW (Fig. 3). The basin lies an altitude of between 1300 and 1520 m, and is located north of the Middyrelva valley. The basin displays classic knob and basin topography, with numerous roches moutonnées and whalebacks interspersed with lakes and pools. There are usually a number of late-lying snowpatches around the margins of the basin and there are numerous periglacial ground forms. The Middyrelva study site (1500 m^2) is located at an altitude of 1150 m in a glacially deepened valley (maximum relief *c.* 500 m), which crosses a drainage divide falling from 1240 m at Middyrvatnet to 980 m at Ulevåvatnet in the SE. The Dyrskarnuten study site (1500 m^2) is similar, comprising a series of small roches moutonnées rising from an altitude of 1080 to 1160 m on a broad col between Votna and Ulevåvatnet to the east. The valleys at Middyrelva and Dyrskarnuten are both moderately vegetated with ericaceous heath. Fingers of vegetated soil spread up the lower slopes of roches moutonnées, and there is also some soil development within weathering pits on the crests. The valleys both support

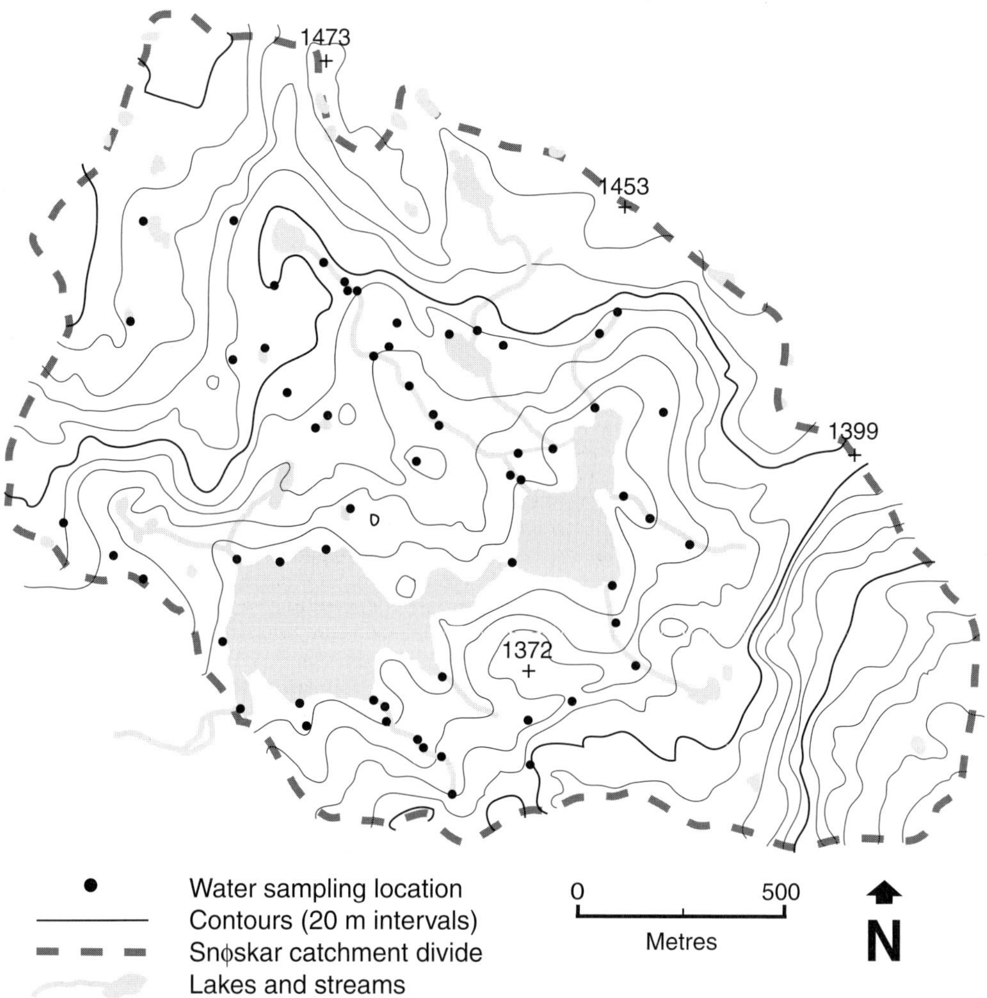

Fig. 3. Solute sampling locations at Snøskar (note some seepages and streams are too small to be shown at this scale).

numerous small tributaries coming off the steep valley sides. The Fjetlandsnuten study site (2000 m^2) is on a small hill rising from a height of from 950 to 1070 m at the west end of Lake Votna. The roches moutonnées studied are particularly exposed and permit views across the Hardangervidda plateau for at least 8 km to the NW. Ericaceous heath is well developed around the fringes of roches moutonnées, including low-growing *Salix arctica* (rock willow), *Rubus chamaemorus* (cloudberry) and *Vaccinium* species (blueberry).

Methods

With the exception of solute analysis described later, all of the other measurements were made on roches moutonnées at the five sites. At the two main sites, Snøskar and Rekkingeskara, the primary criteria for selection of roches moutonnées were bedrock lithology (it was desirable to include all of the major rock types), degree of exposure (particularly sheltered or exposed locations, which might favour weathering activity, were avoided) and spatial coverage. At the remaining three sites, roches moutonnées were selected where they were present. A total of 45 roches moutonnées were used in this study.

Weathering rates

Three methods were used to determine weathering rates. (a) Analysis of solutes in runoff to determine contemporary removal of material in solution from the Snøskar catchment. (b) The use of reference surfaces (Dahl 1967), together with chronological control to determine average weathering rates during the post-glacial period. (c) *In situ* fracture enlargement due to weathering processes.

Determination of weathering rates from water chemistry. A total of 73 late-season water samples were collected from 64 sites (Fig. 3) at Snøskar between 23 August and 5 September 2006. At each sampling location temperature, pH and conductivity (reference 25 °C) were determined. In addition, for 29 of the sampling locations, the metals Al, Ca, Fe, K, Mg, Mn, Na and Si were determined using inductively coupled plasma spectrometry, and anions Cl, NO_3, SO_4 and F were determined using atomic absorption spectrometry. Water samples were taken from a range of sources including snow, snowpatch runoff, lakes, small pools, minor trickles and streams. Total dissolved solids (TDS) were estimated using a conversion factor of 0.7 for conductivity μs to TDS ppm (based on Strömquist & Rehn 1981).

Surface denudation using quartz veins as reference surfaces. Following the method of Dahl (1967), used more recently by André (1995, 1996*a*, *b*, 2002), post-glacial surface lowering was determined by measuring the difference in height between upstanding quartz veins and surrounding bedrock. Measurements were made only on exposed ice-scoured roches moutonnées and whaleback surfaces. The method is based on the principle that being more resistant to surface microweathering, ice-polished quartz vein surfaces represent the pre-weathered, ice-scoured surface at the end of the last glacial maximum.

Measurements, to the nearest 0.1 mm, were usually made with a calliper, although a short section of metal plate was occasionally projected from the vein surface if the surrounding bedrock topography made the use of a calliper difficult. Up to three veins were selected for each roche moutonnée. For each vein, up to 50 measurements were made, always with an approximately equal number of measurements on each side of the vein, and at intervals of 10 mm. Where glacial erosion or subsequent breakdown of vein edges created an arched upper surface (in cross profile), care was taken to record the *maximum* height difference. Following the recommendations of Dahl (1967), quartz veins containing many joints were avoided, as were veins found in bedrock that contained many fractures in close proximity to the veins. Care was taken to ensure that the veins used exhibited a surface sheen, often yellowish in colour, representing glacial polish. The results are reported as a mean for each roche moutonnée. Over 1000 measurements were made at 23 sites, giving an average of 45 measurements at each site.

Fracture enlargement. A calliper was used to measure the width at the top of enlarged fractures to the nearest 0.1 mm, perpendicular to the fracture direction. Care was taken to avoid fractures that had opened as the result of physical displacement of the bedrock (e.g. by tectonic activity, gravitational stress or frost shattering), ensuring that measurements were only taken where enlargement was due to *in situ* microweathering of fracture walls. A total of 1244 measurements were made.

Weathering indices

Weathering rind thickness and rock surface hardness were determined in order to provide a relative estimate of weathering intensity over the post-glacial period.

Weathering rind thickness. The use of weathering rind thickness as a weathering index is based on the widespread assumption that rind development is dependent on the age, or period of exposure, of the surface studied (Thorn 1975), and rind thickness has been used in a variety of ways to study post-glacial bedrock weathering (e.g. Chinn 1981; Dixon *et al.* 2002*b*; Sumner *et al.* 2002). Weathering rind thickness was obtained for exposed bedrock surfaces and also for fracture walls. Surface weathering rind thickness was measured from 25 mm-diameter drilled cores sampled from smooth surfaces on the top of roches moutonnées. Fracture wall weathering rind thickness was measured from rock samples extracted from the intersections of closed joints using a geological hammer. In each case, weathering rind thickness was determined using a clear plastic ruler to the nearest 0.2 mm. A hand-held crack microscope was used to facilitate identification of the boundary between weathering rind and unweathered rock, based primarily on colour change. For drilled cores the maximum weathering rind thickness was obtained from four equally spaced measurements around the core circumference. For rock samples several measurements were made, as necessary, in order to identify the maximum weathering rind thickness. At each site between 15 and 20 measurements were taken, with a total sample number of 667 measurements (525 measurements of surface rind and 142 of fracture wall rind). The data presented represent the mean maximum weathering rind thickness for each roche moutonnée as recommended by Thorn (1975).

Rock surface weakening. Surface hardness was determined using a calibrated 'N'-type Schmidt hammer. This portable field instrument works by measuring the rebound distance of a controlled impact by a piston on a rock surface. A full analysis of the instrument and its widespread usage in geomorphology is given by Goudie (2006). Several researchers, notably Rae *et al.* (2004), Shakesby *et al.* (2006) and Sumner (2004), have used the instrument to investigate post-glacial weathering of bedrock. There has been a great deal of discussion about testing procedures (for example, see Poole & Farmer 1980; Goudie 2006) and a wide range of approaches are in use. A unique approach has been used here. If the primary purpose of using the hammer is to determine the intact strength of rock, then it is appropriate to prepare the surface well to ensure that a fresh face is presented to the hammer. The testing method of Hucka (1965) can then be adopted, which is to use multiple impacts at a single point. However, if the aim is to obtain an index that reflects the relative degree of surface weathering, then the surface, which should be lichen-free and free from loose material, should not be pre-prepared and a single impact at any point is sufficient.

Poole & Farmer (1980) conducted a statistical analysis of the consistency of repeated Schmidt hammer impacts at a series of points on four different rock types. Their results show (Poole & Farmer 1980, fig. 3, p. 170) that the first rebound value is consistently lower than subsequent values. This supports the contention here, that the first rebound value (R_1) can be used to represent the hardness of the weathered surface, and that the second rebound value (R_2) is a much closer approximation (albeit that there is some variability) of intact rock strength. Therefore, in this study, 25 *pairs* of readings were obtained for each roche moutonnée, with each pair of impacts being obtained at a single point (i.e. without moving the hammer). A comparison between the two values allows for some relative estimation of the degree of weakening at sites with contrasting lithological characteristics. More than 2000 measurements were made and all have been adjusted for angle with respect to a horizontal surface (Day & Goudie 1977). Care was taken to avoid sites close to fractures or edges, to select even surfaces and to use the instrument only in dry conditions.

Results

Weathering rates

Chemical erosion from solute runoff. Conductivity of water and snow samples from across the Snøskar catchment yielded an overall mean of 5.5 μs or 3.9 ppm for TDS (Table 1). The highest TDS values recorded were found in some snow samples, and smaller lakes and ponds. The lowest TDS levels recorded were found in streams and small tributaries. A series of conductivity measurements were obtained through a pit dug into snow (Fig. 4). Snow above the strong discontinuity represents net accumulation during the previous 12 months and has a mean TDS concentration lower than mean solute values across the catchment. The slightly elevated solute values at the surface represent the concentration of solutes (derived mainly from precipitation) from ablation of the overlying snow and from summer precipitation. Below the discontinuity the snow is quite old (at least several years) and the higher solute values reflect concentration of solutes arriving from the surface via percolation. While it is not possible to rule out contamination of the snow by runoff of solutes derived directly from bedrock dissolution, this is unlikely, except near the snow–bedrock interface. There are low levels of Cl^-, NO_3^- and SO_4^- in the Snøskar catchment, but F^- was absent. The dominant anion was Cl^- (mean 1.5 ppm), with concentrations being notably higher in surface snow. There are low levels of Al, Ca, Fe, K, Mg, Mn, Na and Si in the

Table 1. *Solute data for the Snøskar catchment*

Location	No.[1]	Cond[2]	pH	Temp[3]	Cl[4]	NO_3	SO_4	Al	Ca	Fe	K	Mg	Mn	Na	Si
West catchment (granitic and andesitic rocks)															
Lake		**7.9**[5]		**9**	**0.3**	**0.6**	**1.7**	**0.030**	**0.33**		**0.09**	**0.06**	**0.015**	**0.01**	**0.25**
	4(4)	3.2–17.1	4.4–5.0	8.1–9.4	0.2–0.4	0.3–0.8	0.5–5.2	0.002–0.113	0.01–1.21	bd	0.05–0.19	0.02–0.19	0.001–0.056	0.07–0.13	0.13–0.51
Pond	1(1)	**19.4**	4.8	**12.7**	**0.3**	**0.61**	**0.4**	**0.044**	bd[6]	**0.007**	**0.17**	**0.01**	**0.001**	**0.10**	**0.28**
Snow		**8.6**		**5.5**	**11.0**	**0.8**	**0.2**				**7.31**	**0.01**	**0.001**	**0.28**	**0.07**
	3(3)	2.8–14.0	4.3–4.3	2.4–8.6	0.4–31.8	0.3–1.3	0.0–0.4	**0.001**	**0.05**	**0.001**	0.18–21.28	0.01–0.02	0.000–0.001	0.10–0.63	0.01–0.18
Stream		**4.4**													
	2(1)	0.5–6.1	5.95	**9.1**	**0.2**	**0.3**	**1.2**	**0.005**	**0.34**	bd	**0.05**	**0.05**	**0.000**	**0.09**	**0.32**
Trickle		**4.2**		**7.3**	**0.2**	**0.3**	**0.5**	**0.010**			**0.01**	**0.02**	**0.001**	**0.05**	**0.21**
	4(4)	2.6–5.7	4.2–4.5	1.9–13.4	0.1–0.2	0.1–0.5	0.3–0.7	0.004–0.028	**0.04**	**0.001**	0.01–0.02	0.01–0.03	0.000–0.002	0.03–0.06	0.10–0.27
Central and south catchment (amphibolite and chlorite mica schist)															
Lake		**3.9**													
	4(1)	3.3–4.7	5.0	**10.7**	**0.4**	**3.2**	**0.8**	bd	**0.13**	bd	**0.20**	**0.031**	**0.001**	**0.08**	**0.21**
Pond		**4.2**													
	3(0)	3.7–4.4	–	–	–	–	–	–	–	–	–	–	–	–	–
Snow	1(1)	**16.7**	3.8	**16.1**	**3.2**	**0.9**	**0.1**	bd	bd	**0.001**	**2.14**	bd	**0.000**	**0.51**	**0.01**
Stream		**4.1**		**9.8**	**0.3**	**0.9**	**0.8**		**0.17**		**0.09**	**0.03**	**0.002**	**0.09**	**0.20**
	13(3)	2.2–6.2	4.1–4.6	8.6–11.3	0.2–0.4	0.5–1.3	0.6–1.0	bd	0.01–0.32	bd	0.04–0.17	0.03–0.05	0.001–0.003	0.07–0.12	0.06–0.26
Trickle		**5.1**		**9.6**		**0.6**	**0.7**				**0.08**	**0.05**		**0.08**	**0.25**
	2(2)	3.6–6.5	4.7–5.4	7.2–12.0	**0.2**	0.5–0.7	0.5–0.9	bd	**0.63**	bd	0.07–0.08	0.02–0.07	**0.001**	0.07–0.09	0.18–0.31
North and east catchment (quartzite and feldspathic schist)															
Lake		**4.4**		**8.6**	**0.2**	**0.8**	**0.8**				**0.06**	**0.03**		**0.07**	**0.24**
	9(2)	2.9–6.4	4.4–4.5	8.0–9.1	0.2–0.2	0.5–1.0	0.6–1.0	**0.003**	**0.29**	bd	0.05–0.07	0.02–0.04	**0.001**	0.07–0.08	0.21–0.26
Pond		**9.2**		**11.6**	**0.5**	**0.4**	**1.5**				**0.20**	**0.05**	**0.001**	**0.10**	**0.26**
	4(2)	5.3–12.9	4.3–5.8	8.0–15.2	0.5–0.6	0.2–0.6	0.8–2.2	**0.005**	**1.44**	bd	0.15–0.25	0.01–0.09	0.001–0.002	0.07–0.12	0.08–0.44
Stream		**4.3**		**9.9**											
	10(1)	3.3–7.5	4.0–4.4	8.8–11.0	**0.3**	**0.3**	**0.5**	bd	bd	bd	**0.09**	**0.01**	**0.001**	**0.06**	**0.20**
Trickle	1(1)	**4.5**	4.2	**7.1**	**0.2**	**0.4**	**0.6**	**0.002**	bd	bd	**0.01**	**0.02**	**0.001**	**0.07**	**0.21**
Seepage		**5.0**		**13.2**	**0.2**	**0.4**	**0.6**				**0.07**	**0.02**	**0.002**	**0.05**	**0.23**
	3(3)	2.7–6.4	4.0–4.5	11.6–14.6	0.2–0.3	0.1–0.6	0.4–0.9	**0.011**	**0.01**	**0.002**	0.03–0.14	0.01–0.03	0.001–0.004	0.04–0.06	0.22–0.24
MEAN	**64**	**5.51**	–	–	**1.48**	**0.64**	**0.78**	**0.014**	**0.332**	**0.002**	**0.934**	**0.032**	**0.003**	**0.111**	**0.208**

[1]Value indicates number of samples used to obtain conductivity, pH and temperature (value in parenthesis indicates number of samples used in elemental analysis); [2]conductivity in μS; [3]temperature in °C; [4]all elements in ppm; [5]mean given in bold, other values give range; [6]bd, measurement below instrument detection levels.

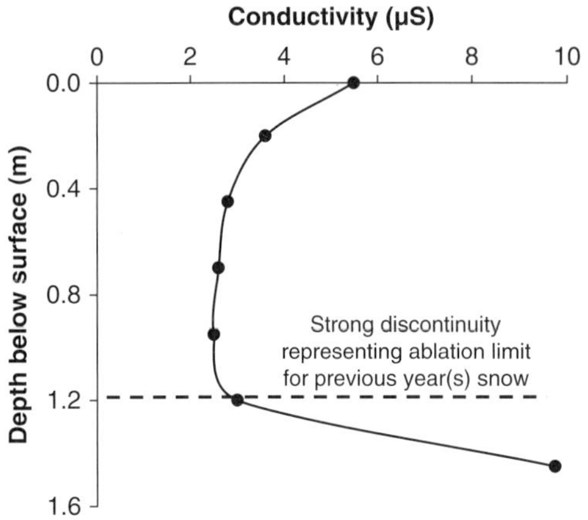

Fig. 4. Conductivity depth profile for one snowpatch (all snow above the dashed line is from last winter).

catchment. The dominant cation was K^+ (mean 0.9 ppm), also notably higher in snow. Ca (0.3 ppm) and Si (0.2 ppm) are also relatively important and appear to have slightly greater concentrations in streams and ponds. Observation of the spatial distribution of solute concentrations suggests that there are slightly elevated levels of nitrate in the central, lowest part of the basin and on a low-level plateau on the east side. These areas support greater vegetation cover than is found elsewhere in the catchment and this is, therefore, an indication that nitrate has its origin in the very limited sediments and immature soils with organic matter. Silica levels are also slightly lower in the centre of the basin where the rocks are much less dominated by quartz.

Overall, solute concentrations for the catchment are extremely low and do not appear to relate to spatial variations in bedrock lithology. This would be difficult to determine statistically in the absence of a much more detailed sampling regime, since there are frequent variations in rock type and mineralogical composition. Generally, there is little spatial variation in the slightly below mean solute concentrations for streams where water is moving. Where water collects, in lakes, ponds and snow, solute concentrations are a little higher.

Surface lowering. Hardangervidda is thought to have been largely ice-covered until the end of the Younger Dryas (Mangerud *et al.* 1979). Evidence presented by Anderson (1980) indicates that Hardangervidda was deglaciated about 8750 ± 250 ^{14}C years BP. This correlates with 10 000 calendar years BP (Stuiver *et al.* 1998). A similar date was found by Dixon *et al.* (2002*a*) for deglaciation of the Riksgränsen area in northern Norway, which although 9° further north in latitude, has a comparable climatic regime to Hardangervidda by virtue of its much lower elevation. Therefore, for the purposes of calculating post-glacial weathering rates in this study a date of 10 000 years BP is used. The overall mean rate of post-glacial lowering as determined from the measurement of quartz vein reference surfaces is 0.55 mm ka^{-1}, and individual measurements for different roches moutonnées range from 0.05 to 2.2 mm ka^{-1} (Table 2). The range and mean rates of surface lowering are broadly comparable with those determined by André working on a similar range of rocks in the Abisko-Riksgränsen region of north Sweden (André 1995, 1996*a*, *b*, 2002) and by Dahl working at Narvik in north Norway (Dahl 1967).

Fracture enlargement. At the surface, although many joints have been opened by post-glacial weathering, there are many that remain tightly closed. The overall mean fracture enlargement is 2.4 mm (Table 2), which compares well with values obtained at Riksgränsen by André (1995) for a similar range of rocks and comparable environmental conditions. This single value masks huge variation from fractures that have barely opened up to those with 75 mm of enlargement. Most commonly, fractures are either incipient (e.g. 0.1–0.2 mm) or opened to approximately 5–20 mm. The majority of fractures contain some infilling, commonly vegetative material such as moss, and small fragments of rock that probably originate from fracture walls. Relatively few fractures contained an accumulation of fine sediment. This indicates either that fine sediment has not been produced or that it has been flushed away. Likewise, there is little accumulation of organic soil in fractures. Several distinctive cross-profiles of enlarged fractures can be identified (Fig. 5), which may be indicative of the weathering processes at work.

Weathering indices

Weathering rind thickness. The overall mean maximum weathering rind thickness for exposed surfaces was 3.4 mm, with an individual maximum value of 26.2 mm (Table 2). The overall mean maximum weathering rind thickness for fracture walls was 2.5 mm, with an individual maximum of 20.7 mm (Table 2). These results show that weathering rind is generally thicker on exposed surfaces than on fracture walls (Fig. 6).

Rock surface weakening. Using the Schmidt hammer the mean values for R_1 and R_2, respectively, are 54 and 64 (Table 2). The range of mean values for each roche moutonnée are 38–71 for R_1 and 55–73 for R_2. The scattergraph (Fig. 7) shows the correlation between mean R_1 and R_2 site

Table 2. *Summary of microweathering data for the Røldal area. Mean value given in bold, other values give range of individual measurements*

Study location	Rock types	Surface lowering (mm ka^{-1})	Fracture enlargement (mm)	WRT (mm) exposed surfaces	WRT (mm) fracture walls	Surface hardness R_1	Surface hardness R_2
Rekkingeskara (n = 22)	Quartzite, feldspathic and biotite gneiss, hornblende–biotite gneiss and biotite–epidote schist	**0.40** 0.05–1.28 n = 422	**2.4** 0.1–75.0 n = 449	**2.8** 0.1–12.0 n = 86	**2.9** 0.1–20.7 n = 86	**54.8** 24.6–75.0 n = 555	**65.1** 36.2–77.2 n = 555
Snøskar (n = 15)	Quartzo-feldspathic schist, mica schist, chlorite mica schist and amphibolite	**0.56** 0.07–2.20 n = 379	**2.3** 0.1–46.0 n = 536	**3.7** 0.2–26.2 n = 70	**1.7** 0.1–11.1 n = 44	**54.3** 17.7–77.2 n = 339	**64.0** 36.4–79.3 n = 339
Dyrskarnuten (n = 3)	Chlorite-rich amphibolitic schist	**0.89** 0.28–2.09 n = 119	**3.9** 0.1–35.0 n = 78	**2.9** 0.4–7.8 n = 10	**3.1** 1.0–5.6 n = 4	**43.8** 27.5–54.5 n = 50	**59.4** 53.4–67.4 n = 50
Middyrelva (n = 2)	Chlorite mica schist	**0.91** 0.10–1.98 n = 72	**1.1** 0.1–4.7 n = 48	**4.7** 1.5–12.0 n = 8	**0.7** 0.1–1.0 n = 3	**55.2** 36.2–75.0 n = 40	**62.4** 47.1–77.1 n = 40
Fjetlandsnuten (n = 3)	Actinolite amphibolitic gneiss	**0.40** 0.10–1.05 n = 76	**2.7** 0.1–17.0 n = 132	**3.6** 0.9–8.9 n = 13	**2.2** 0.3–9.2 n = 12	**47.3** 34.1–64.3 n = 25	**60.2** 49.2–68.6 n = 25
All sites (n = 45)	–	**0.55** 0.05–2.20 n = 1068	**2.4** 0.1–75.0 n = 1244	**3.4** 0.1–26.2 n = 187	**2.5** 0.1–20.7 n = 149	**53.8** 17.7–77.2 n = 1009	**64.1** 36.2–79.3 n = 1009

WRT, weathering rind thickness.

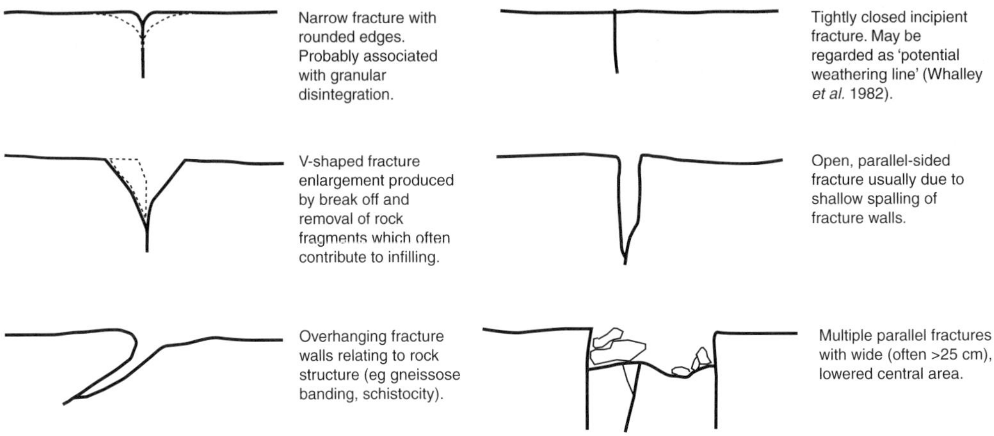

Fig. 5. Observed cross-profiles of enlarged fractures.

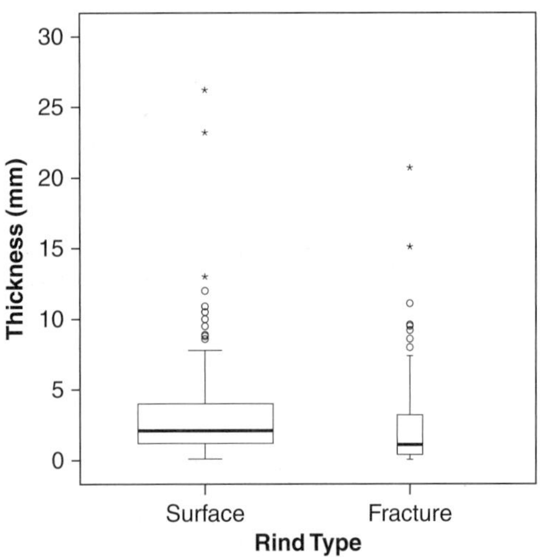

Fig. 6. Box plots comparing weathering rind thickness for exposed surfaces and fracture walls.

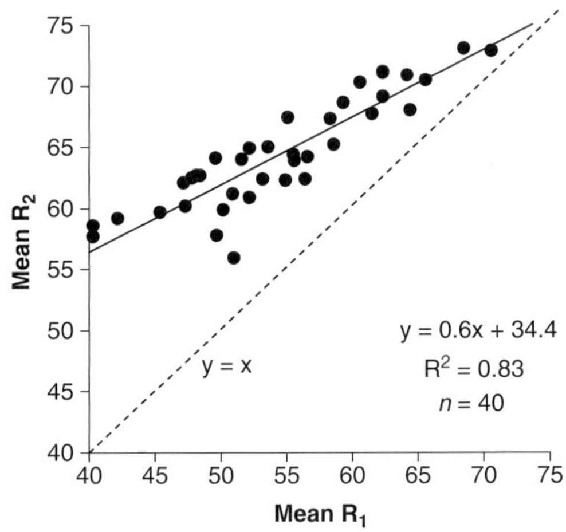

Fig. 7. Correlation between mean R_1 and R_2 site values.

values and demonstrates: (i) a significant increase in rebound value for the second measurement (R_2); and (ii) the difference between R_1 and R_2 is significantly greater at lower values of R_1. This indicates that rocks which are weaker in their fresh, unweathered state show a greater reduction in surface hardness due to weathering than rocks which are inherently stronger in their fresh state.

Weathering morphology

Many ice-smoothed rock surfaces display a small-scale morphology indicative of weathering processes (Figs 8, 9 and 10a–d). Shallow, small-scale surface flaking or spalling is particularly ubiquitous. Occasionally the lifted flakes remain intact (Fig. 8), but mostly shallow scars, typically 1–5 mm deep, remain on the rock surface. These scars are often visible because they have a fresh, lichen-free appearance in comparison with the surrounding bedrock. Flaking is extremely common on quartzo-feldspathic bedrock, and generally occurs on surfaces parallel with structural controls such as foliation and banding. A particular form of flaking very common on thinly banded gneiss is *stepped* flaking or spalling giving the effect of edges having been rounded. This occurs where thin foliations in the rock peel off, deepest at upper surfaces and less so at the sides of a block (Fig. 9). This form may represent weathering modification of glacially rounded surfaces. The photograph also shows macro-breakdown of

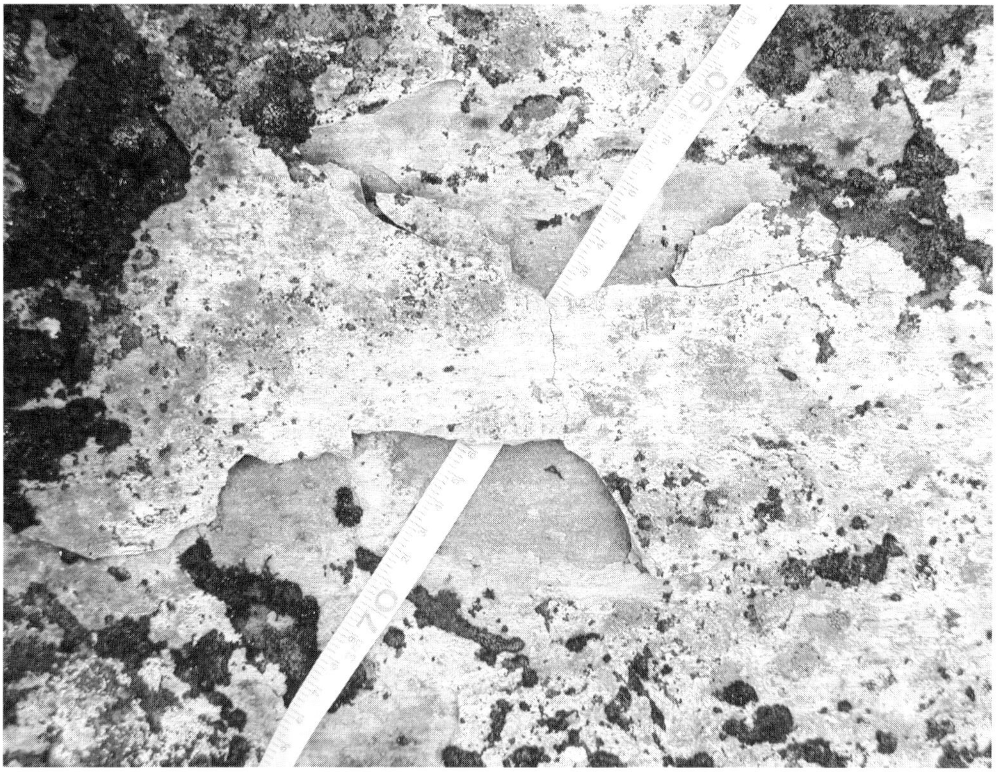

Fig. 8. Shallow surface spalling in quartzitic schist.

roches moutonnées via small-scale fracturing. The loose blocks are slowly moving away from the main rock mass in the adjacent solifluction.

Weathering pits are similar to those reported by André (2002), typically irregular but rounded in plan form and shallow with fairly smooth surfaces (Fig. 10a). Lengths range from 20 to 100 cm and widths from 10 to 50 cm. Pit depth is usually up to 5 cm, but depths of up to 15 cm also occur. Most pits are elongate to some degree, often oriented in alignment with fractures or geological structure. These pits occur infrequently, and generally only on the amphibolites and bedrock containing a high proportion of biotite. Pits also tend to be much more prevalent on roches moutonnées that have a moderate lichen cover. A second type of much smaller, deeper, weathering hollow occurs very commonly in the amphibolites and on the chlorite mica-schist at Middyrelva. These hollows occur particularly on vertical, rather than horizontal surfaces, and are elongated along foliation or banding. Hollows are typically up to 20 mm deep and 50 mm in diameter, but there are some larger forms up to 300 mm in length. The hollows at Middyrelva (Fig. 10b) are fully developed into a dense cover of tafoni. They typically occur on lichen-free surfaces and have a rough texture indicative of granular breakdown.

A particularly unusual weathering form observed in the Røldal area are highly irregular, very deep, and often undercut, hollows (Fig. 10c). These have the appearance of karren forms produced by limestone dissolution in karst terrain and for that reason they are referred to here as pseudokarren. They are also very similar in appearance to the 'weathering pits' described by Dahl (1966). Pseudokarren only occur at three locations: on a group of three amphibolitic roches moutonnées at Fjetlandsnuten; on a single roche moutonnée in amphibolite at Snøskar; and on a group of granitic gneisses at Rekkingeskara. Hollows are characteristically very rough textured and pitted on a small scale. They either occur as isolated features (Fig. 10c), as coalesced hollows or as crenulations or indentations along edges (Fig. 10d). Pseudokarren vary in shape and depth, but single hollows are commonly up to 80 cm in length and of the order of 10–20 cm deep.

In amphibolitic and mica-rich bedrock there are many examples of weathering-modified glacial erosional forms such as striations, crescentic gouges, P-forms and Nye channels. For example,

Fig. 9. Fracturing and downslope movement of loose blocks. Stepped spalling (see text) associated with foliation in granitic schist.

judging from their distinctive plan form, some weathering pits have almost certainly developed from the enlargement of crescentic gouges. Striations are ubiquitous, but are often difficult to locate on lichen-covered bedrock. Coarser striations can sometimes be found on exposed surfaces, but well-preserved fine striations are generally only found through minor excavation of sediment at the lower margins of roches moutonnées. Other indicators of weathering activity include: surface discoloration, seen quite spectacularly in some of the feldspathic schist at Snøskar; fracture enlargement and rounding of fracture cross-profiles (Fig. 5); weathering rind; and the presence of upstanding quartz veins, previously discussed.

Discussion

Weathering and erosion rates in the Røldal area

Using a mean TDS value for the catchment of 3.85 ppm (derived from the mean conductivity

Fig. 10. (**a**) Typical shallow weathering pit in actinolite amphibolite. (**b**) Honeycomb weathering pits in chlorite mica-schist.

Fig. 10. (*Continued*) (**c**) An isolated pseudokarren in actinolite amphibolite. (**d**) Crenulated and undercut edges of pseudokarren 'solution' forms in amphibolite.

of 5.5 μs) and an effective precipitation of 1275 mm (subtracting 250 mm from mean annual precipitation for evapo-transpiration), the mean rate of chemical erosion for the Snøskar catchment equates to 4.9 t km^{-2} year^{-1}. Table 3 shows a range of erosion rates obtained for comparable environments and it can be seen that those obtained for Snøskar are similar to rates at Latnjavagge (5.4 t km^{-2} year^{-1}) in Swedish Lapland (Beylich *et al.* 2004). These rates are considerably lower than the range of values of 19.2–46 t km^{-2} year^{-1} determined for Kärkevagge (Darmody *et al.* 2000; Campbell *et al.* 2002*a*). However, the erosion rate for Snøskar does not include any allowance for atmospheric inputs, which were not obtained but could be of the order of the mean conductivity value for the Snøskar catchment (a mean precipitation conductivity of 9 μs was found for Kärkevagge by Darmody *et al.* 2000). For this reason, the erosion rate obtained is like to be an overestimate. Further reasons to suspect that the rate may be an overestimate are: (a) the data represent summer activity when one might expect chemical processes to be more active; (b) data were collected during a dry period following 3 weeks of very wet weather and thus solute concentrations may be relatively higher than if discharge rates had been greater; and (c) the calculated rate assumes uniform contact between surface runoff and the bedrock from which solutes are derived. In reality, during the peak snowmelt period, meltwater moving through the snowpack will have relatively poor contact with bedrock. The calculated erosion rate may also lack rigour because of the limited temporal period over which the data were collected.

Using the overall mean surface lowering rate of 0.55 mm ka^{-1} (obtained using quartz veins as references surfaces) and assuming a mean rock density of 2650 kg m^{-3}, surface lowering equates to the removal of 1.5 t km^{-2} year^{-1} of material from the Røldal area. The erosion rate derived from the *maximum* of site mean values for surface lowering (2.2 mm ka^{-1}) is 5.8 t km^{-2} year^{-1}. These rates are similar to those calculated from surface lowering at Abisko-Riksgränsen in Sweden (André 1995, 1996*a*, *b*, 2002). At its simplest level, surface lowering will result from a range of weathering and erosion processes, including dissolution. One might therefore expect the solutional erosion rate to be less than the denudation rate calculated from measurements of surface lowering.

Table 3. *Rates of chemical erosion and surface lowering for the Røldal area. Values in parentheses are ranges, other values are means. Values in italics are erosion rates (t km^{-2} year^{-1}) calculated from direct measurement of surface lowering (mm ka^{-1})*

Location	Method	Surface lowering (mm ka^{-1})	Erosion rate (t km^{-2} year^{-1})	Lithology	Source
Kärkevagge, Sweden	Solute load		19–46	Pyrite-rich granitic schists	Rapp (1960); Darmody *et al.* (2000); Campbell *et al.* (2002*a*)
Latnjavagge, Sweden	Solute load		5.4	Micaschist	Beylich *et al.* (2004)
Abisko-Riksgränsen, Sweden	Surface lowering	0.5 (0.2–1.2)	*1.3* *(0.5–3.2)*	Amphibolite Biotite-rich granite and syenite Carbonate sedimentary rocks	André (1995, 1996*a*, *b*, 2002)
		5.4	*14.3*	Carbonate sedimentary rocks	
Narvik, north Norway	Surface lowering	1.0		Granite	Dahl (1967)
Sognjefjell, SW Norway	Solute load	15.5		Calcitic schists	Owen *et al.* (2006)
Global	Solute load		18–19	Crystalline rocks	Meybeck (1987)
Snøskar, southern Norway	Solute load		4.9	Feldspathic schists, biotite-rich schist and amphibolite	Nicholson (this paper)
Røldal, Norway	Surface lowering	0.55 (0.05–2.20)	*1.5* *(0.3–5.8)*	Feldspathic schists, biotite-rich schist and amphibolite	Nicholson (this paper)

However, the two estimates of total denudation rate are not strictly comparable because they are derived from two independent sets of measurements (bedrock lowering and solute runoff) that reflect a very different range of micro-environments and processes: on the one hand, measurements of surface lowering reflect weathering of exposed bedrock surfaces; on the other hand, solute runoff will include weathering and erosion taking place on bedrock surfaces and beneath blockfields, within and on top of snowpatches, within rock joints and through leaching from soils. It is likely that the denudation rate derived from bedrock surface lowering is an underestimate of total denudation in the Røldal area.

Controls on cold environment weathering processes

Analysis of relationships between variables (Table 4) indicates that there is a statistically significant inverse correlation between surface lowering and surface hardness (R_1 and R_2). This demonstrates, not unexpectedly, that weaker bedrock is more susceptible to surface lowering than stronger bedrock. That the inverse correlation with surface lowering is stronger for R_1 than for R_2 may also be an indication that a weakened bedrock surface is a prerequisite for surface lowering. These relationships also point to the fact that surface lowering is achieved through some mechanism that results in mechanical weakening of the rock. Previous studies indicate that this could be through biophysical disruption of bedrock by lichen thalli (e.g. André 1995; Carter & Viles 2004) or fungal activity (e.g. Etienne & Dupont 2002; Arocena *et al.* 2003), biochemical dissolution and alteration by lichen and fungi (e.g. Etienne & Dupont 2002; Arocena *et al.* 2003; Hall *et al.* 2005), development of mineral grain porosity due to chemical dissolution (e.g. Dixon *et al.* 2002*a*), or through entirely physical processes such as thermal shock (e.g. Hall *et al.* 2002). In the Røldal area, the very low levels of chemical erosion derived from analysis of solute runoff suggest very limited chemical weathering in this high mountain plateau environment. However, this is not supported by the ubiquitous presence of weathering rind or from observations of weathering morphology.

Weathering rind is ubiquitous on bedrock surfaces in the Røldal area, although its thickness varies. This is clear evidence that chemical and/or biochemical processes are an important component of microweathering in this area. There is increasing evidence to suggest that rind formation is strongly dependent on moisture availability (Etienne 2002; Dixon *et al.* 2006). There is also evidence that rind formation is associated with an increase in porosity. For example, field experiments at Kärkevagge, using buried granite disks, demonstrated that early development of rind coincides with a significant increase in rock porosity (Dixon *et al.* 2006). Oguchi & Matsukura (2000) also noted a coincidence of higher porosity in andesites with greater rind thickness. The fundamental cause of the increase in porosity associated with rind formation is chemical or biochemical in nature. Dixon *et al.* (2006) argue that intra- and inter-grain dissolution and the formation of microcracks bring about these changes at the rock surface. Etienne (2002) also argued the case for the role of organic acids from fungal growth. The common presence of iron oxides in weathering rind (e.g. Dixon *et al.* 2006) indicates the role of oxidation of ferromagnesian minerals. Dixon *et al.* (2006) indicated that the incidence of increased porosity was greater in feldspar and quartz, and this concurs with the finding in the Røldal area that weathering rind is thickest in the quartzo-feldspathic rocks (Nicholson 2008). It is suggested that the widespread occurrence of weathering rind on bedrock in the Røldal area is also indicative that chemical weathering processes are active.

It is useful to consider the relationship between weathering rind thickness and rock surface hardness. Several authors have noted the importance of micro-erosion in the formation and evolution of weathering rind (Etienne 2002; Gordon & Dorn 2005). It is highly probable that susceptibility of rind to micro-erosion is related to rock surface hardness, particularly since the latter is partially dependent on rock porosity (e.g. Nicholson 2001).

Table 4. *Correlation matrix of weathering variables for the Røldal area. Values in bold indicate that the correlation is significant at the 99% confidence level*

		R_1	R_2	SL	WRT_S	FE	WRT_F
Surface hardness	R_1	1.0					
Surface hardness	R_2	**0.93**	1.0				
Surface lowering (mm ka^{-1})	SL	**−0.60**	**−0.51**	1.0			
Surface rind thickness (mm)	WRT_S	0.30	0.26	0.10	1.0		
Fracture enlargement (mm)	FE	**−0.49**	−0.40	**0.55**	−0.17	1.0	
Fracture rind thickness (mm)	WRT_F	0.06	−0.03	−0.17	0.05	−0.17	1.0

However, the relationship between the two indices is not straightforward. In this study there is a weak positive correlation between surface rind thickness and surface hardness (Table 4). The coincidence of greater porosity with rind formation found in other studies would lead one to expect the opposite (i.e. that thicker rinds would coincide with weaker bedrock). However, it is suggested that in relatively weak rocks or rocks with relatively high porosity the rate of rind formation may be outpaced by rind removal due to micro-erosion, thus giving the false impression that there is little absolute rind development.

The common coincidence of shallow surface flaking with quartzo-feldspathic rocks and their occurrence parallel with metamorphic foliation and banding suggest that there might be some lithological control on their formation. However, an alternative explanation is possible, analogous to the cause of exfoliation envisaged by Etienne (2002) in relation to the evolution of weathering rind. Etienne observed microcracks parallel to the surface at depths of 1–3 mm, which broadly coincide with the thickness of flakes observed in the Røldal area. Etienne (2002) proposed that these microcracks, representing structural heterogeneity between weathered rind and the unaltered bedrock beneath, might coalesce to form larger flakes. Furthermore, he envisaged that lifting, or exfoliation of flakes, could come about through biological activity (e.g. growth of fungi or lichen) or due to physical stresses (e.g. ice lens growth or thermal stress). It seems reasonable that this range of processes (Etienne 2002) could be responsible for the ubiquitous presence of shallow surface flakes in the Røldal area.

That weathering pits are often oriented in alignment with metamorphic structure or local fractures, and occur primarily on biotite-rich bedrock, is an indication that their development is at least partially geologically controlled. Observations also suggest a strong coincidence with lichen-covered bedrock, supporting the contention of previous studies (André 1995, 2002) that biological weathering (e.g. of biotite crystals) could be key. However, given that late-lying snow inhibits lichen colonization, it could be argued that weathering pits favour snow-free sites where frequent fluctuations in temperature and moisture conditions are more likely to occur.

The honeycomb-like weathering tafoni observed are similar to those observed by French & Guglielmin (1999) on meta-granites in Antarctica, which were ascribed to granular disintegration associated with frost action in the presence of salts. French & Guglielmin (1999) believed that these features developed rapidly over a period of 2000–3000 years. The cause of the deep pitting found in the Røldal area is unknown. However, given the predominance of these features in the amphibolitic and chlorite mica-schist, it seems likely that selective and differential weathering of ferromagnesian minerals leading to granular disintegration could be key (Campbell & Claridge 1987).

The deepest of the pseudokarren observed represents a weathering rate at least two orders of magnitude greater than that obtained from measurements of bedrock surface lowering, and so it is clear that they represent locally intense weathering conditions. One possibility that needs to be considered is that pseudokarren represent a pre-glacial weathering surface that has survived glaciation. However, given that two of the three sites where they are found display numerous striations, and at Fjetlandsnuten there are also Nye channels and other P-forms of glacial meltwater erosion origin, this explanation seems highly unlikely. A more promising explanation is that pseudokarren are sites favouring intense biotic and/or chemical activity concentrated around flaws such as cracks or shallow exfoliation scars. The former might explain the elongate nature of most of these features. Pseudokarren have the appearance of having been formed through dissolution, although it is quite likely that vegetative material has contributed to dissolution through the provision of organic acids (Dahl 1966; Dixon *et al.* 2002*a*).

The common preservation of glacial striations in the Røldal area is a clear indication that weathering has had relatively limited impact on landform denudation since the last period of glaciation. Although glacial striations were found on all rock types, they are almost an order of magnitude more abundant on amphibolite despite surface lowering, and several examples of their presence were observed in close proximity to upstanding quartz bands. This may indicate that the original striae were larger and persist despite bedrock surface lowering. Alternately, it may simply indicate that surface lowering has perpetuated the original shape and size of striations.

The role of fractures in weathering

Much of the literature on the role of fractures in cold-environment weathering has focused on the traditional concept of freeze–thaw as the dominant process (e.g. Walder & Hallett 1985; Murton *et al.* 2006). Field-based studies have included the role of fractures and microcracks in the formation of blockfields (e.g. Boelhouwers 2004; Whalley *et al.* 2004), the macro-breakdown of rockwalls (e.g. Matsuoka & Sakai 1999) and fractures as structural controls on macro-breakdown (e.g. Gordon 1981; Glasser *et al.* 1998; Olvmo & Johansson 2002). However, there has been relatively little field-based study of the direct role of fractures in cold-environment microweathering. One exception is the work of André (e.g. 2002) who measured fracture enlargement and obtained similar values to those obtained in the present study. If the mean fracture enlargement of 2.4 mm obtained in this study is assumed to reflect bi-directional recession of the fracture walls, then uni-directional recession can be assumed to be 1.2 mm. Given that measurements of fracture enlargement were made at the intersection of fractures with the surface, for which a chronological reference exists, it is reasonable to accept that this value represents uni-directional widening over a period of 10 000 years. The rate per thousand years is therefore 0.12 mm ka^{-1}. Given a mean surface lowering rate of 0.55 ka^{-1}, this value represents only one fifth of the equivalent lowering achieved at the surface over the same period of time.

There is a highly significant inverse correlation between fracture enlargement and R_1, and an equally significant positive correlation with surface lowering (Table 4). Given the correlation between surface lowering and surface hardness (see earlier), this suggests that, although fracture enlargement occurs at a much slower rate than removal of material from exposed bedrock surfaces, the processes and controls involved are very similar. In previous studies (André 2002) it has been proposed that post-glacial granular disintegration is the primary cause of fracture enlargement. However, a two-stage process of fracture enlargement is envisaged here, in which initial modification of fractures occurs entirely *in situ* by chemical and physical microweathering. This is clearly indicated by the presence of weathering rind on fracture walls and results in weakening and micro-erosion (e.g. through transport of solutes and fine particles), leading to the initial enlargement of fractures. With void space now created between fracture walls there is scope for further enlargement through multiple processes operating at a larger scale. Evidence from fracture cross profiles (Fig. 5) indicates that there are three main processes involved: (a) granular disintegration producing rounding of fracture edges; (b) spalling of fracture walls producing thin flakes or wedge-shaped rock fragments; and (c) more general surface break-up in association with multiple fracture intersections and shattered zones. Some fracture cross-profiles (e.g. overhanging fractures, Fig. 5) reflect the interaction of weathering processes with rock structure.

The fact that weathering rind is thicker on exposed surfaces than on fracture walls (Fig. 6) is perhaps contrary to what might have been expected. That there is no correlation between fracture wall rind thickness and any of the other variables measured (Table 4) also indicates that there are different controls on the development of fracture rind and a different set of processes involved. One might have expected fracture wall rind to develop well, given that fractures are sites of potential moisture accumulation, thus aiding chemical activity. Furthermore, it is unlikely that significant rind has been removed from fracture walls since samples were taken from closed fractures. This is an interesting finding and there are several possible explanations. It is possible that, in reality, moisture does not easily penetrate closed fractures and thus weathering activity is inhibited by the limited quantity of moisture available. Alternately, moisture penetrates but travels downwards rapidly and thus its residence time, and the opportunities for fracture wall weathering, are limited. It is also likely that biotic weathering processes are inhibited in cracks by the lack of photic activity. This would lend support to the view that bacteria and fungi are important in rind development (Etienne 2002). It may also help to explain why rind thickness is greater in pale, quartz-rich rocks (Nicholson 2008) that probably allow greater penetration of solar radiation. A further explanation for the contrast between surface and fracture wall rind thickness is that development of rind at the surface is particularly enhanced by the presence of moisture at the interface between bedrock and late-lying snow, a view shared by others (Thorn 1975; Ballantyne *et al.* 1989; Nyberg 1991; Thorn & Hall 2002). An alternate view is that fractures have only opened during the Holocene and therefore there has simply been less time for fracture wall rind to develop.

The finding that fracture enlargement is much less than lowering of the surface broadly concurs with the observation that weathering rind found on fracture walls is thinner than that on exposed surfaces and suggests that fractures have less influence on microweathering than might have been expected. However, it is important to note that there is substantial observational evidence from Rekkingeskara and Snøskar that fractures play a significant role in large-scale landform development. In particular, there are numerous examples of roches moutonnées that have experienced significant breakdown, the products of which are being assimilated into surrounding blockstreams.

Conclusions

In this study multiple weathering indices demonstrate that, for the Røldal area, rates of periglacial microweathering are generally low, although surface morphology indicates that there is locally intense weathering activity. Rates are similar to those obtained for comparable regions indicating that, at a regional level, climatic conditions are a

The role of buoyancy in palsa formation

MATTI SEPPÄLÄ[1]* & KAUKO KUJALA[2]

[1]*Department of Geography, P.O. Box 64, FI-00014 University of Helsinki, Helsinki, Finland*

[2]*Water Resources and Environmental Engineering Laboratory, P.O. Box 4300, FI-90014 University of Oulu, Oulu, Finland*

**Corresponding author (e-mail: matti.seppala@helsinki.fi)*

Abstract: The formation of a palsa is based on the thermal properties of peat. Frozen wet peat has a high thermal conductivity, and therefore cold can penetrate deep into peat layers if the snow cover is thin; while the dry peat in summer insulates the frozen core of a palsa, so that the permafrost core is preserved. The volumetric growth of the palsa is based on the buoyancy effect of the frozen core, which lifts it, causing some water to accumulate under the core, where it freezes during the next winter and forms thin ice layers. Only when the frozen peat core touches the frost-susceptible silt or silty till layer at the bottom of the mire does ice segregation start to play an important role in the formation of the palsa.

Palsas, defined as peat mounds with a permanently frozen peat and mineral soil core, are typical phenomena in the circumpolar zone of discontinuous permafrost (Lundqvist 1969; Seppälä 1972, 1988*a*). They can be up to 150 m in diameter and can reach a height of 12 m (Lagarec 1982). The summits of palsas are free of snow even in the middle of winter, however, because the wind carries the snow off and deposits it on the slopes and elsewhere on the flat mire surface (Seppälä 1990, 1994). Permafrost is found on palsa mires only in the palsas themselves, its formation being based on the physical properties of peat. Dry peat is a good insulator, but wet peat conducts heat better (Seppälä 1986, 1988*a*) and frozen peat better still (Brown 1966; Kujala *et al.* 2008). This means that heat is extracted from deeper layers in winter, whereas the dry peat on the palsa surface insulates the frozen core and prevents thawing of deeper, frozen layers during the short summer. Without a covering peat layer the permafrost would disappear in palsa regions, where the mean annual air temperature is close to −2 °C.

When the frost susceptibility of peat is studied under laboratory conditions, frost heave tests show that the surface peat samples are non-susceptible to frost according to all the test parameters, and that no formation of ice lenses is observed. Ice lens formation due to water intake within peat material is not probable (Kujala *et al.* 2008). This means that frozen peat does not form segregated ice lenses by physical frost heave processes (e.g. according to secondary frost heave theory). According to An & Allard (1995, p. 236), almost no segregation ice is formed in most peat types, particularly fibrous peat, during freezing. However, ice layers are present within peat palsas, as has been observed in many drillings. This is not normal ice segregation of the sort observed in freezing silt, for example, but rather a perched saturated layer that is formed at the thawing front as the active layer in a palsa thaws and some water migrates into the frozen layer and eventually into the permafrost underneath (cf. Mackay 1983; Smith 1985), along the thermal gradient. This downward migration of water from the thawing active layer is physically similar to the upward migration of water to the freezing front (An & Allard 1995, p. 236).

The main purpose of this work is to explain the formation of ice layers in palsas in Finnish Lapland. There have been some earlier attempts at modelling palsa formation, but they are not really valid for the understanding of palsas observed in the field. Outcalt & Nelson (1984*a*) made a computer simulation of buoyancy and snow cover effects on palsa dynamics. Buoyancy plays a certain role in the early stage of palsa formation, and this characteristic has also been shown experimentally (Seppälä 1982). A rise of the palsa hummock takes place when the surrounding mire surface has thawed. Outcalt & Nelson (1984*a*) used randomly changing annual snow depths ranging from 0 to 99 cm, however, which is not the case anywhere on present-day palsa mires. On the contrary, the snow depth is shallow from one year to the next at the time when the hummock forms, and the higher the hummock becomes the smaller are the changes in the snow depth covering the mound.

Outcalt & Nelson (1984*b*) and Outcalt *et al.* (1986) presented a different model for frost

From: KNIGHT, J. & HARRISON, S. (eds) *Periglacial and Paraglacial Processes and Environments.*
The Geological Society, London, Special Publications, **320**, 51–56.
DOI: 10.1144/SP320.4 0305-8719/09/$15.00

mounds in the High Arctic, formed by hydrostatic pressure in a closed groundwater system. These features were studied in the zone of continuous permafrost in northern Alaska and were described not as palsas but as palsa-like frost mounds above a thick permafrost layer, which are like pingos with a peat cover. Their model of formation does not depend on the properties of peat, as is the case in palsas. There are also many other types of frost mounds observed (cf. Seppälä 1988*a*), which again should not be called palsas.

The mathematical one-dimensional heat and mass transfer model, designed by An & Allard (1995) to simulate the long-term aggradation of permafrost and the formation of palsas, takes account of the variable phase-change temperatures, the build-up of a discrete ice-lens, frost heave and thaw consolidation in order to consider freeze–thaw cycles near the soil surface and annual cyclic temperature variations. The model starts out from an unfrozen site with 1 m of peat overlying silt and assumes that the site remains uncovered by snow. Present-day mean annual air temperature conditions (of −5.6 °C) and an annual temperature cycle integrating freezing and thawing indices were used (An & Allard 1995, p. 237). Here, we try to explain the palsa formation on mires with much thicker peat layers (>2 m) without silt, which is a frost-susceptible material. The most difficult matter in the modelling of palsas is to simulate the natural conditions, which change throughout the winter. Air temperatures typically fluctuate, freezing of the active layer can be fast or slow and the snow depth increases around the palsa but its surface is free of snow for most of the winter, although it may be covered at first and will always be covered for a while after a snowfall, before the wind blows the snow off.

Internal structure of palsas

The permanently frozen core of palsas contains frozen peat with small ice crystals and thin layers of ice. Also thicker (*c*. 15 cm) layers of ice have been found in core drillings, but mainly in the silt-cored palsas (Åhman 1977; Allard & Rousseau 1999). Salmi (1972) cut a palsa in Finnish Lapland with a chainsaw and described thick layers of ice in the frozen core formed in peat. Many 1–10 cm-thick ice layers were also observed in several peat palsas with core drillings in Finnish Lapland.

Buoyancy

Since peat is non-frost susceptible (Kujala *et al.* 2008), we have tried to find some mechanism other than ice segregation to explain the formation of ice layers in palsas. Field observations (Seppälä 1982, 2006) of new palsa embryos show that they rise above the mire surface during the summer when the surrounding seasonal frost layer has thawed. The frozen core bubbles up like a floating cork above the water-saturated peat, which is fibrous and stays more or less in position. This process causes a void space in water-saturated peat under the upheaved frozen core, which is filled immediately by water from more densely saturated loose peat around it. In the succeeding winter, this water layer freezes when heat is extracted from the deeper frozen core (Seppälä 1982) and a new ice layer is formed. In the succeeding summer, buoyancy uplifts the thicker frozen core, the palsa rises some more above the mire surface and a new water layer is formed under the frozen core.

The density of most liquids and solids varies slightly with changes in temperature and pressure. Density of ice is 0.917×10^3 kg m^{-3} and that of water is 1.00×10^3 kg m^{-3} at standard atmospheric temperature and pressure, defined as 0 °C and 1 atm. Because the density of ice is less that that of water, the ice has a net upward force.

The formation of ice layers in palsas is based on Archimedean law, as shown in the following simple calculations based on some assumptions (Fig. 1). If the frost penetrates to the depth of 1 m and the water content of saturated peat is 80% and its density 1100 kg m^{-3}, then we can calculate the uplift force caused by the frozen palsa. We can also calculate the thickness of the active layer where the uplift force is in balance with the peat layer, when the water content of active layer is 55% and its density is 467 kg m^{-3}. With these hypotheses the uplift pressure calculated with equation (1) would be 81 N m^{-2}. This pressure would lift up the active layer by a total of 17 cm (equation 2). This uplift corresponds well with the thickness of ice layers observed in palsas in Finnish Lapland.

$$
\begin{aligned}
F &= \rho_{\text{sat. unfrozen}} \cdot g \cdot V - \rho_{\text{sat. frozen}} \cdot g \cdot V \\
&= 1100 \frac{\text{kg}}{\text{m}^3} \cdot 9.81 \frac{\text{m}}{\text{s}^2} \cdot 1\,\text{m}^3 - 1019 \frac{\text{kg}}{\text{m}^3} \\
&\quad \cdot 9.81 \frac{\text{m}}{\text{s}^2} \cdot 1\,\text{m}^3 \\
&= 81\,\text{N} \qquad (1)
\end{aligned}
$$

where F is uplift force (N), $\rho_{\text{sat. unfrozen}}$ is the density of saturated unfrozen peat (kg m^{-3}), g is acceleration due to gravity (m s^{-2}) and V is volume (m^3).

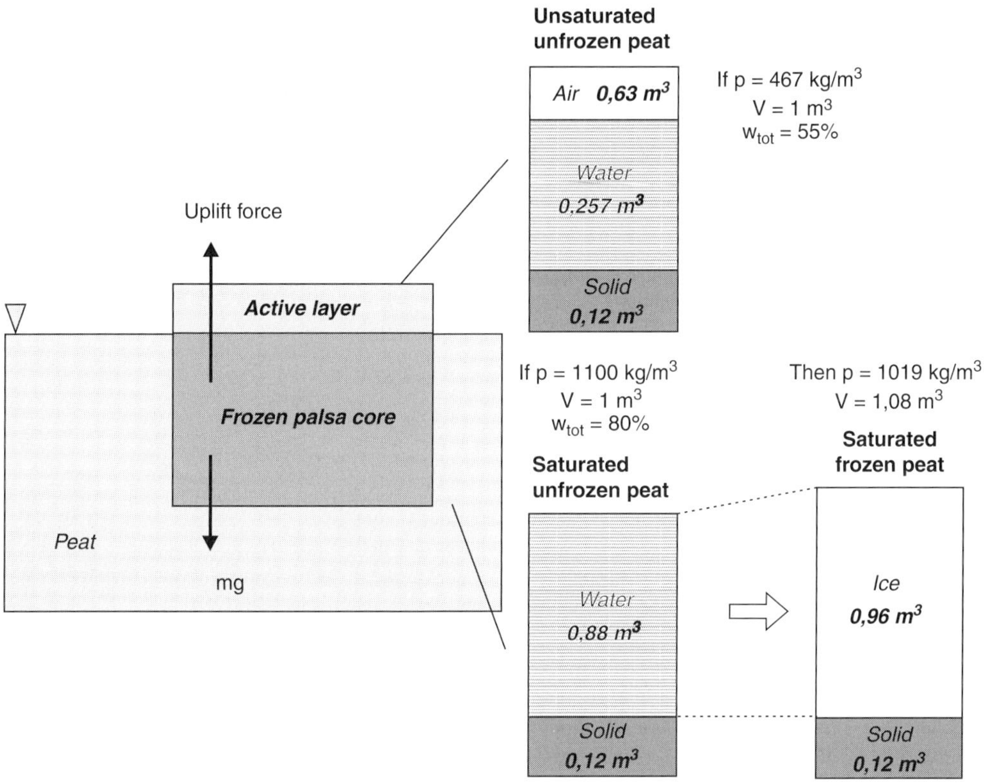

Fig. 1. Force balance of palsa and basic properties of frozen and unfrozen peat layers.

$$x \cdot 1\,\mathrm{m}^2 \cdot 467\frac{\mathrm{kg}}{\mathrm{m}^3} \cdot 9.81\frac{\mathrm{m}}{\mathrm{s}^2} = 81\frac{\mathrm{kg}}{\mathrm{m}^3} \cdot 9.81\frac{\mathrm{m}}{\mathrm{s}^2} \cdot 1\,\mathrm{m}^3 \Rightarrow 17\,\mathrm{cm} \quad (2)$$

where x is amount of uplift (cm).

If the density of the active layer is lower, then the uplift will be much more. In the field we have noticed up to 30 cm of uplift of new palsas during the first summer.

Model of the formation of palsas

Palsa formation starts when the snow cover present on a mire surface is so thin that the winter frost (Fig. 2a) penetrates sufficiently deep to prevent the summer heat from thawing it completely (Seppälä 1982, 1986, 1988*b*). During the first summer the small embryo palsa rises above the mire surface when the seasonal frost in surrounding peat layers has thawed (Fig. 2b). This is caused by the buoyancy of the freezing core (cf. Outcalt & Nelson 1984*a*). This stage of palsa formation has been seen in the field several times during the last few years (e.g. Seppälä 1986, 2003, 2006). The new palsa embryo is dry; *Sphagnum* moss on its surface has a very pale colour and is dying out as the result of desiccation. Vegetation assemblages on a growing palsa change from that on the surrounding mire, reflecting the changed drainage status (Seppälä 1982, 1988*a*). The dry moss cover insulates the frozen core, and the active layer on it is only some 25–30 cm thick in Finnish Lapland. The upheaval process probably also causes some void to develop under the frozen body, which is floating on very wet peat or almost pure water. At this point in the cycle, the thermal characteristics of peat start to play an important role.

In autumn the evaporation decreases and rains make the peat wet again. This gives a chance for frost to penetrate still deeper into the saturated and frozen peat. In the following winter the small and developing hump on the mire surface has even less snow because of snow removal by wind, and it is exposed to cold air. Frost penetrates deeper. In the succeeding summer the hump shows further

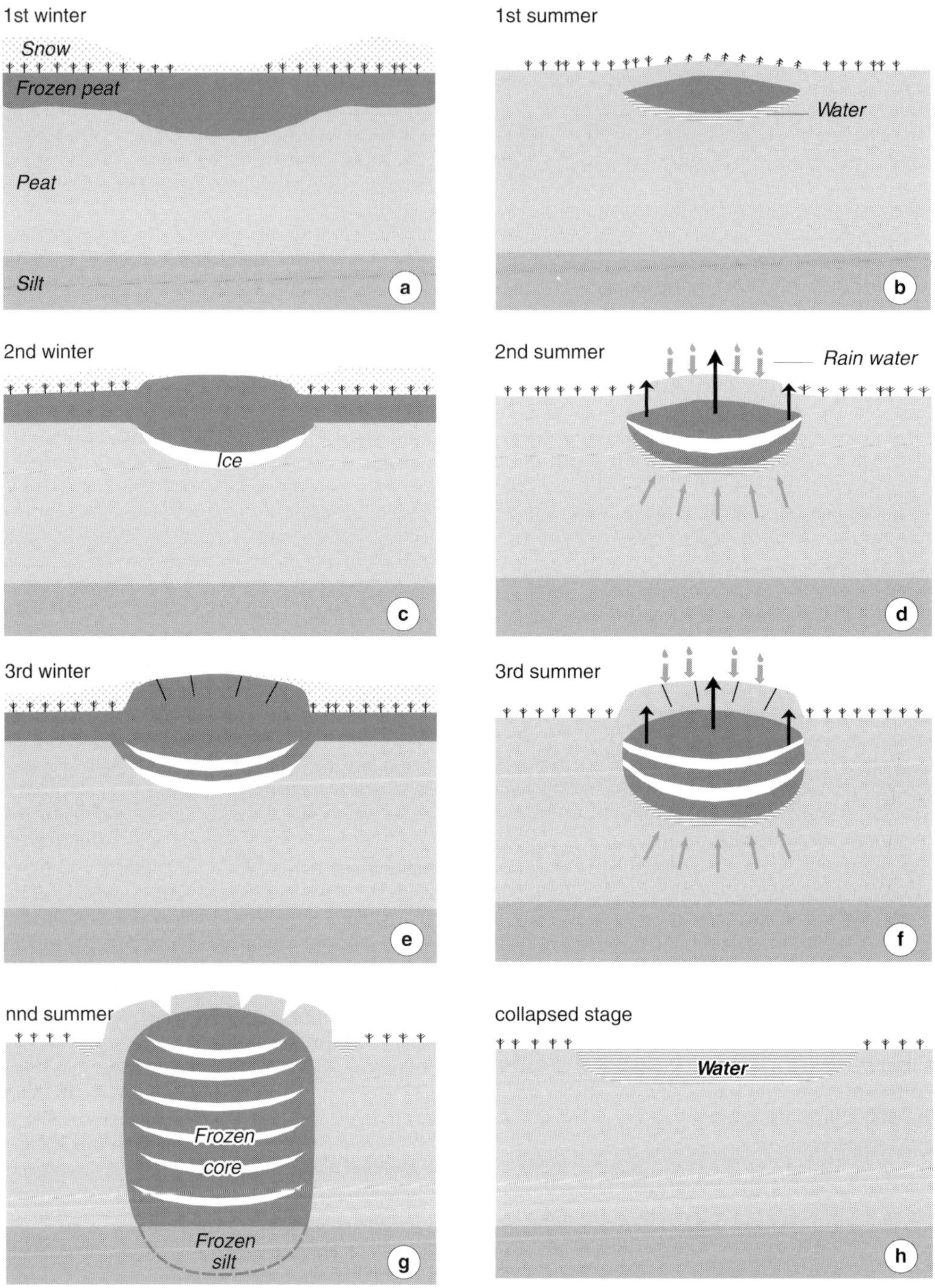

Fig. 2. Principal phases of palsa formation. Formation of ice-rich layers (white) under the freezing core by buoyancy. Water movement is indicated with grey/blue arrows and upheaval by black arrows. Cracking of active layer marked in (e) and (f).

upheaval (Fig. 2d). As the surface rises, the wind becomes even more effective in drying the surface peat during the summer and keeping it clear of snow in winter. During the summer, when the active layer thaws, water migrates to the permafrost table and keeps the peat above it saturated. A high proportion of water in the frozen core is found just below the permafrost table, indicating this migration of water from the surface. But also, deeper inside the frozen core of palsa, we find thin ice layers that are probably formed by the buoyancy of the frozen core and frost that sucks water from below to the freezing front when the frozen core of palsa grows downwards (Fig. 2d, f).

When the freezing of the palsa core reaches the till or silt layers at the base of the mire, the mature stage of palsa development begins (Fig. 2g). The palsa surface has already cracked during earlier stages and its edges are steep, which enables snow to collect on the slopes of the palsa. Peat blocks can also collapse down along open cracks into the pool that have been formed surrounding the mature palsa. Abrasion can remove peat from the palsa surface (Seppälä 2003) and degradation of the palsa has started. The palsa loses its insulating peat layer and its frozen core cannot stay frozen. The old palsa is destroyed by thermokarst, and becomes scarred by pits and collapse forms. As a sign of former palsas we find small water bodies, ponds or open peat surfaces without vegetation developed on mires (Fig. 2h). These forms can be surrounded by low rim ridges, indicating former peat blocks that have collapsed along the frost table. From such pools a new palsa may ultimately emerge after a renewed phase of peat formation. The cyclic palsa formation recommences from the beginning.

This is a descriptive model of the formation of palsas based on field observations and laboratory measurements, but it is very difficult to put this into a mathematical model because so many factors change at the same time, and they change in different ways in different stages of development. At first the crucial factor is the snow depth and thermal characteristics of snow and peat. Then the water content of peat and air temperature are important. Migration of free water in peat is the least known factor in the formation of the large volume of frozen core. Some 80–90% of the palsa core volume is frozen water, but the freezing expansion of water does not explain its great height. Water-saturated peat 2 m thick can produce a palsa 5–7 m in height above the mire surface in Finnish Lapland.

Discussion and conclusions

The thermal conductivity of peat (Kujala *et al.* 2008) is a fundamental factor for the freezing of a palsa, as it increases as the water content of the peat increases and as the temperature drops below freezing point. It is well known that the formation and growth of a palsa cannot be explained only by means of thermal calculations. The growth in the volume of a palsa is in close connection with the balance of its mechanical forces, which means that the palsa has a substantial tendency to rise upwards. Buoyancy therefore explains the formation of thick ice layers in peat. Buoyancy in palsa formation was pointed out first by Outcalt & Nelson (1984*a*), but they did not explain the ice layers in palsa cores with such a buoyancy mechanism.

According to our studies (Kujala *et al.* 2008) original palsa peat is not a frost-susceptible material. However, palsas contain ice layers, which cannot be explained with ice segregation. Buoyancy creates a void beneath the frozen palsa core when it becomes buoyant within the unfrozen mire. This void space with less peat is below the water table and so becomes filled with water. In some drillings water under pressure has been found when penetrating through the frozen core. The buoyancy of the palsa leads to a rupture or series of ruptures within the peat beneath the palsa core as it is dilated. This leads to a zone beneath the palsa core where the density of the peat is much reduced and filled with water, which then freezes during the coming winters when cold penetrates deeper into the mire. Hydrostatic pressure is not able to feed the growth of an ice lens in a way similar to that proposed for hydrostatic pingos, because no permafrost is found surrounding palsas.

The temperature inside a palsa is controlled by weather conditions, and the refreezing of the active layer also depends on the depth of the overlying snow cover. The lower parts of the active layer can remain unfrozen for fairly long periods in autumn, and we can often find an unfrozen layer of peat above the permafrost table throughout a mild winter.

Palsas are very reliable indicators of changing environmental conditions, but before we are able to interpret such effects more widely we should monitor physical conditions, such as temperature and moisture inside the palsas. Increasing summer temperatures do not necessarily mean decay for our palsas if the peat is dry during the hot period, as this acts to insulate the frozen core.

Thanks are due to anonymous referee who provided constructive comments and Dr J. Knight who revised the language of the manuscript.

References

ÅHMAN, R. 1977. Palsar I Nordnorge. *Meddelanden från Lunds Universitets Geografiska Institution, Avhandlingar*, **78**, 1–165.

ALLARD, M. & ROUSSEAU, L. 1999. The internal structure of a palsa and a peat plateau in the Rivière Boniface region, Québec: inferences on the formation of ice segregation mounds. *Géographie physique et Quaternaire*, **53**, 373–387.

AN, W. & ALLARD, M. 1995. A mathematical approach to modelling palsa formation: Insights on processes and growth conditions. *Cold Regions Science and Technology*, **23**, 231–244.

BROWN, R. J. E. 1966. Influence of vegetation on permafrost. *In*: *Proceedings of the Permafrost International Conference November 1963, Lafayette, Indiana*. National Academy of Science, National Research Council, Washington, DC, Publication, **1287**, 20–25.

KUJALA, K., SEPPÄLÄ, M. & HOLAPPA, T. 2008. Physical properties of peat and palsa formation. *Cold Regions Science and Technology*, **52**, 408–414.

LAGAREC, D. 1982. Cryogenic mounds as indicators of permafrost conditions, northern Québec. *In*: *Proceedings of the Fourth Canadian Permafrost Conference, Calgary*. National Rescarch Council of Canada, Ottawa, 43–48.

LUNDQVIST, J. 1969. Earth and ice mounds: a terminological discussion. *In*: PÉWÉ, T. L. (ed.) *The Periglacial Environment: Past and Present*. McGill-Queen's University Press, Montreal, 203–215.

MACKAY, J. R. 1983. Pingo growth and subpingo water lenses, Western Arctic Canada. *In*: *Proceedings of the 4th International Conference on Permafrost, Fairbanks, Alaska*. National Academy Press, Washington, DC, 762–766.

OUTCALT, S. I. & NELSON, F. 1984*a*. Computer simulation of buoyancy and snow-cover effects in palsa dynamics. *Arctic and Alpine Research*, **16**, 259–263.

OUTCALT, S. I. & NELSON, F. 1984*b*. Growth mechanisms in aggradation palsas. *Zeitschrift fur Gletscherkunde und Glazialgeologie*, **20**, 65–78.

OUTCALT, S. I., NELSON, F. E., HINKEL, K. M. & MARTIN, G. D. 1986. Hydrostatic-system palsas at Toolik Lake, Alaska: field observations and simulation. *Earth Surface Processes and Landforms*, **11**, 79–94.

SALMI, M. 1972. Present developmental stages of palsas in Finland. *In*: *Proceedings of the 4th International Peat Congress Helsinki, Finland*, Volume 1, 121–141.

SEPPÄLÄ, M. 1972. The term 'palsa'. *Zeitschrift für Geomorphologie, N.F.*, **16**, 463.

SEPPÄLÄ, M. 1982. An experimental study of the formation of palsas. *In*: *Proceedings of the Fourth Canadian Permafrost Conference, Calgary*. National Research Council of Canada, Ottawa, 36–42.

SEPPÄLÄ, M. 1986. The origin of palsas. *Geografiska Annaler*, **68A**, 141–147.

SEPPÄLÄ, M. 1988*a*. Palsas and related forms. *In*: CLARK, M. J. (ed.) *Advances in Periglacial Geomorphology*. Wiley, Chichester, 247–278.

SEPPÄLÄ, M. 1988*b*. Frozen peat mounds in continuous permafrost, northern Ungava, Québec, Canada. *Zeitschrift für Geomorphologie N.F., Supplement Band*, **71**, 107–116.

SEPPÄLÄ, M. 1990. Depth of snow and frost on a palsa mire, Finnish Lapland. *Geografiska Annaler*, **72A**, 191–201.

SEPPÄLÄ, M. 1994. Snow depth controls palsa growth. *Permafrost and Periglacial Processes*, **5**, 283–288.

SEPPÄLÄ, M. 2003. Surface abrasion of palsas by wind action in Finnish Lapland. *Geomorphology*, **52**, 141–148.

SEPPÄLÄ, M. 2006. Palsa mires in Finland. *The Finnish Environment*, **23**, 155–162.

SMITH, M. W. 1985. Observations of soil freezing and frost heaving at Inuvik, Northwest Territories, Canada. *Canadian Journal of Earth Sciences*, **22**, 283–290.

Basal glacier ice and massive ground ice: different scientists, same science?

RICHARD I. WALLER[1]*, JULIAN B. MURTON[2] & PETER G. KNIGHT[1]

[1]*Research Institute for the Environment, Physical Sciences & Applied Mathematics, Keele University, Keele, Staffordshire ST5 5BG, UK*

[2]*Department of Geography, University of Sussex, Brighton, BN1 9QJ, UK*

Corresponding author (e-mail: r.i.waller@esci.keele.ac.uk)

Abstract: Whilst glaciologists and permafrost researchers investigate ice bodies using similar techniques, there has been surprisingly little collaboration between the two communities. This paper examines the potential benefits of interdisciplinary research into the formation of basal ice beneath glaciers and the origin of massive ice in glaciated permafrost regions. Active collaboration in these areas has already improved our understanding of the formation of basal ice beneath cold-based glaciers, the critical role played by basal freezing in controlling the dynamic behaviour of stagnating ice streams and the significance of glacier–permafrost interactions at the margins of Pleistocene ice sheets. However, in order to promote future collaboration certain obstacles need to be overcome. The contrasting ice-classification schemes employed by glaciologists and permafrost scientists, for example, need to be unified in order to allow detailed comparisons of ice-rich sequences in both environments. This could, in turn, enable exciting research advances, most notably by facilitating the identification of preserved remnants of Pleistocene ice sheets within permafrost regions that provide a potentially invaluable and currently largely untapped source of palaeoglaciological information.

The study of ice, liquid water and sediment mixtures is a major research focus within both glaciology and permafrost science (e.g. Christoffersen *et al.* 2006; Rempel 2007). Glaciologists have long recognized the importance of debris-rich ice at the base of glaciers and ice sheets, and have made rapid progress over the last 50 years in identifying its key characteristics, processes of formation and glaciological significance (for relevant reviews see Hubbard & Sharp 1989; Knight 1997; Cook *et al.* 2006). Similarly, permafrost scientists have elucidated the nature, origin and palaeoenvironmental significance of massive bodies of ground ice (e.g. Mackay 1971, 1989; Mackay & Dallimore 1992; Murton 2005). Not surprisingly, researchers in both fields use similar techniques, ranging from the simple description and characterization of ice–sediment facies (e.g. Lawson 1979; Murton & French 1994; Hubbard & Sharp 1995), to the application of stable-isotope analyses (e.g. Lorrain & Demeur 1985; Knight 1989; Cardyn *et al.* 2007), geophysics (e.g. Dallimore & Davis 1992; Arcone *et al.* 1995; Kneisel *et al.* 2008) and drilling (Mackay 1971; Murray & Porter 2001). Nonetheless, interdisciplinary collaboration remains limited (Harris & Murton 2005).

This paper argues that greater collaboration between glaciology and permafrost science is essential to elucidate the processes of ice formation and the palaeoenvironmental significance of debris-bearing ice. It reviews recent research about the formation of basal ice, and the origin and palaeoenvironmental significance of massive ice. Three case studies are presented to illustrate the ways in which progress in these related fields can be facilitated by the combined knowledge and efforts of glaciologists and permafrost scientists. Finally, we identify obstacles to collaboration and highlight future research possibilities.

The origin and glaciological significance of basal ice

Many glaciers have a distinctive basal horizon that differs markedly from the overlying firnified glacier ice due to its interaction with the glacier bed (Fig. 1). Such basal-ice layers can reach tens of metres in thickness, and are usually distinguished from the overlying glacier ice by an abrupt increase in debris content and an anisotropic physical structure (Hubbard & Sharp 1989). Basal ice is also characterized by a distinctive isotopic signature owing to the fractionation that occurs on freezing. This results in basal ice being enriched in the heavier isotopes of oxygen (less negative $\delta^{18}O$ ratios) relative to the water from which it is derived. In a closed-system setting, basal ice also

From: KNIGHT, J. & HARRISON, S. (eds) *Periglacial and Paraglacial Processes and Environments*. The Geological Society, London, Special Publications, **320**, 57–69.
DOI: 10.1144/SP320.5 0305-8719/09/$15.00

Fig. 1 Debris-rich basal ice exposed at the base of the Matanuska Glacier, Alaska (61°46′31″N, 147°45′33″W). The ice axe is given for scale.

displays a co-isotopic relationship that differs from that associated with firnified ice (Knight 1997). Basal-ice layers have been described at temperate glaciers in Alaska and the Alps (e.g. Hubbard & Sharp 1995), subpolar glaciers in Greenland and Svalbard (e.g. Boulton 1970), polar glaciers in the high Arctic and Antarctica (e.g. Tison *et al.* 1993), and glaciers descending from tropical ice caps (Knight 1988). The presence of basal ice can influence the behaviour of glaciers or ice sheets with, for example, its enhanced debris content affecting its crystal size and fabric, rheology and, consequently, the dynamic response of the entire ice body (e.g. Lawson 1996; Fitzsimons 2006). In addition, the presence of debris-rich basal ice can influence rates of glacier erosion, transport and deposition, and, therefore, sediment budgets (e.g. Rea & Whalley 1994). Finally, recent work has suggested that basal freeze-on associated with the development of thick sequences of basal ice can cause ice-stream stagnation by increasing till consolidation and shear strength (Christoffersen & Tulaczyk 2003*a*, *b*).

Most research on basal ice has focused on its mechanisms of formation. Work on this subject has continued for almost five decades and a broad range of mechanisms have now been advocated (for lengthier reviews see Hubbard & Sharp 1989; Knight 1997). A large motivation behind this research is the belief that subglacial processes and conditions can be reconstructed from examining the distinctive characteristics of basal ice, and, therefore, that the conditions beneath inaccessible ice-sheet interiors can be determined via observations at easily accessible ice margins (e.g. Sugden *et al.* 1987). Consequently, basal-ice researchers have attempted to develop genetic classification schemes in which the appearance of distinctive basal-ice facies can be related to the occurrence of specific basal processes and conditions (e.g. Sharp *et al.* 1994; Hubbard & Sharp 1995).

Whilst initial research on the origin of basal ice advocated debris entrainment via shearing (e.g. Goldthwait 1951), this hypothesis was soon criticized by both Weertman (1961) and Boulton (1970), who argued that shearing alone was incapable of entraining debris. They suggested that thermally controlled processes involving the freeze-on of water onto the glacier sole provide the dominant mechanism of debris entrainment and basal-ice formation. Subsequent research has identified two main thermally controlled processes responsible for basal-ice formation: regelation and basal adfreezing.

Regelation, as originally described by Weertman (1957), involves the melting of basal ice on the up-glacier side of an obstacle due to the development of excess pressure and an accompanying reduction in the pressure melting point. The resulting meltwater then flows to the low-pressure area on the lee side of the obstacle, where it refreezes, releasing latent heat of fusion. This heat is then conducted back to the stoss face where it contributes to further melting. As a result the mechanism involves the mass transport of water in a thin basal layer and the bulk transport of the whole ice mass above. Kamb & La Chappelle (1964) subsequently observed a layer of ice heavily laden with debris consisting of both mud and rock fragments 1–2 mm in diameter beneath the Blue Glacier in Washington, USA. They related this debris-rich layer to Weertman's process of regelation, and suggested that debris accumulates on the stoss side of obstacles and is then entrained into the regelation ice when the meltwater refreezes, producing banners of sediment that extend from the crests of the bedrock obstacles. The operation of this process has subsequently been advocated by many workers in the field, notably Lawson (1979) and Sugden *et al.* (1987).

Whilst regelation was originally conceived of as operating beneath rigid-bed glaciers, the process has now been extended to soft-bed glaciers. Iverson & Semmens (1995) have suggested that regelation may occur downwards into porous sediments as a result of regelation past individual grains. They argued that this mechanism may represent the dominant process of entrainment in soft-bedded environments, particularly where effective pressures are high. This process has subsequently been used by Christoffersen *et al.* (2006) in combination with congelation to explain the development of a variety of basal-ice facies (see later). Iverson *et al.* (2007) have recently provided the first field measurements of the process of 'regelation infiltration' by installing a prism of simulated till within a bedrock trough in an artificial cavity beneath Engabreen, a temperate glacier in Norway. During a 12-day experiment in 2001, when the cavity was allowed to close, the glacier bed was subsequently found to have infiltrated 50–80 mm into the upper part of the till prism at a rate close to that predicted by theoretical models (Philip 1980).

Boulton (1970), amongst others, pointed out that whilst regelation may explain the presence of debris within basal-ice, as it merely redistributes existing ice it is unable to explain the observed thickness of some basal-ice layers in subpolar regions, which can reach tens of metres in thickness (e.g. Knight *et al.* 1994). The process of basal adfreezing provides an alternative mechanism of basal-ice formation capable of generating thicker basal-ice sequences as it involves the net addition of ice.

Basal adfreezing was initially proposed by Weertman (1961), and has been widely advocated as the principal cause of thick and laterally extensive basal-ice sequences beneath polythermal glaciers (e.g. Knight *et al.* 1994; Zdanowicz *et al.* 1996). Weertman (1961) proposed his 'freezing model' after concluding that the shearing hypothesis of Goldthwait (1951) failed to provide a tenable explanation for a series of ice-cored moraines observed near Thule, Greenland. Weertman argued that under the thick ice of the ice-sheet interior, the temperature gradient would be insufficient to drain away the heat generated at the bed. This excess heat would lead to basal melting, producing water that flows along the local pressure gradient towards the margin. As the ice thins nearer the margin, the temperature gradient steepens, removing all heat generated at the bed and returning the base to freezing conditions, where the water refreezes. As a result, there is a net transport of mass from the ice-sheet interior to a location closer to the margin, allowing the accretion of thick basal-ice sequences.

Glaciohydraulic supercooling provides an alternative mechanism of basal adfreezing and basal-ice formation that has recently been described to explain the formation of thick basal-ice beneath temperate glaciers (Alley *et al.* 1998; Lawson *et al.* 1998, see Cook *et al.* 2006 for a detailed review). The process is associated with the pressure-dependence of the melting point of ice and the flow of subglacial water through topographic basins or 'overdeepenings'. At the base of the overdeepening, the melting temperature is depressed because of the overburden pressure, and water can remain liquid at temperatures below the pressure melting point. As water is forced by the pressure gradient to ascend the adverse slope out of the overdeepening, the pressure-dependent melting temperature gradually rises towards 0 °C as the overburden pressure decreases. If the adverse bed slope is sufficiently steep (>1.2–1.7 times the gradient of the ice-surface slope) the heat generated and the rate of increase in water temperature will be insufficient to match the changing pressure melting point and the water will become supercooled (Alley *et al.* 2003).

Whilst the theoretical basis for glaciohydraulic supercooling has been established for some time (e.g. Röthlisberger 1968), field evidence of the process has not been identified until recently. Work at the Matanuska Glacier, Alaska, reported by Lawson *et al.* (1998) has provided evidence for the supercooling of subglacial discharge via direct temperature measurements and the observation of features such as anchor ice terraces around supercooled outlets. Owing principally to physical, sedimentological and isotopic similarities between the basal ice and ice accreting around vents discharging supercooled water, the authors have also related the process to the formation of a thick basal-ice layer exposed at the glacier margin (Fig. 1). Subsequent work in Iceland by Roberts *et al.* (2002) has suggested that the process comprises the dominant mechanism of basal-ice formation under normal flow conditions and that it is also capable of entraining large quantities of sediment during extreme flow conditions (i.e. jökulhlaups). However, additional research has either failed to find evidence for the operation of glaciohydraulic supercooling where large overdeepenings occur (e.g. Swift *et al.* 2006), or has concluded that the process cannot explain the observed variability in basal-ice facies (Cook *et al.* 2007). Consequently, the relative importance of the process versus other established mechanisms of basal-ice formation remains the subject of ongoing research.

Substantial progress has clearly been made in identifying the processes and products of basal-ice formation. However, the focus on thermally controlled processes that involve melting and refreezing has led to the question of how basal ice can be formed beneath cold-based glaciers on permafrost where temperatures remain below freezing. This issue is explored further in a later section.

The origin and palaeoenvironmental significance of massive ice

Many permafrost regions contain ground ice, a term used to refer to any type of ice formed in freezing or frozen ground (French 2007). Large bodies of massive ice constitute one of the most spectacular forms of ground ice, and have been widely documented in both the western Canadian Arctic (e.g. French & Harry 1990; Mackay & Dallimore 1992) and western Siberia (e.g. Vtyurin & Glazsovskiy 1986; Astakhov & Isayeva 1988; Astakhov *et al.* 1996). Massive ice is defined by an ice content in excess of 250% by weight (Harris *et al.* 1988) and ordinarily occurs as large, tabular bodies that can reach over 10 m in thickness. Whilst the distribution, stratigraphic setting, and physical and chemical characteristics of massive ice have been increasingly well constrained, its mode of origin and palaeoenvironmental significance is subject to debate.

There are two principal theories for the origin of massive ice. First, massive ice represents exceptionally large bodies of segregated intrusive ice created by subaerial permafrost aggradation and high sub-permafrost porewater pressures (e.g. Mackay 1971, 1989; Mackay & Dallimore 1992). Once aggrading permafrost reaches or generates a pressurized aquifer, a sustained water flow to the freezing front allows the development of a thick, tabular body of 'intrasedimental ice'. Supporting evidence has come from the analysis of borehole logs in the western Canadian Arctic which indicates that massive ice frequently occurs at the boundary between clay-rich materials and underlying sands and gravels, i.e. at the boundary between materials of high and low frost-susceptibility (Mackay 1971). In addition, coastal exposures at sites like Peninsula Point near Tuktoyaktuk indicate a prominent subhorizontal layering and a conformable contact with the overlying sediments (Fig. 2). Finally, ice dykes extending upwards from the top of massive ice into the overburden demonstrate the existence – during massive ice formation – of porewater pressures sufficiently high to hydrofracture the overlying frozen sediments (Mackay & Dallimore 1992). Rampton (1988*a*, 1991) has argued that high porewater pressures and massive-ice formation relate to a glacially imposed hydraulic gradient driving subglacial meltwater to an aggrading proglacial permafrost table during deglaciation (see also Lacelle *et al.* 2004).

An alternative explanation for the origin of massive ice is that it represents buried glacier ice that has been preserved within permafrost following ice retreat. The preservation of glacier ice for centuries within ice-marginal sediments under non-permafrost conditions is well established (e.g. Everest & Bradwell 2002), whilst Sugden *et al.* (1995) have suggested that under permafrost conditions glacier ice can be preserved for millions of years. This hypothesis is the preferred explanation for massive ice in western Siberia by many Russian scientists, who suggest that the widespread,

Fig. 2 Massive intrasedimental ice with prominent sub horizontal stratification exposed at Peninsula Point, western Canadian Arctic (69°24′34″N, 133°07′46″W). The ice face is approximately 5 m high.

prolonged preservation of glacier ice within the permafrost indicates 'incomplete' or 'retarded' deglaciation (e.g. Astakhov & Isayeva 1988; Grosvald *et al.* 1986; Kaplyanskaya & Tarnogradskiy 1986; Astakhov *et al.* 1996). Key lines of evidence for this mode of origin include the stratigraphic setting and structural characteristics of the massive ice. In contrast to intrasedimental ice, massive ice of glacial origin has an unconformable upper contact and is often entirely surrounded by diamicton, whose low permeability precludes an open-system water supply to an aggrading freezing front. Consequently, such ice is frequently described within moraine complexes (e.g. French & Harry 1988; Dyke & Savelle 2000; Lacelle *et al.* 2007). In addition, the ice frequently contains both large- and small-scale tectonic structures such as thrust faults and recumbent folds (e.g. French & Harry 1990; Astakhov *et al.* 1996) (Fig. 3). Such features are widely considered indicative of shear deformation within the basal layers of an ice sheet and to be inconsistent with an origin through segregation following deglaciation, although Rampton (1988*a*, 1991) has argued that such features might be imparted on intrasedimental ice by glacier re-advance. Finally, some workers have suggested that buried glacier ice displays a distinctive isotopic signature (e.g. Lorrain & Demeur 1985; Lacelle *et al.* 2007).

Distinguishing between these modes of origin is crucial to our understanding of their significance. If they represent intrasedimental ice then they provide important information on the climatic regime at the time of their formation. Alternatively, if they represent buried glacier ice then they provide an unparalleled source of palaeoglaciological information. As the latter hypothesis involves a glacial origin, it is clear that its rigorous appraisal requires a thorough knowledge and understanding of the physical, chemical and isotopic characteristics of the full range of glacial-ice facies. An example of how this can assist in the interpretation of massive ice is provided in the following sections.

Interdisciplinary research

Given that basal-ice and permafrost researchers examine fundamentally identical materials using similar techniques, it is unsurprising that there are significant overlapping areas of research that would benefit from more collaboration. The following section examines three such examples before considering the current barriers to and future possibilities afforded by a more interdisciplinary approach.

Examples

Basal-ice formation in permafrost environments

With the majority of proposed mechanisms of basal-ice formation involving melting and

Fig. 3 Debris-rich massive ice with folded clear ice layers exposed at Mason Bay, western Canadian Arctic (69°32′31″N, 134°02′56″W). The ice face is approximately 3 m high.

refreezing over a variety of spatial scales, the development of basal-ice sequences beneath cold-based glaciers resting on permafrost where there is negligible meltwater available for refreezing has provided an ongoing glaciological problem. This is symptomatic of a broader issue within glaciology concerning the plausibility of the assumption that basal processes cease at temperatures below the pressure melting point (e.g. Waller 2001).

Shaw (1977) suggested that cold-based glaciers entrain debris when they override frontal aprons of ice and sediment. Many polar and subpolar glaciers terminate in a vertical cliff, the foot of which is masked by an apron of ice blocks, refrozen meltwater, supraglacial and englacial debris, and wind-blown sediment. This apron may be entrained by glacier advance, and subsequently folded, attenuated and foliated by flow metamorphism. This model has subsequently been applied by Evans (1989), who described the dominance of apron entrainment in the formation of basal-ice layers in glaciers in the Canadian High Arctic. The shearing hypothesis originally proposed by Goldthwait (1951) has also experienced a renaissance as a potential explanation for basal-ice formation beneath cold-based glaciers, with Tison *et al.* (1993) concluding that the overriding of active ice over stagnant ice and subsequent shearing was the only tenable hypothesis of debris entrainment at a site at the margin of the Antarctic ice sheet.

Continued research on basal-ice formation beneath cold-based glaciers has emphasized the importance of glacier–permafrost interactions. The excavation of a 26 m-long tunnel beneath the Suess Glacier in the Dry Valleys of Antarctica (e.g. Fitzsimons 2006) has revealed the presence of a heterogeneous basal-ice layer up to 3.8 m thick containing large blocks of sediment with well-preserved sedimentary structures thought to reflect the erosion and entrainment of subglacial permafrost.

Observations by the authors at the margin of the Russell Glacier in western Greenland suggest that glacier–permafrost interactions play a significant role in the development of the basal-ice layer at this site. This outlet glacier discharges ice from the southwestern sector of the Greenland ice sheet, approximately 15 km east of Kangerlussuaq (67°06′N, 50°09′W), and terminates in a region of continuous permafrost where permafrost thicknesses are estimated at 127 ± 31 m (van Tatenhove 1995). The Russell Glacier has experienced significant oscillations during the last century, with a retreat phase between 1943 and 1968, and a subsequent re-advance between the early 1970s and 1999 of more than 200 m in places (Knight *et al.* 2000). The basal-ice layer at this site has been extensively studied (e.g. Knight 1989; Knight *et al.* 1994; Waller *et al.* 2000) and comprises two main facies: a conspicuously debris-rich stratified facies and an overlying dispersed facies.

Knight *et al.* (1994) suggested that the stratified facies is formed by the entrainment and deformation of blocks of old snow, frozen till and laminated ice–debris layers. Observations associated with the recent re-advance of the glacier margin suggest that glacier–permafrost interactions play an important role in the formation of the stratified

Fig. 4 Frozen moraine sediment adhering to the base of the Russell Glacier, western Greenland (67°06′12″N, 50°13′30″W) and being ripped-up by continued ice motion. The metre rule is given for scale.

facies basal ice with frozen moraine sediments being ripped up by the advancing ice margin (Fig. 4), with the result that the entrained basal ice and moraine sediments display very similar particle-size distributions.

It is suggested that in the absence of freezing or regelation, entrainment occurs in response to the transmission of basal shear stress from the glacier bed into the frozen subglacial sediment, resulting in the traction of a layer of subglacial sediment and its entrainment as a layer of stratified facies ice. In other words, the glacier couples with the underlying permafrost and the effective bed of the glacier (the plane beneath which glacier-induced motion ceases) shifts from the ice–bed interface to a plane of weakness within the underlying substrate. In permafrost environments the base of the permafrost layer may be characterized by high porewater pressures and, therefore, provides a potential plane of weakness that can allow the entrainment and subsequent deformation of the overlying frozen material (Mathews & Mackay 1960; Astakhov *et al.* 1996). Alternatively, planes of localized weakness may occur within the permafrost in the form of ice-rich or clay-rich layers (Astakhov *et al.* 1996; Fitzsimons *et al.* 1999). Whilst this provides a viable hypothesis of basal-ice entrainment in the absence of freezing and thawing, it does generate an additional set of uncertainties. Fitzsimons (2006) has pointed out that in such situations it can be difficult, if not impossible, to identify the exact location of the glacier bed. Whilst this is traditionally viewed as being texturally defined by a clear ice rock or ice sediment interface, in circumstances where a glacier with basal ice overrides ice-rich sediment then no such boundary is evident. Fitzsimons (2006) argues that the glacier bed is more accurately viewed as a heterogeneous zone in which the lower boundary of glacier-induced flow is dependent on thermal and mechanical processes, and likely, therefore, to vary spatially and temporally.

Glaciologists have started to acknowledge that cold-based glaciers can couple with frozen substrates (e.g. Echelmayer & Zhongxiang 1987; Cuffey *et al.* 2000). In addition, permafrost researchers have identified evidence for the widespread and deep-seated deformation of permafrost in response to it being overridden by Pleistocene ice sheets (e.g. Astakhov *et al.* 1996; Murton *et al.* 2004). Therefore, in the same way in which glacial geologists have worked in close collaboration with glaciologists to constrain the mechanisms and glaciological significance of subglacial sediment deformation beneath warm-based ice, closer collaboration between glaciologists working on modern-day cold-based glaciers, and permafrost researchers examining glacier–permafrost interactions beneath former ice sheets, is required to fully understand the mechanisms of sediment entrainment and mobilization beneath cold-based glaciers.

Frost heave, basal-ice formation and ice-stream dynamics

Recent work on basal freezing and ice accretion beneath ice streams (Christoffersen & Tulaczyk 2003*a*, *b*) has already benefitted from the use of existing models of frost heave originally devised by permafrost engineers. Their findings have major implications both for our understanding of the dynamic behaviour of ice streams (Bougamont *et al.* 2003) and the glaciological significance of distinctive basal-ice facies (Christoffersen *et al.* 2006).

In comparing borehole observations from two ice streams of the West Antarctic Ice Sheet, the active Whillans Ice Stream (formerly called Ice Stream B) and the slow-moving Kamb Ice Stream (formerly called Ice Stream C), Christoffersen & Tulaczyk (2003*a*) observed that the latter was characterized by: (1) comparatively steep basal temperature gradients; (2) comparatively thick layers of debris-rich basal ice approximately 12–25 m thick; (3) unfrozen till displaying supercooled temperatures (up to -0.35 °C); and (4) a decrease in till porosity close to the ice–sediment interface. They hypothesized that whilst the rapid motion of the Whillans Ice Stream is promoted by active basal melting and a reduction in effective pressure and shear strength within the subjacent till, the stagnation of the Kamb Ice Stream was caused by basal freeze-on and an associated consolidation and strengthening of the subglacial till.

In attempting to develop a numerical, vertical-column model of the freezing process in such a situation, Christoffersen & Tulaczyk (2003*b*, p. 2) noted that 'there is a paucity of theoretical and empirical investigations of heat, water, and solute flow during basal freeze-on'. As a result, recognizing the general similarity of ice formed by basal freezing and by near-surface frost heave, the authors utilized theoretical treatments of frost heave originally developed by permafrost engineers. In particular, they used the Clapeyron equation that provides a means for coupling pressure and temperature in a freezing porous medium, and which provides the fundamental basis for frost-heave models (e.g. Fowler & Krantz 1994). In addition, whilst ice–water phase changes are often considered solely the result of temperature and pressure, Christoffersen & Tulaczyk (2003*b*) also considered the roles played by both solute concentration and interfacial effects. The latter are particularly important in fine-grained porous media such as the clay-rich till sampled beneath the nearby Whillans Ice Stream (Tulaczyk *et al.* 1998), in which the curvature of ice–water interfaces can lead to high interfacial pressures, a substantial reduction in the freezing point and the presence of supercooled water (e.g. Willliams & Smith 1989).

Christoffersen & Tulaczyk (2003*b*) provided model results for both high interfacial pressures (i.e. fine-grained till) and low interfacial pressures (i.e. coarse-grained till) under two hydrogeological settings (closed- and open-water systems). In the case of a closed-water system, freeze-on extracts porewater resulting in till consolidation, an increase in shear strength and a reduction in ice velocity. For both high and low interfacial pressures, till strength matches the driving stress after 65 years and the ice stream shuts down. Subsequently, the freezing front migrates into the subglacial till causing further till consolidation and the development of segregated ice lenses, whose thickness and spacing are determined by depth and by surface tension (i.e. particle size). In the case of an open-water system, the timing of ice-stream shutdown and the spacing of ice lenses are similar, although the individual ice lenses are thicker and the degree of till consolidation is more limited.

The model's prediction of the development of intercalated layers of debris-rich ice (frozen till) and debris-poor ice (segregation ice) is consistent with borehole observations of a debris-rich basal ice layer 10–14 m thick described by Vogel *et al.* (2005). They described a basal-ice layer comprising alternating layers of clear ice and debris-rich ice layers, with the thickness and sediment content of debris-rich layers generally increasing towards the ice–sediment interface. A rapid transition from relatively clean ice to sediment-rich ice is argued to reflect a change from abundant water availability to limited water availability, in turn associated with a shutdown of the Kamb Ice Stream approximately 300 years ago. In addition, the presence of decimetre-thick layers of clear ice close to the ice–sediment interface and a basal water layer is thought to suggest imminent ice-stream reactivation, with a water supply in excess of that consumed by basal freeze-on leading to a relubrication of the bed.

Christoffersen *et al.* (2006) have subsequently used the same model to quantitatively define the development of four types of basal ice beneath soft-bedded glaciers whose formation is associated with subglacial frost-heave processes and controlled by the delivery of water to the freezing front. Clear ice (type I) forms when the influx of water matches or exceeds the freezing rate, such that the heat budget is satisfied with no extraction of water from the underlying till. In the other three ice types, a progressive reduction in the water supply results in enhanced ice–bed coupling and debris entrainment through regelation. Laminated ice (type II) is produced by periodic regelation events associated with an influx of groundwater slightly lower than the freezing rate. Brief periods of soft-bed regelation entrain thin layers of debris-rich ice, whilst intervening longer periods of congelation entrain thicker layers of clear ice. Massive regelation ice (type III) is created when the influx of water is roughly half that required to satisfy the freezing rate. Regelation events are still episodic, but the increase in effective pressure results in an increase in the regelation rate and the thickness of the regelation ice layers, ultimately generating a facies of approximately 50% debris by volume. Finally, solid dirty regelation ice (type IV) is associated with a closed system whereby all the water is drawn from the till pore spaces. In this case, very little congelation ice is generated and the effective stress increases rapidly, producing a rapidly thickening regelation layer with a sediment content of more than 60% by volume (i.e. lacking excess ice). Rempel *et al.* (2007) identified some problems with the modelling framework used by Christoffersen *et al.* (2006), notably concerning the spatial distribution of ice saturation. Further discussion on this topic is given by Christoffersen *et al.* (2007).

The origin of massive ice in glaciated permafrost terrains

Massive ice within glaciated permafrost terrains has usually been interpreted either as buried glacier ice or as intrasedimental ice formed in subaerial permafrost regions. Distinguishing between them is fundamental to understanding the interactions between glaciers and permafrost, and the characteristics and behaviour of former ice sheets. However, interpretation of massive ice has often been problematic (Vtyurin & Glazovskiy 1986; French & Harry 1990) because intrasedimental ice is difficult to distinguish from basal glacier ice, since both ice types form by the same freezing processes operating in porous media (cf. Mackay 1989; Rempel 2008) and, therefore, display similar physical and chemical properties. Indeed, it seems increasingly likely that the only real distinction between them is spatial, rather than genetic, because massive ice can form within sedimentary sequences (i.e. intrasedimentally) beneath, at the margin or in front of glaciers, or in permafrost regions unrelated to glacial activity. Thus, the more useful distinctions amongst massive-ice occurrences in permafrost regions are between: (1) subglacial or submarginal ice (e.g. Murton *et al.* 2005); (2) proglacial ice (e.g. Lacelle *et al.* 2004); and (3) non-glacial ice (e.g. Lawson 1983). The lack of a clear genetic distinction between basal glacier ice and massive intrasedimental ice may also explain why rigorous hypothesis testing to discriminate them has proved difficult. Instead, the genetic discrimination is clearer between firnified glacier ice and other ice types (Cardyn *et al.* 2007; Lacelle *et al.* 2007).

An example of a collaborative research programme bringing together glaciologists and permafrost scientists in order to interpret the origin and spatial significance of massive ice has been carried out in the Tuktoyaktuk Coastlands of western Arctic Canada (Murton *et al.* 2004, 2005). Here many exposures of massive ice and icy sediments at former ice-marginal sites of the Laurentide Ice Sheet (LIS) contain features characteristic of both so-called basal and intrasedimental ice. Basal-ice features comprise: (1) ice facies and facies groupings similar to those from the basal-ice layers of contemporary glaciers and ice sheets in Alaska, Greenland and Iceland (Fig. 3); (2) ice-crystal fabrics similar to those from basal ice in Antarctica and ice-cored moraines on Axel Heiberg Island, Canada; and (3) a thaw or erosional unconformity along the top of the massive ice and icy sediments, buried by glacigenic or aeolian sediments. Intrasedimental ice consists of pore ice and segregated ice formed within Pleistocene sands deposited before glacial overriding.

The co-existence of basal and intrasedimental ice within massive ice and icy sediments suggests that they formed within the basal ice layer of the LIS. This layer is thought to have developed by accretion of both new and existing ice, its formative mechanisms including: (a) large-scale freeze-on of meltwater and sediment at the transition from warm- to cold-based ice; (b) permafrost aggradation beneath thinning, stagnant basal ice; and (c) porewater expulsion in ice-free areas prior to glacial overriding. Stagnation of the ice sheet and melt-out of till from the ice surface allowed burial and preservation of the basal-ice layer on a regional scale in the Tuktoyaktuk Coastlands. At other sites close to but within the Laurentide ice limit, buried glacier ice beneath frozen glacitectonite that contains subglacially eroded clasts of ground ice indicates that near-surface permafrost did not degrade beneath some parts of the ice margin (cf. Mackay *et al.* 1972; Mackay & Matthews 1983; Rampton 1988*b*). Interestingly, permafrost has also been inferred beneath the margin of Pleistocene ice sheets in Europe (Haeberli 1981).

To integrate all of the studies of massive ice in the Tuktoyaktuk Coastlands, a two-stage model of ground-ice development has been proposed based on the cryostratigraphic distinction of two generations of ground ice: pre- and post-deformation ice (Murton 2005). Pre-deformation ice was glacially deformed or eroded beneath the cold-based margin of the LIS during Marine Isotope Stage 2 (Murton *et al.* 2007), and comprises: (1) buried basal ice; (2) massive segregated ice; and (3) ice clasts subglacially eroded from pre-existing ground ice. In contrast, post-deformation ice has not been glacially disturbed because it formed during or after deglaciation, and includes; (4) dykes and sills of intrusive ice; (5) massive segregated intrusive ice; (6) ice wedges and composite wedges; (7) segregated ice; and (8) pool ice. The superimposition of post-deformation ice into permafrost containing older, pre-deformation ice indicates that substantial quantities of overpressurized water were injected into ice-marginal permafrost during or after deglaciation. The required external water source for the post-deformation intrusive ice was probably overpressurized subpermafrost groundwater in front of the retreating margin of the LIS. Injection of this water into proglacial permafrost hydraulically fractured the permafrost and formed ice dykes, ice sills and massive segregated intrusive ice. This model may have wider application to the development of massive ice wherever ice sheets advanced and retreated across lowland regions of continuous permafrost.

Current problems

A number of problems currently act as barriers to more effective collaboration between the glaciological and permafrost research communities. One of the most fundamental problems involves the contrasting nature of the schemes employed to describe and classify ice–sediment mixtures. With the classification schemes used within basal ice (e.g. Lawson 1979; Sharp *et al.* 1994) and permafrost research (e.g. Murton & French 1994; Shur & Jorgenson 1998) involving different parameters (e.g. volumetric ice content, particle size, ice-crystal size, bubble content, distribution of ice and sediment, etc.), it is currently difficult to translate between the resulting classifications. The resolution of this problem through the development of a more comparable classification scheme appears straightforward in theory but may prove harder to achieve in practice. Within the field of basal-ice research, for example, despite concerns that 'any attempt to compare field descriptions from different publications is hampered by a babel of classifications as diverse as the number of workers in the field' (Knight 1993, p. 352), there is still no universally accepted classification scheme (see however Hubbard *et al.* in press). Until this situation is resolved through deliberate and determined collaboration between the two communities, any attempt to compare massive-ice facies with basal-ice facies as a starting point when determining the origin of massive ice will continue to prove problematic.

An additional concern involves the contrasting foci of researchers within the two communities. Whilst glaciologists are principally concerned with processes and their products, permafrost scientists are often required to carefully examine the lithostratigraphic and cryostratigraphic context of the ice–sediment mixture being studied. Identification of the origin of massive ice, for example, often involves the resolution of problematic stratigraphic issues such as dating and facies correlation (e.g. Harry *et al.* 1988). The requirement to engage with such issues may deter the active involvement of some glaciologists.

Future research possibilities

Powerful incentives to overcome these problems are the future research possibilities offered by the remnants of ice sheets buried in Arctic and Antarctic permafrost. Such remnants are thought to be widespread in, for example, western Siberia (e.g. Kaplyanskaya & Tarnogradskiy 1986), northern European Russia (Henriksen *et al.* 2003), the western Canadian Arctic (Dyke & Savelle 2000; Dyke & Evans 2003) and East Antarctica (Sugden *et al.* 1995). If these ice masses can be reliably distinguished from non-glacial ice types, then these products of 'incomplete' or 'retarded' deglaciation provide a palaeoglaciological archive of immense value. For example, Grosvald *et al.* (1986) estimated that approximately 10 000 km^3 of relict glacier ice remains preserved within western Siberia alone. Whilst this is likely to represent an overestimate, even if only a fraction of this is related to Pleistocene ice sheets then its detailed investigation still has the potential to revolutionize our knowledge and understanding of the stratigraphy, physical properties and dynamic behaviour of continental ice sheets within high-latitude regions. Grosvald *et al.* (1986) consequently argued that this ice should be the subject of special glaciological studies. Such an invitation is particularly pressing in the twenty first century as the dramatic rise in temperatures within the Arctic threatens the loss of this archive before its importance was ever appreciated.

Conclusions

It is clear that there are a variety of interdisciplinary research problems within glacial and permafrost environments, whose resolution requires greater collaboration between glaciologists and permafrost researchers. Whilst such interdisciplinary research has to date been rather limited, the examples discussed earlier illustrate the benefits that such an approach could bring to both disciplines. In particular, the application of frost-heave models to subglacial settings in order to elucidate the influence of ice accretion on till properties and the flow of ice streams has potentially major implications for our understanding of the dynamic behaviour of ice sheets.

Greater co-operation is also required to examine the nature of glacier–permafrost interactions beneath cold-based, soft-bedded glaciers. Whilst glaciologists working on modern-day glaciers have traditionally assumed that basal processes cease to operate at temperatures below the pressure melting point, permafrost researchers have long recognized the ability of both sediments to retain liquid water at subfreezing temperatures and of glaciers to actively couple with permafrost. Further examination of the operation of these processes in both past and present subglacial environments is required to evaluate their potential influence on glacier flow and sediment transport.

There are some obstacles to effective collaboration that need to be targeted and resolved before real progress can be made. The resolution of such problems, however, has the potential to generate a wide range of new research possibilities. The most obvious includes the development of an ice–sediment-classification scheme that will allow detailed comparisons to be made between the ice facies observed at modern-day glaciers with those observed in exposures of massive ice. This would greatly assist in the accurate identification of buried glacier ice with permafrost environments which, by unlocking a new palaeoglaciological archive, could revolutionize our understanding of past ice sheets.

The authors gratefully acknowledge the financial and logistical support that enabled a series of field expeditions to the western Canadian Arctic and Greenland where these ideas were developed. Financial support was provided by the University of Greenwich (RW), the Leverhulme Trust (PGK & JM), the Tyrell Fund of the Geological Society (JM) and the Royal Society (RW & PGK). The authors would also like to thank the referees (S. Harrison and one anonymous) for helpful comments that led to substantial improvements to this review.

References

ALLEY, R. B., LAWSON, D. E., EVENSON, E. B., STRASSER, J. C. & LARSON, G. J. 1998. Glaciohydraulic supercooling: a freeze-on mechanism to create stratified, debris-rich basal ice. II. Theory. *Journal of Glaciology*, **44**, 563–569.

ALLEY, R. B., LAWSON, D. E., LARSON, G. J., EVENSON, E. B. & BAKER, G. S. 2003. Stabilizing feedbacks in glacier-bed erosion. *Nature*, **424**, 758–760.

ARCONE, S. A., LAWSON, D. E. & DELANEY, A. J. 1995. Short-pulse wavelet recovery and resolution of dielectric constants within englacial and basal ice of Matanuska Glacier, Alaska, U.S.A. *Journal of Glaciology*, **41**, 68–86.

ASTAKHOV, V. I. & ISAYEVA, L. L. 1988. The 'Ice Hill': an example of 'retarded deglaciation' in Siberia. *Quaternary Science Reviews*, **7**, 29–40.

ASTAKHOV, V. I., KAPLYANSKAYA, F. A. & TARNOGRADSKIY, V. D. 1996. Pleistocene permafrost of West Siberia as a deformable glacier bed. *Permafrost and Periglacial Processes*, **7**, 165–191.

BOUGAMONT, M., TULACZYK, S. & JOUGHIN, I. 2003. Response of subglacial sediments to basal freeze-on 2. Application in numerical modeling of the recent stoppage of Ice Stream C, West Antarctica. *Journal of Geophysical Research*, **108**(B4), 2223, doi: 10.1029/2002JB001936.

BOULTON, G. S. 1970. On the origin and transport of englacial debris in Svalbard glaciers. *Journal of Glaciology*, **9**, 213–229.

CARDYN, R., CLARK, I. D., LACELLE, D., LAURIOL, B., ZDANOWICZ, C. M. & CALMELS, F. 2007. Molar gas ratios of air entrapped in ice: a new tool to determine the nature and origin of relict massive ground-ice bodies in permafrost. *Quaternary Research*, **68**, 239–248.

CHRISTOFFERSEN, P. & TULACZYK, S. 2003*a*. Thermodynamics of basal freeze-on: predicting basal and subglacial signatures of stopped ice streams and interstream ridges. *Annals of Glaciology*, **36**, 233–243.

CHRISTOFFERSEN, P. & TULACZYK, S. 2003*b*. Response of subglacial sediments to basal freeze-on. 1. Theory and comparison to observations from beneath the West Antarctic Ice Sheet. *Journal of Geophysical Research*, **108**(B4), 2222, doi: 10.1029/2002JB001935.

CHRISTOFFERSEN, P., TULACZYK, S., CARSEY, F. D. & BEHAR, A. E. 2006. A quantitative framework for interpretation of basal ice facies formed by ice accretion over subglacial sediment. *Journal of Geophysical Research*, **111**, F01017, doi: 10.1029/2005JF000363.

CHRISTOFFERSEN, P., TULACZYK, S., BEHAR, A. & CARSEY, F. 2007. Reply to comment by A. W. Rempel *et al.* on 'A quantitative framework for interpretation of basal ice facies formed by ice accretion over subglacial sediment'. *Journal of Geophysical Research*, **112**, F02037, doi: 10.1029/2006JF000732.

COOK, S. J., KNIGHT, P. G., WALLER, R. I., ROBINSON, Z. P. & ADAM, W. G. 2007. The geography of basal ice and its relationship to glaciohydraulic supercooling: Svínafellsjökull, southeast Iceland. *Quaternary Science Reviews*, **26**, 2309–2315.

COOK, S. J., WALLER, R. I. & KNIGHT, P. G. 2006. Glaciohydraulic supercooling: the process and its significance. *Progress in Physical Geography*, **30**, 577–588.

CUFFEY, K. M., CONWAY, H. *ET AL.* 2000. Entrainment at cold glacier beds. *Geology*, **28**, 351–354.

DALLIMORE, S. R. & DAVIS, J. L. 1992. *Ground Penetrating Radar Investigations of Massive Ice*. Geological Survey of Canada, Paper, **90-4**, 41–48.

DYKE, A. S. & EVANS, D. J. A. 2003. Ice-marginal terrestrial landsystems: northern Laurentide and Innuitian ice sheet margins. *In*: EVANS, D. J. A. (ed.) *Glacial Landsystems*. Arnold, London, 143–165.

DYKE, A. S. & SAVELLE, J. M. 2000. Major end moraines of Younger Dryas age on Wollaston Peninsula, Victoria Island, Canadian Arctic: implications for paleoclimate and for formation of hummocky moraine. *Canadian Journal of Earth Sciences*, **37**, 601–619.

ECHELMEYER, K. & ZHONGXIANG, W. 1987. Direct observation of basal sliding and deformation of basal drift at subfreezing temperatures. *Journal of Glaciology*, **33**, 83–98.

EVANS, D. J. A. 1989. Apron entrainment at the margins of sub-polar glaciers, North-west Ellesmere Island, Canadian High Arctic. *Journal of Glaciology*, **35**, 317–324.

EVEREST, J. & BRADWELL, T. 2002. Buried glacier ice in southern Iceland and its wider significance. *Geomorphology*, **52**, 347–358.

FITZSIMONS, S. J. 2006. Mechanical behaviour and structure of the debris-rich basal ice layer. *In*: KNIGHT, P. G. (ed.) *Glacier Science & Environmental Change*. Blackwell, Oxford, 329–335.

FITZSIMONS, S. J., MCMANUS, K. J. & LORRAIN, R. D. 1999. Structure and strength of basal ice and substrate of a dry-based glacier: evidence for substrate deformation at sub-freezing temperatures. *Annals of Glaciology*, **28**, 236–240.

FOWLER, A. C. & KRANTZ, W. B. 1994. A generalized secondary frost heave model SIAM. *Journal of Applied Mathematics*, **54**, 1650–1675.

FRENCH, H. M. 2007. *The Periglacial Environment*, 3rd edn. Wiley, Chichester.

FRENCH, H. M. & HARRY, D. G. 1988. Nature and origin of ground ice, Sandhills Moraine, southwest Banks Island, Western Canadian Arctic. *Journal of Quaternary Science*, **3**, 19–30.

FRENCH, H. M. & HARRY, D. G. 1990. Observations on buried glacier ice and massive segregated ice, Western Arctic Coast, Canada. *Permafrost and Periglacial Processes*, **1**, 31–44.

GOLDTHWAIT, R. P. 1951. Development of end moraines in east-central Baffin Island. *Journal of Geology*, **59**, 567–577.

GROSVALD, M. G., VTYURIN, B. I., SUKHODROVSKIY, V. L. & SHISHORINA, Z. G. 1986. Underground ice in western Siberia: origin and geological significance. *Polar Geography and Geology*, **10**, 173–183.

HAEBERLI, W. 1981. Ice motion on deformable sediments. *Journal of Glaciology*, **27**, 365–366.

HARRIS, C. & MURTON, J. B. (eds). 2005. *Cryospheric Systems: Glaciers and Permafrost*. Geological Society, London, Special Publications, **242**.

HARRIS, S. A., FRENCH, H. M., HEGINBOTTOM, J. A., JOHNSTON, G. H., LADANYI, B., SEGO, D. A. & VAN EVERDINGEN, R. O. 1988. *Glossary of Permafrost and Related Ground-ice Terms*. Permafrost Subcomittee on Geotechnical Research, National Research Council of Canada, Ottawa, Technical Memorandum, **142**.

HARRY, D. G., FRENCH, H. M. & POLLARD, W. H. 1988. Massive ground ice and ice-cored terrain near Sabine Point, Yukon Coastal Plain. *Canadian Journal of Earth Sciences*, **25**, 1846–1856.

HENRIKSEN, M., MANGERUD, J., MATIOUCHKOV, A., PAUS, A. & SVENDSEN, J. I. 2003. Lake stratigraphy implies an 80 000 yr delayed melting of buried dead ice in northern Russia. *Journal of Quaternary Science*, **18**, 663–679.

HUBBARD, B. & SHARP, M. J. 1989. Basal ice formation and deformation: a review. *Progress in Physical Geography*, **13**, 529–558.

HUBBARD, B. & SHARP, M. J. 1995. Basal ice facies and their formation in the Western Alps. *Arctic and Alpine Research*, **27**, 301–310.

HUBBARD, B., COOK, S. & COULSON, H. in press. Basal ice facies: a review and unifying approach. *Quaternary Science Reviews*, doi: 10.1016/j.quascirev.2009.03.005.

IVERSON, N. R. & SEMMENS, D. J. 1995. Intrusion of ice into porous media by regelation: mechanism of sediment entrainment by glaciers. *Journal of Geophysical Research*, **100**(B7), 10,219–10,230.

IVERSON, N. R., HOOYER, T. S. *ET AL.* 2007. Soft-bed experiments beneath Engabreen, Norway: regelation infiltration, basal slip and bed deformation. *Journal of Glaciology*, **53**, 323–340.

KAMB, B. & LA CHAPPELLE, E. 1964. Direct observations of the mechanism of glacier sliding over bedrock. *Journal of Glaciology*, **5**, 159–172.

KAPLYANSKAYA, F. A. & TARNOGRADSKIY, V. D. 1986. Remnants of the Pleistocene ice sheets in the permafrost zone as an object for paleoglaciological research. *Polar Geography and Geology*, **10**, 257–266.

KNEISEL, C., HAUCK, C., FORTIER, R. & MOORMAN, B. 2008. Advances in geophysical methods for permafrost investigations. *Permafrost and Periglacial Processes*, **19**, 157–178.

KNIGHT, P. G. 1988. The basal ice and debris sequence at the margin of an equatorial ice-cap; El Cotopaxi, Ecuador. *Geografiska Annaler*, **70A**, 9–13.

KNIGHT, P. G. 1989. Stacking of basal debris layers without bulk freezing-on: isotopic evidence from West Greenland. *Journal of Glaciology*, **35**, 214–216.

KNIGHT, P. G. 1993. Glaciers (progress report). *Progress in Physical Geography*, **17**, 349–353.

KNIGHT, P. G. 1997. The basal ice layer of glaciers and ice sheets. *Quaternary Science Reviews*, **16**, 975–993.

KNIGHT, P. G., SUGDEN, D. E. & MINTY, C. D. 1994. Ice flow around large obstacles as indicated by basal ice exposed at the margin of the Greenland ice sheet. *Journal of Glaciology*, **40**, 359–367.

KNIGHT, P. G., WALLER, R. I., PATTERSON, C. J., JONES, A. P. & ROBINSON, Z. P. 2000. Glacier advance, ice-marginal lakes and routing of meltwater and sediment: Russell Glacier, Greenland. *Journal of Glaciology*, **46**, 423–426.

LACELLE, D., BOJORNSON, J., LAURIOL, B., CLARK, I.D. & TROUTET, Y. 2004. Segregated-intrusive ice of subglacial meltwater origin in retrogressive thaw flow headwalls, Richardson Mountains, NWT, Canada. *Quaternary Science Reviews*, **23**, 681–696.

LACELLE, D., LAURIOL, B., CLARK, I. D., CARDYN, R. & ZDANOWICZ, C. 2007. Nature and origin of a Pleistocene-age massive ground-ice body exposed in the Chapman Lake moraine complex, central Yukon Territory, Canada. *Quaternary Research*, **68**, 249–260,

LAWSON, D. E. 1979. *Sedimentological Analysis of the Western Terminus Region of the Matanuska Glacier, Alaska*. US Army, Cold Regions Research and Engineering Laboratory, Report, **79-9**.

LAWSON, D. E. 1983. Ground ice in perennially frozen sediments, Northern Alaska. *In*: *Permafrost: Fourth International Conference, Proceedings, Fairbanks, Alaska*. National Academy Press, Washington, DC, 695–700.

LAWSON, D. E., STRASSER, J. C., EVENSON, E. B., ALLEY, R. B., LARSON, G. J. & ARCONE, S. A. 1998. Glaciohydraulic supercooling:a freeze-on mechanism to create stratified, debris-rich basal ice: I. Field evidence. *Journal of Glaciology*, **44**, 547–562.

LAWSON, W. 1996. The relative strengths of debris-laden basal ice and clean glacier ice: some evidence from Taylor Glacier, Antarctica. *Annals of Glaciology*, **23**, 270–276.

LORRAIN, R. D. & DEMEUR, P. 1985. Isotopic evidence for relic Pleistocene glacier ice on Victoria Island, Canadian Arctic archipelago. *Arctic & Alpine Research*, **17**, 89–98.

MACKAY, J. R. 1971. The origin of massive icy beds in permafrost, western Arctic Coast, Canada. *Canadian Journal of Earth Sciences*, **8**, 397–422.

MACKAY, J. R. 1989. Massive ice: some field criteria for the identification of ice types. Current Research, Part G, Geological Survey of Canada, Paper, **89-1G**, 5–11.

MACKAY, J. R. & DALLIMORE, S. R. 1992. Massive ice of the Tuktoyaktuk area, western Arctic coast, Canada. *Canadian Journal of Earth Science*, **29**, 1235–1249.

MACKAY, J. R. & MATTHEWS, J. V. 1983. Pleistocene ice and sand wedges, Hooper Island, Northwest Territories. *Canadian Journal of Earth Sciences*, **20**, 1087–1097.

MACKAY, J. R., RAMPTON, V. N. & FYLES, J. G. 1972. Relic Pleistocene permafrost, Western Arctic, Canada. *Science*, **176**, 1321–1323.

MATHEWS, W. H. & MACKAY, J. R. 1960. Deformation of soils by glacier ice and the influence of pore pressures and permafrost. *Transactions of the Royal Society of Canada*, **54**, 27–36.

MURRAY, T. & PORTER, P. R. 2001. Basal conditions beneath a soft-bedded polythermal surge-type glacier: Bakaninbreen, Svalbard. *Quaternary International*, **86**, 103–116.

MURTON, J. B. 2005. Ground-ice stratigraphy and formation at North Head, Tuktoyaktuk Coastlands, western Arctic Canada: a product of glacier–permafrost interactions. *Permafrost and Periglacial Processes*, **16**, 31–50.

MURTON, J. B. & FRENCH, H. M. 1994. Cryostructures in permafrost, Tuktoyaktuk Coastlands, Western Arctic Canada. *Canadian Journal of Earth Sciences*, **31**, 737–747.

MURTON, J. B., FRECHEN, M. & MADDY, D. 2007. Luminescence dating of Mid to Late Wisconsinan aeolian sand as a constraint on the last advance of the Laurentide Ice Sheet across the Tuktoyaktuk Coastlands, western Arctic Canada. *Canadian Journal of Earth Sciences*, **44**, 857–869.

MURTON, J. B., WALLER, R. I., HART, J. K., WHITEMAN, C. A., POLLARD, W. & DALLIMORE, S. R. 2004.

Stratigraphy and glaciotectonic structures of permafrost deformed beneath the northwest margin of the Laurentide ice sheet, Tuktoyaktuk Coastlands, Canada. *Journal of Glaciology*, **50**, 399–412.

MURTON, J. B., WHITEMAN, C. A., WALLER, R. I., POLLARD, W. & DALLIMORE, S. R. 2005. Basal ice facies and supraglacial melt-out till of the Laurentide Ice Sheet, Tuktoyaktuk Coastlands, western Canadian Arctic. *Quaternary Science Reviews*, **24**, 681–708.

PHILIP, J. R. 1980. Thermal fields during regelation. *Cold Regions Science & Technology*, **3**, 193–203.

RAMPTON, V. N. 1988*a*. Origin of massive ground ice on Tuktoyaktuk Peninsula, Northwest Territories, Canada: a review of stratigraphic and geomorphic evidence. *In*: SENNESET, K. (ed.) *Proceedings of the 5th International Conference on Permafrost, 2–5 August 1988*, Trondheim, Norway, Volume 1. Tapir, Trondheim, 850–855.

RAMPTON, V. N. 1988*b*. *Quaternary Geology of the Tuktoyaktuk Coastlands, Northwest Territories*. Geological Survey of Canada.

RAMPTON, V. N. 1991. Observations on buried glacier ice and massive segregated ice, Western Arctic Coast: Discussion (Short Communication). *Permafrost and Periglacial Processes*, **2**, 163–165.

REA, B. R. & WHALLEY, W. B. 1994. Subglacial observations from Oksfjordjokelen. *Earth Surface Processes and Landforms*, **19**, 659–673.

REMPEL, A. W. 2007. Formation of ice lenses and frost heave. *Journal of Geophysical Research*, **112**, pF02S21, doi: 10.1029/2006JF000525.

REMPEL, A. W. 2008. A theory for ice-till interactions and sediment entrainment beneath glaciers. *Journal of Geophysical Research*, **113**, F01013, doi: 10.1029/2007JF000870.

REMPEL, A. W., WETTLAUFER, J. S. & WORSTER, M. G. 2007. Comment on 'A quantitative framework for interpretation of basal ice facies formed by ice accretion over subglacial sediment' by Poul Christoffersen *et al*. *Journal of Geophysical Research*, **112**, F02036, doi: 10.1029/2006JF000701.

ROBERTS, M. J., TWEED, F. S. *ET AL*. 2002. Glaciohydraulic supercooling in Iceland. *Geology*, **30**, 439–442.

RÖTHLISBERGER, H. 1968. Erosive processes which are likely to accentuate or reduce the bottom relief of valley glaciers. *International Association of Hydrological Sciences Publication*, **79**, 87–97.

SHARP, M. J., JOUZEL, J., HUBBARD, B. & LAWSON, W. 1994. The character, structure and origin of the basal ice layer of a surge-type glacier. *Journal of Glaciology*, **40**, 327–340.

SHAW, J. 1977. Till body morphology and structure related to glacier flow. *Boreas*, **6**, 189–201.

SHUR, Y. L. & JORGENSON, M. T. 1998. Cryostructure development on the floodplain of the Colville River delta, northern Alaska. *In*: LEWKOWICZ, A. G. & ALLARD, M. (eds) *Proceedings of the 7th International Permafrost Conference, Yellowknife, Canada, 23–27 June 1998*. Collection Nordicana, Centre d'études nordiques, Université Laval, 993–999.

SUGDEN, D. E., KNIGHT, P. G., LIVESEY, N., LORRAIN, R. D., SOUCHEZ, R. A., TISON, J. L. & JOUZEL, J. 1987. Evidence for two zones of debris entrainment beneath the Greenland ice sheet. *Nature*, **328**, 238–241.

SUGDEN, D. E., MARCHANT, D. R., POTTER, N., JR, SOUCHEZ, R. A., DENTON, G. H., SWISHER, C. C. III, & TISON, J. L. 1995. Preservation of Miocene glacier ice in East Antarctica. *Nature*, **376**, 412–414.

SWIFT, D. A., EVANS, D. J. A. & FALLICK, A. E. 2006. Tranverse englacial debris-rich ice bands at Kvíárjökull, southeast Iceland. *Quaternary Science Reviews*, **25**, 1708–1718.

TISON, J. L., PETIT, J. R., BARNOLA, J. M. & MAHANEY, W. C. 1993. Debris entrainment at the ice–bedrock interface in sub-freezing temperature conditions (Térre Adelie, Antarctica). *Journal of Glaciology*, **39**, 303–315.

TULACZYK, S., KAMB, B., SCHERER, R. P. & ENGELHARDT, H. F. 1998. Sedimentary processes at the base of a West Antarctic ice stream: constraints from textural and compositional properties of subglacial debris. *Journal of Sedimentary Research*, **68**, 487–496.

VAN TATENHOVE, F. G. M. 1995. *The Dynamics of Holocene Deglaciation in West Greenland with Emphasis on Recent Ice-marginal Processes*. PhD thesis, Universiteit van Amsterdam.

VOGEL, S. W., TULACZYK, S. *ET AL*. 2005. Subglacial conditions during and after stoppage of an Antarctic Ice Stream: Is reactivation imminent? *Geophysical Research Letters*, **32**, L14502, doi: 10.1029/2005GL022563.

VTYURIN, B. I. & GLAZOVSKIY, A. F. 1986. Composition and structure of the Ledyanaya Gora tabular ground ice body on the Yenisey. *Polar Geography and Geology*, **10**, 273–285.

WALLER, R. I. 2001. The influence of basal processes on the dynamic behaviour of cold-based glaciers. *Quaternary International*, **86**, 117–128.

WALLER, R. I., HART, J. K. & KNIGHT, P. G. 2000. The influence of tectonic deformation on facies variability in stratified debris-rich basal ice. *Quaternary Science Reviews*, **19**, 775–786.

WEERTMAN, J. 1957. On the sliding of glaciers. *Journal of Glaciology*, **3**, 33–38.

WEERTMAN, J. 1961. Mechanism for the formation of inner moraines found near the edge of cold ice caps and ice sheets. *Journal of Glaciology*, **3**, 965–978.

WILLIAMS, P. J. & SMITH, M. V. 1989. *The Frozen Earth: Fundamentals of Geocryology*. Cambridge University Press, Cambridge.

ZDANOWICZ, C. M., MICHEL, F. A. & SHILTS, W. W. 1996. Basal debris entrainment and transport in glaciers of southwestern Bylot Island, Canadian Arctic. *Annals of Glaciology*, **22**, 107–113.

Proglacial, periglacial or paraglacial?

OLAV SLAYMAKER

Department of Geography, University of British Columbia, 1984 West Mall, Vancouver, British Columbia, Canada V6T 1Z2 (e-mail: olav@geog.ubc.ca)

Abstract: The terms proglacial and periglacial are well-understood descriptors of contemporary and past environments, but the paraglacial concept is more controversial and has prompted vigorous debate. Definitions are reviewed and the paraglacial concept is considered critically. It is argued that the term 'paraglacial' defined as 'non-glacial processes conditioned by glaciation' describes landscapes that are adjusted neither to Last Glacial Maximum nor to contemporary geomorphic processes. Where a landscape is paraglacial it can be characterized in terms of rate of change and trajectory of that change. It cannot be defined in relation to glaciers (as in proglacial) or by cold-climate processes (as in periglacial). Almost all paraglacial landforms and all paraglacial landscapes are transient and transitional. An interesting challenge of paraglacial landscapes is then to determine their rates of change; how far they have advanced along the trajectory from glacial to non-glacial; and how to recognize empirically the temporal and spatial relationships between proglacial, periglacial, paraglacial and fluvial landscapes. Implications of this approach to paraglacial landscapes are discussed in relation to historical and dynamic geomorphology.

The terms 'periglacial' and 'proglacial' are clearly defined; but the term paraglacial, coined as recently as 1971 (Ryder 1971*a*, *b*), is more controversial (Ballantyne 2002*a*). It is argued here that paraglaciation is concerned primarily with the long-term disturbance regime imposed by glaciation on landforms and landscapes, and opens up new questions of transience and transition in Earth systems (Schumm 1973; Brunsden 1980, 1993; Hewitt 2006; Slaymaker & Kelly 2007). The main purpose of the present paper is to cast light on the concept of paraglaciation, as compared with the well-established understanding of 'periglacial' and 'proglacial'.

Introduction

There are two definitions of 'periglacial' that are in common use: (a) an environment of frequent freeze–thaw cycles and deep seasonal freezing (encompassing about 35% of the Earth's continental surface); and/or (b) a permafrost environment (which covers only 20%) (French 2007). The word connotes distinctive processes, landforms and landscapes (Worsley 2004). The literal meaning of the term proglacial is 'in front of the glacier' (Penck & Bruckner 1909), and emphasis is placed on processes and landforms in close proximity to the ice margin. The word describes distinctive processes, sediment–landform associations and landform assemblages in glacifluvial, glacilacustrine and glacimarine environments (Benn & Evans 1998). In contrast to these, the term 'paraglacial', defined as 'non-glacial processes conditioned by glaciation' (Church & Ryder 1972), is more contentious, as discussed in a major paper on paraglacial geomorphology (Ballantyne 2002*a*), and does not depend either on unique processes or unique location. In fact, paraglacial environments may include both proglacial and periglacial environments, as well as fluvial and mass movement landforms and processes.

The term 'paraglacial' is now being used without careful distinction between it and the long-established traditional term 'proglacial' (e.g. Iturrizaga 1999). Indeed, and in part as a result of Ballantyne's magisterial papers (Ballantyne 2002*a*, *b*), the term paraglacial is now also being used to cover such a variety of circumstances, including aeolian and coastal environments, that the word could well become redundant. It is the lack of clarity of the use of the term 'paraglacial' that forms the motivation for this paper. We review briefly the definitions of periglacial and proglacial processes, landforms and environments. But the substance of the paper considers paraglaciation, ways in which it differs from periglacial and proglacial, and implications of the paraglacial concept for our understanding of geomorphology in general.

Periglacial systems, processes and landform assemblages

The presence of permafrost and intense frost activity defines a periglacial environment uniquely and, classically, such environments are adjusted to a certain average regime of permafrost development and freeze–thaw cycle frequency and magnitude.

Proximity to glaciers is irrelevant to the definition. Periglacial systems can be found in: (a) the

From: KNIGHT, J. & HARRISON, S. (eds) *Periglacial and Paraglacial Processes and Environments*.
The Geological Society, London, Special Publications, **320**, 71–84.
DOI: 10.1144/SP320.6 0305-8719/09/$15.00

polar desert and semi-deserts of the high Arctic; (b) the tundra zone; (c) the boreal forest zone; (d) the maritime and continental sub-arctic; and (e) mid- and low-latitude alpine areas (French 2007). The most distinctive periglacial landforms are associated with permafrost, and include tundra polygons, pingos and palsas, and thermokarst features where permafrost is melting. In intense frost areas, coarse, angular rock debris, tors and a wide variety of patterned ground are formed. Periglacial processes are effective in breaking down bedrock and in forming distinctive local-scale forms, particularly through heaving and sorting processes *in situ*. They are not, however, notably effective in evacuating those sediments to make them available for fluvial transport and deposition. Consequently, periglacial landforms are somewhat subdued in the landscape and their preservation in the geological record is limited. Two notable exceptions to this generalization are the extensive periglacial 'head' and stratified slope deposits of northwestern Europe (DeWolf 1988), and the remarkable, still poorly explained, phenomenon of the cryoplain (Washburn 1979).

Proglacial systems, sediment–landform associations and landform assemblages

Location immediately in the front of glaciers defines a proglacial environment uniquely, and, classically, such environments are adjusted to the average regime of fluvial, lacustrine and marine processes that occur immediately adjacent to the ice. The hydrology of proglacial rivers exhibits a singular pattern of flow, both seasonally and diurnally. Moderate flood flows are common and extraordinary jökulhlaup floods occur in front of many glaciers. Sediment is therefore frequently entrained, and fluvial sedimentary features evolve rapidly (Church & Gilbert 1975). There is very little overlap in the appearance of proglacial and periglacial features, although it is common that proglacial and periglacial environments may overlap.

Glacifluvial proglacial erosional forms include drainage diversions and spillways. Depositional forms include sandar (outwash plains), valley trains and braided outwash fans. Braided river facies are characterized by Miall (1978) as one of four types (Trollheim, Scott, Donjek and Platte types). In addition, jökulhlaup-dominated facies have also been recognized (Benn & Evans 1998). Glacilacustrine proglacial forms include strandlines (erosional) and deltas and delta moraines (depositional) (Hicks *et al.* 1990; Hasholt *et al.* 2000). Facies are either deltaic or lake bottom sediments. Glacimarine proglacial forms are generally considered in two settings: (1) the fjord environment where sedimentation is influenced by tide water or floating glaciers, rivers, slope and marine processes; and (2) the continental shelf and deep ocean where sedimentation is dominated by grounded ice margins, ice shelves and open-marine processes.

Periglacial and proglacial systems as equilibrium systems

It is commonly the case that periglacial and proglacial processes and forms are analysed under the assumption that they represent process–response

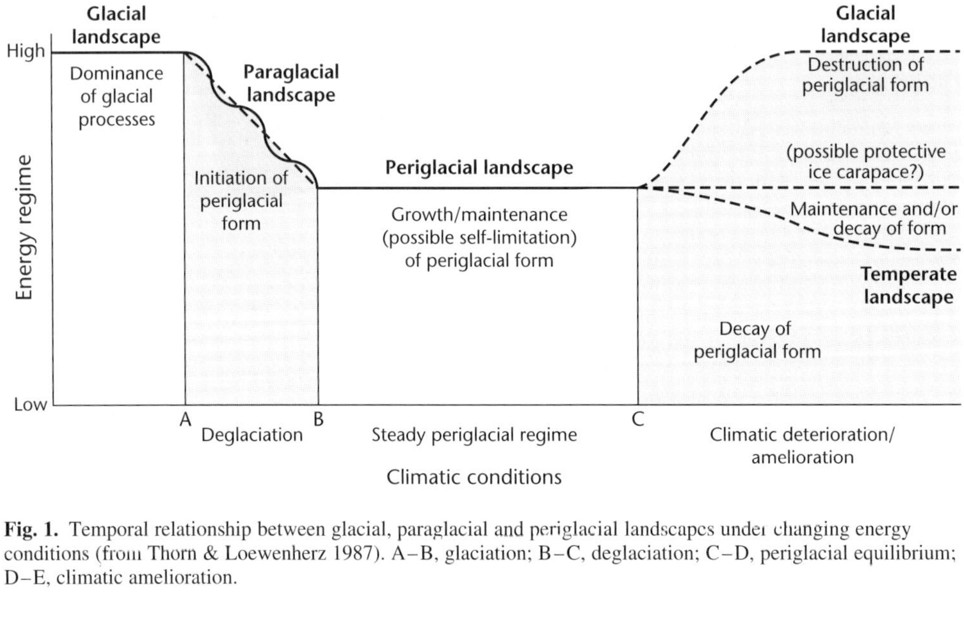

Fig. 1. Temporal relationship between glacial, paraglacial and periglacial landscapes under changing energy conditions (from Thorn & Loewenherz 1987). A–B, glaciation; B–C, deglaciation; C–D, periglacial equilibrium; D–E, climatic amelioration.

system adjustments to governing conditions that are well known and understood, as in the preceding paragraphs. In fact, French (2007) pointed out that most periglacial landscapes have recently emerged from under continental ice sheets and that they are still in the process of adjustment to contemporary thermal and precipitation regimes. In order to characterize 'never glaciated periglacial terrain' he identified the landforms of Banks Island, N.W.T. (Canada) and the Barn Mountains, Yukon Territory (Canada), and demonstrated the large role of fluvial processes in those landscapes. Thorn & Loewenherz (1987) (Fig. 1) proposed a model of temporal relationships between glacial, paraglacial and periglacial landscapes under changing energy conditions. They emphasized the extent to which the evolution of periglacial forms at local scale is superimposed on the longer term evolution of periglacial landscapes, which, at any given time, may have differing proportions of glacial, fluvial and periglacial elements.

Paraglacial systems: origin and evolution

The evolution of the word paraglacial, literally meaning 'beyond the glacier', has been described by Slaymaker (2004). Ballantyne (2002*a*) identified the slow process of adoption of the term outside North America between 1971 and 1983, and the dramatic increase in its usage post-1984 (Fig. 2). Since 1984 he saw four trends: (1) an extension of the geomorphic contexts in which the paraglacial concept has been explicitly used; (2) a focusing of research on present-day paraglacial processes and land systems; (3) use of the paraglacial concept as a framework for research across a wide range of contrasting deglacial environments; and (4) a growing awareness of the palaeoenvironmental significance of paraglacial facies in Quaternary stratigraphic studies.

The term paraglacial was introduced by Ryder (1971*a*, *b*) to describe alluvial and colluvial fans that had accumulated through the reworking of glacial sediments by rivers and debris flows following late Wisconsinan deglaciation in the interior of British Columbia (Canada) (Figs 3 and 4). She showed that fan accumulation had been initiated soon after valley floors became ice free and continued until shortly after the deposition of Mazama tephra (6600 years BP). The paraglacial concept was formalized by Church & Ryder (1972). The circumstance that generated the idea was their recognition of the analogue between contemporary proglacial sandur development on Baffin Island (Fig. 5) and the post-glacial evolution of presently inactive alluvial fans in the semi-arid interior of British Columbia. Church & Ryder (1972) defined the paraglacial environment as one that is characterized by 'non-glacial processes that are directly

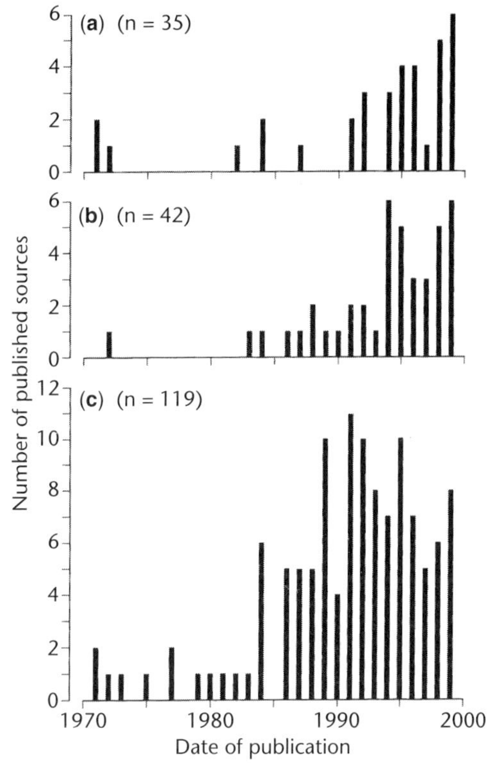

Fig. 2. Annual increase in the number of publications dealing with paraglacial geomorphology (after Ballantyne 2002*a*). (**a**) Publications with 'paraglacial' in the title; (**b**) publications using the term 'paraglacial' in the context of processes operating on recently deglaciated terrain; (**c**) publications using the term 'paraglacial' in the context of landscape adjustment following Pleistocene or Early Holocene deglaciation.

conditioned by glaciation' and identified three aspects of the influence of paraglacial glacigenic sediment supply on fluvial transport: (1) the dominant component of reworked sediment may shift from till to secondary sources, such as alluvial fans and valley fills; (2) regional uplift patterns will condition the timing of changes in the balance between fluvial deposition and erosion; and (3) consequently, the total period of paraglacial effect is prolonged beyond the period of initial reworking of glacigenic sediments. This concept was refined further by Clague (1986) in his discussion of the rhythm of geomorphic activity in British Columbia; by Slaymaker (1987) in identifying the role of the intermediate-scale basins as the sites of maximum specific sediment yield; and by Church & Slaymaker (1989) (Fig. 6) in their formulation of the sediment wave model and estimation of the relaxation time of paraglaciation.

Fig. 3. Paraglacial alluvial fan in Thompson River valley, B.C., NTS Sheet 92I/12 (photograph courtesy of J. M. Ryder). This is one of the fans that inspired the paraglacial concept.

Fig. 4. Paraglacial valley fill, central Fraser River valley (air photograph 1087–46 BC by the Province of B.C.). Fraser River trenched the valley fill in the early Holocene.

Fig. 5. Ekalugad valley, Baffin Island, containing a classic sandur (after Church 1972). This is the site that suggested an analogy between contemporary sandur evolution and the early Holocene paraglacial deposition of valley fill in Fraser valley.

Taking into account the full range of paraglacial research in the previous 30 years, Ballantyne (2002*a*) proposed a new working definition of paraglacial as 'non-glacial earth surface processes, sediment accumulations, landforms, land systems and landscapes that are directly conditioned by glaciation and deglaciation' (p. 1938). This definition added two things to the Church & Ryder definition, namely: (1) the fact that paraglaciation is a spatial scale-dependent concept, in that different paraglacial forms occur at different scales; and (2) that the process of deglaciation also conditions paraglaciation. The ensuing section will elaborate some of these ideas.

Components of the paraglacial concept

Landforms and land systems involved

Geomorphic contexts in which the term paraglacial is now being used include: (1) debris-cone, alluvial-fan and valley-fill deposits (e.g. Ryder 1971*a*, *b*;

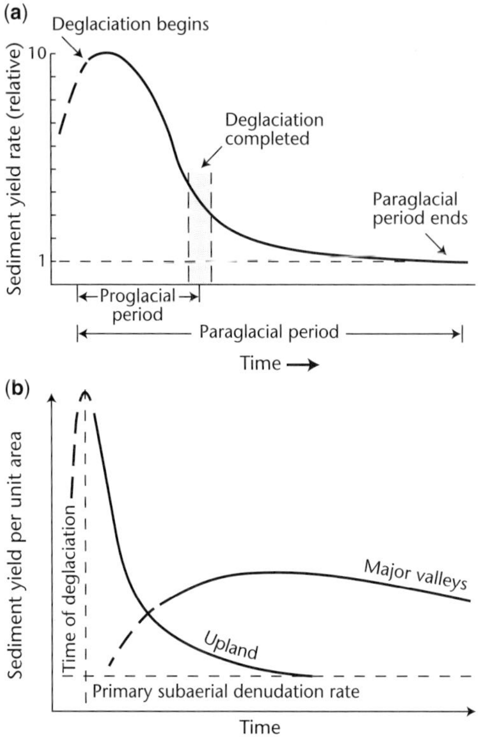

Fig. 6. The paraglacial environment according to: (**a**) Church & Ryder (1972), sediment yield as a function of time since deglaciation; and (**b**) Church & Slaymaker (1989), the sediment wave model.

Beaudoin & King 1994); (2) rock slopes (e.g. Wyrwoll 1977; Johnson 1984; Bovis 1990); (3) sediment-mantled slopes (Curry 2000); (4) glacier forefields (Matthews *et al.* 1998); (5) glacilacustrine systems (Shaw & Archer 1979); and (6) coastal systems (Forbes & Syvitski 1994).

Non-glacial Earth surface processes

All non-glacial processes, including mass movement, fluvial, lacustrine, aeolian, coastal and marine processes, are legitimately incorporated into paraglaciation on the original Church & Ryder (1972) definition of 'non-glacial processes conditioned by glaciation'.

Relation to proglacial

One way of clarifying the distinction between proglacial and paraglacial is in terms of distance from the glacier. Ice contact, proglacial and paraglacial are terms used for materials and landforms located at progressively greater distances from the ice front (Embleton-Hamann 2004). Pedagogically, this is a helpful approach, but there remains the problem of establishing meaningful thresholds between these categories (Table 1). The original Church & Ryder (1972) solution was to subsume the proglacial environment under the more general term paraglacial, and to identify the proglacial period as the earliest stage of the paraglacial period. Eyles & Kocsis (1988) objected to the latter strategy on the two grounds that: (1) the proglacial environment has a clear and separate definition in geology, and extending the term paraglacial to such processes is superfluous; and (2) the timescale of paraglacial sediment reworking was completely open-ended.

In response to the first objection, the concept of paraglaciation does no harm to the traditional understanding of proglacial forms and processes. The concept simply expands the context into which the local-scale characteristics of proglacial environments fit. In response to the second objection, see the discussion in the next subsection on 'Scale issues'.

Scale issues

Both Benn & Evans (1998) and Ballantyne (2002*a*) make the case that because there are no processes unique to paraglacial environments it would be better to think of paraglacial as referring to a period of time. Benn & Evans (1998) proposed that paraglacial should be defined as 'the period of rapid environmental adjustment following glacier retreat' (p. 261) and Ballantyne (2002*b*) has proposed to recast the definition of paraglacial to 'the time scale over which glacially conditioned sediment stores are either exhausted or attain stability' (p. 371).

These are two possible responses to the second of Eyles & Kocsis' objections. Benn & Evans (1998) consider paraglacial activity under: (1) terrestrial ice-marginal environments (local scale); (2) paraglacial associations of sediment and landforms (meso-scale paraglacial fans, terraces and slope deposits); and (3) the paraglacial land system (regional scale). Their typology elaborates both the temporal (time to sediment exhaustion) and the spatial (landforms, associations and systems) scale-dependence of the concept.

Although these contributions, from both Benn & Evans and from Ballantyne, are helpful clarifications and, indeed, extensions of the Church & Ryder definition, the writer is not persuaded that their solutions express the most innovative contribution of the paraglacial concept.

Sediment exhaustion v. sediment wave model

One specific issue that affects the understanding of the paraglacial environment, and hence directly

Table 1. *Classification of proglacial landforms (after Embleton-Hamann 2004)*

Environment	Process	Landform
Terrestrial, ice-marginal	Meltwater erosion	Ice-marginal meltwater channels
	Mass movement/meltwater deposition	Ice-marginal ramps and fans
		Dump and push moraines
		Recessional moraines
	Glacitectonics	Composite ridges and thrust block moraines Hill-hole pairs and cupola hills
		Hill-hole pairs and cupola hills
	Meltwater deposition	Kame and kettle topography
Subaquatic, ice-marginal	Mass movement/meltwater deposition	Morainal banks
		De Geer moraines
	Meltwater deposition	Grounding-line fans
		Ice-contact (kame-) deltas
	Debris flows	Grounding-line wedge
Transitional from ice-marginal to fluvial	Meltwater erosion	Scabland topography
		Spillways
	Meltwater deposition	Outwash plain (sandur)
		Outwash fan
		Valley train
		Pitted outwash
		Kettle hole/pond
Transitional from ice-marginal to lacustrine and marine	Meltwater deposition/ mass movement	Deltas
	Deposition from suspension settling and iceberg activity	Cyclopels, cyclopsams, varves
		Dropstone mud and diamicton
		Iceberg dump mounds
		Iceberg scour marks

impacts on the way in which it is modelled, concerns the way in which the unstable glacigenic sediment is released from storage. Release of sediment from storage has been considered in two main ways: (1) the release of the *in situ* glacigenic sediment; and (2) the re-entrainment of sediment previously reworked as, for example, in the entrenchment of paraglacial alluvial fans and valley fills (Church & Ryder 1972). Two competing models are provided in the literature: the cascade of sediment wave model (Church & Slaymaker 1989) (Fig. 6); and the sediment exhaustion model (Cruden & Hu 1993).

The sediment wave model. This model was suggested by Church & Slaymaker (1989) and was derived from a set of observations of specific sediment yield as a function of drainage basin area in British Columbia. Slaymaker (1987) had noted that a plot of specific sediment yield v. drainage basin area for the Canadian Cordillera departed significantly from the normally accepted relationship. That normal relationship for major river basins of the world shows maximum specific sediment yield in the smallest tributaries and a monotonic decrease with basin area (e.g. Milliman & Syvitski 1992). The conventional explanation of this relationship is that small basins are steeper, often experiencing more intense rainstorms and therefore higher erosion rates, and tended to produce higher specific sediment yield. But in the Canadian Cordillera (Slaymaker 1987), and more specifically within the province of British Columbia (Church & Slaymaker 1989), it was noted that specific sediment yield for basins between 10 km^2 and 30 000 km^2 increased monotonically and then beyond 30 000 km^2 the relation became negative, as is the 'normal case'. Apparently here, sediment is being entrained disproportionately rapidly by larger rivers in a downstream direction (it is, of course, perfectly normal for larger rivers to carry more sediment than smaller rivers, but the specific sediment yield normally decreases). Church & Slaymaker (1989) inferred that there must be a sediment wave that had already passed its peak in the smaller basins (hence the smaller basins have lower specific sediment yields at the present time), reaching a maximum specific sediment yield at around 30 000 km^2. A similar conclusion was reached by Ashmore (1993) in relation to the paraglacial landscapes of the Prairie provinces of Canada, and Church *et al.* (1999) subsequently showed that a similar situation exists throughout much of glaciated Canada. Harbor & Warburton (1993) extended the interpretation of the western Canadian data by showing that the pattern of paraglacial fluvial sediment transport can be described by a family of curves (Fig. 7). Over millennial timescales, a 'wave' of reworked sediment of primary or secondary glacigenic origin passes progressively from steep tributary basins into trunk valleys. Ballantyne (2002*a*) critiqued this model in that it suggests that the paraglacial reworking of glacigenic sediment was very limited before the wave of reworked sediment from upstream began to arrive. 'This seems unrealistic', he says, 'particularly in valleys where postglacial trunk streams inherited a thick and erodible valley fill or till, glacifluvial deposits or glaci-lacustrine sediments' (p. 1997). The point is well taken and is worth further investigation. However, there are two interim responses to Ballantyne's critique: (1) that the sediment exhaustion model has difficulty in reproducing the observed patterns in Canada; and (2) that the disaggregated paraglacial sediment waves can take into account the variable distribution of paraglacial sediment stores, as demonstrated by Harbor & Warburton (1993).

The sediment exhaustion model. This model, which is favoured by Ballantyne (2002*b*), takes the form:

$$S_t = S_0 e^{-Kt} \quad (1)$$

where t is time elapsed since deglaciation, S_t is the proportion of 'available' sediment remaining for

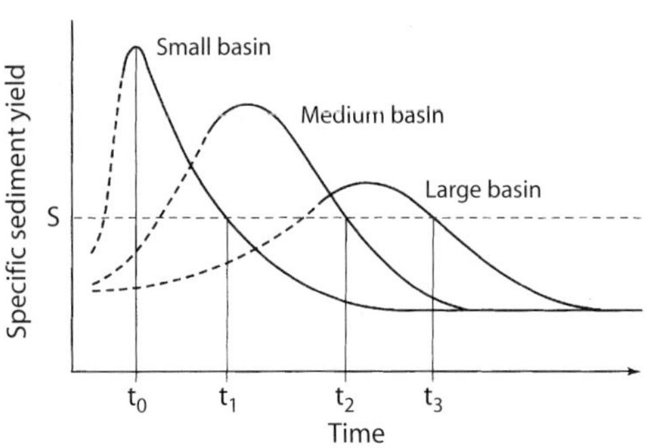

Fig. 7. The paraglacial sediment wave for drainage basins of different size according to Harbor & Warburton (1993). Specific sediment yield diminishes in amplitude and becomes progressively lagged behind deglaciation as drainage basin size increases.

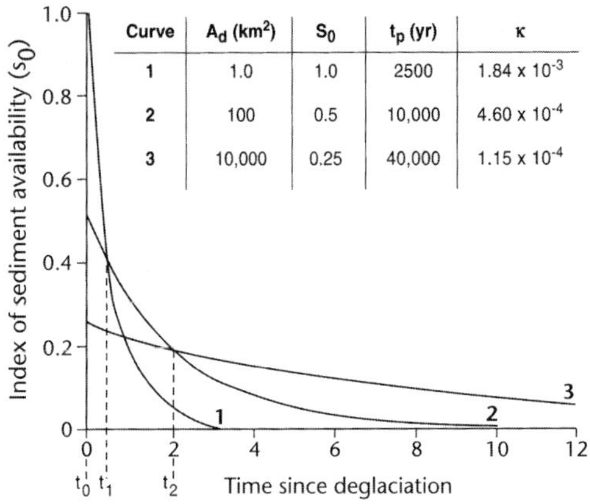

Fig. 8. Exponential sediment exhaustion model of paraglacial sediment release for drainage basins of different size. The model assumes that initial sediment availability is inversely related to basin size. The rationale for this assumption is that small basins are well coupled and larger basins are progressively less well coupled (after Ballantyne 2002*b*).

reworking at time t; S_0 is the total 'available' sediment at $t = 0$; and K is the rate of loss of 'available' sediment by either reworking or stabilization. If we assume that $S_0 = 1$ at $t = 0$, then the rate of loss of 'available' sediment (K) may be expressed as:

$$K = \ln(S_t)/-t. \quad (2)$$

The value of this model is that it allows calculation of the 'half-life' of the paraglacial system under investigation (Fig. 8). The model also provides an objective solution to the question of the total duration of the paraglacial period.

Complications to the orderly evolution of both the sediment wave model and the exhaustion model are identified below, but few field calibrations are available. The effects of episodic reduction in base level (Church & Ryder 1972), paraglacial coasts affected by marine transgression through a drumlin field providing transient periods of sediment availability (Forbes & Taylor 1987), extreme climatic events (Ballantyne & Whittington 1999) and many other circumstances may interrupt the steady-state conditions postulated by the exhaustion model (Fig. 9). A hypothetical sediment wave model for the Lillooet River valley in the Coast Mountains of British Columbia (Fig. 10), which includes interruptions from new sediment waves from volcanic, Neoglacial and human impact sources, was developed by Jordan & Slaymaker (1991). In 2005 Friele *et al.* produced independent calibration of the contribution of volcanic sediments to the architecture of the valley fill. Sediment from two major collapse events of the debuttressed slopes of Mt Meager was dated and traced along

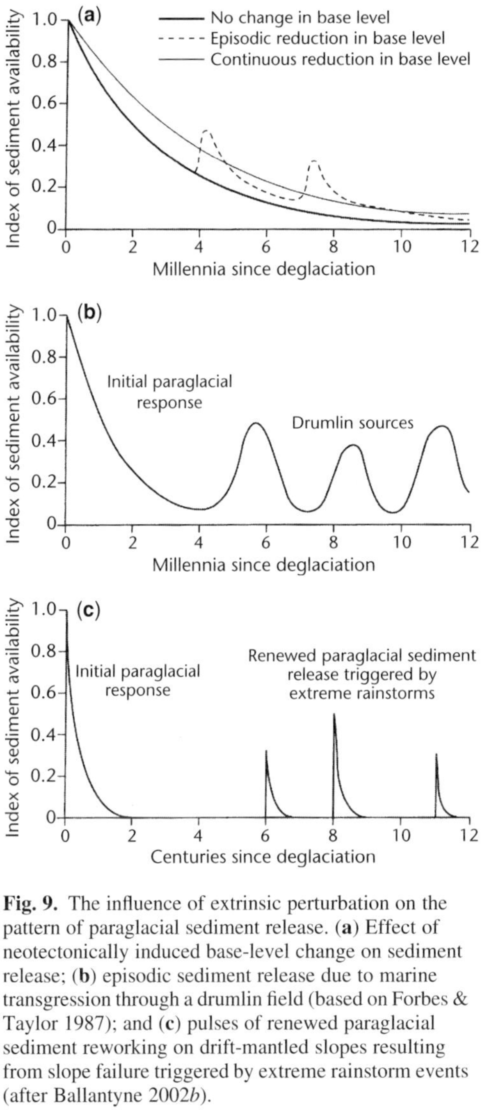

Fig. 9. The influence of extrinsic perturbation on the pattern of paraglacial sediment release. (**a**) Effect of neotectonically induced base-level change on sediment release; (**b**) episodic sediment release due to marine transgression through a drumlin field (based on Forbes & Taylor 1987); and (**c**) pulses of renewed paraglacial sediment reworking on drift-mantled slopes resulting from slope failure triggered by extreme rainstorm events (after Ballantyne 2002*b*).

the length of the valley in a series of drilling sites. The Neoglacial and human activity influences hypothesized by Jordan and Slaymaker have not been confirmed because it would seem that the signal of the very large mass movement events on the debuttressed volcano overwhelms the signal from other sources.

Transient landforms and transitional landscapes: implications for geomorphology

It seems probable that most of the world's glaciated landscapes are transitional or transient. If this is the case, then paraglaciation has major implications for our understanding of geomorphic change. The Church & Ryder (1972) definition of paraglacial as 'non-glacial processes conditioned by glaciation' remains a necessary condition for recognizing paraglaciation. But it may not be a sufficient condition to express the innovativeness of the concept. A paraglacial environment is an environment that is in the process of recovering from disturbance. It is not adjusted to any average hydroclimate or runoff conditions. It is explicitly in the process of becoming adjusted in the aftermath of the massive upheaval of the Last Glacial Maximum.

Because of these differences between paraglacial and other geomorphic environments, it seems to us to be preferable to define paraglacial as a descriptor of landforms and landscapes that are in transition from glacial to non-glacial conditions. This transitional landscape lasts until the glacially conditioned sediment stores are either removed or attain stability (Schumm & Rea 1995). Paraglaciation and its associated processes and forms are therefore, we suggest, neither a unique process nor a location-based or a temporal concept, but they have a dynamic systems definition involving a rate of change and a trajectory.

Two other questions follow: (i) does the process of onset of continental glaciation also define a paraglacial context (i.e. the transition from non-glacial to glacial)?; and (ii) what is the best way to characterize the progress of the process of adjustment of any specified paraglacial landscape? Neither of these two questions has been explored in depth in the literature, although Clague (1986) introduced a discussion of the first question through his model of the rhythm of geomorphic work (Fig. 11b). He noted the evidence of thick accumulations of advance and recessional outwash in the lower Fraser River valley as signalling periods of maximum geomorphic work. In relation to the second question, research is ongoing in mapping process domains in glaciated river basins in the southern Coast Mountains of British Columbia (Brardinoni & Hassan 2006). They have proposed a metric that characterizes the extent to which contemporary hydroclimate and runoff have imposed a fluvial signal on this rugged glaciated landscape.

Implications for historical geomorphology

Although there has been huge progress in developing the chronology of the Quaternary (Bowen 1978), temperature fluctuations over the past 800 000 years have been calibrated (EPICA 2004), and the advance and retreat of the major ice sheets during the Wisconsinan is better understood than ever before (Siegert 2001); there has been limited progress in developing new theoretical frameworks

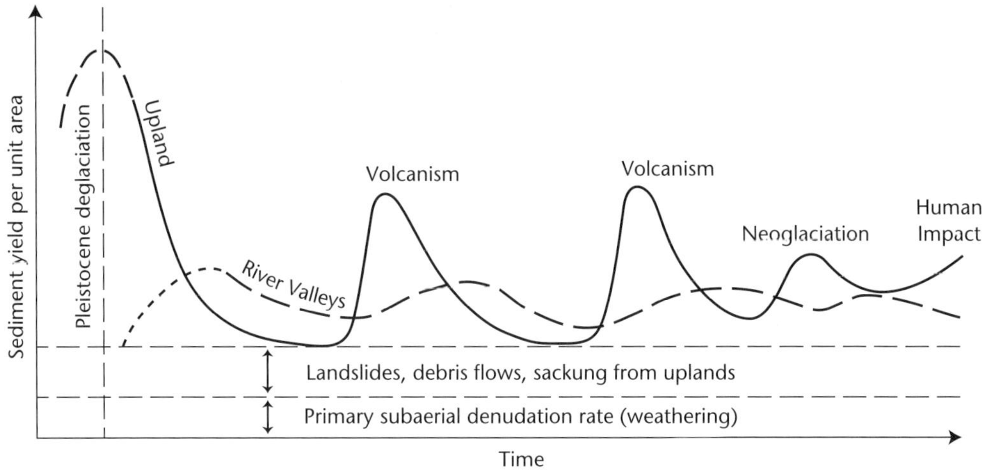

Fig. 10. Theoretical pattern of paraglacial sediment release in Lillooet River valley, B.C. (Jordan & Slaymaker 1991). The pulses labelled 'volcanic' refer to large, Holocene mass movement events on debuttressed slopes of a Quaternary volcanic complex (field calibration provided by Friele *et al.* 2005); the Neoglacial and human-disturbance pulses are presumed but not confirmed empirically.

for the historical evolution of landscapes. In this context, the paraglacial model as defined by Ryder (1971*a*, *b*) and Church & Ryder (1972) constitutes one of the more interesting advances. Knox (1984) and Clague (1986) in North America, and Starkel (1987) in Europe, have pursued the topic of rhythmicity, periodicity and predictability in landform evolution (Fig. 11). More recently, Church (1996), Ballantyne (2002*a*) and Hewitt (2002, 2006) have reflected on the nature of transitions in landscape history. That which is underlined in all of these studies is the alternation of periods of rapid change with periods of comparative quiescence, a concept which is consistent with that of 'punctuated equilibrium' advanced by Eldredge & Gould (1972) in palaeobiology.

Implications for dynamic geomorphology

Schumm's (1973) 'complex response to disturbance' model emphasized erosion following threshold exceedance at specific locations within a landscape that is otherwise largely inactive. Attention was thereby focused on disturbances at the micro- or local scale. Brunsden (1980, 1993) enlarged the idea to landscape scale and over geological time, and explored the relation between impulse and form response. He described landscape reaction time and relaxation time followed by 'characteristic form' time (Fig. 12).

The interesting question, which is raised by Hewitt (2006) and Slaymaker & Kelly (2007), is whether glaciated landscapes, for example, can achieve 'characteristic form' during the short interglacial periods of the order of 20 000–30 000 years. The answer from our work in British Columbia is scale-dependent. Small headwater systems of less than 1 km^2 have exported the original and reworked glacigenic sediments; larger systems have not yet achieved characteristic form (Church *et al.* 1999). The implications for dynamic geomorphology are that the focus of much research on glaciated landscapes needs to be on transitional landscapes or transient systems (Brardinoni & Hassan 2006). The fact that 'the landscape is imprisoned in its history' (Church & Slaymaker 1989, p. 454) means that there is a priority to assess contingencies in the landscape alongside the energy and mass fluxes that have figured so prominently in so-called process geomorphology. Both contingency and immanence (*sensu* Simpson 1963) have to be taken into account.

Conclusions

We conclude that the most fruitful use of the term paraglacial is radically different from the way in which the apparently related terms periglacial and proglacial are used. The paraglacial concept is an idea that has the potential to change the direction of a subfield of Earth science by focusing attention on the ways in which glaciated landscapes respond to non-glacial conditions. Whereas Earth systems of small scale (perhaps <1 km^2) may be treated as equilibrium responses to a variety of non-glacial processes, the vast majority of glaciated landscapes is still incompletely adjusted to non-glacial conditions, and should be regarded as transient

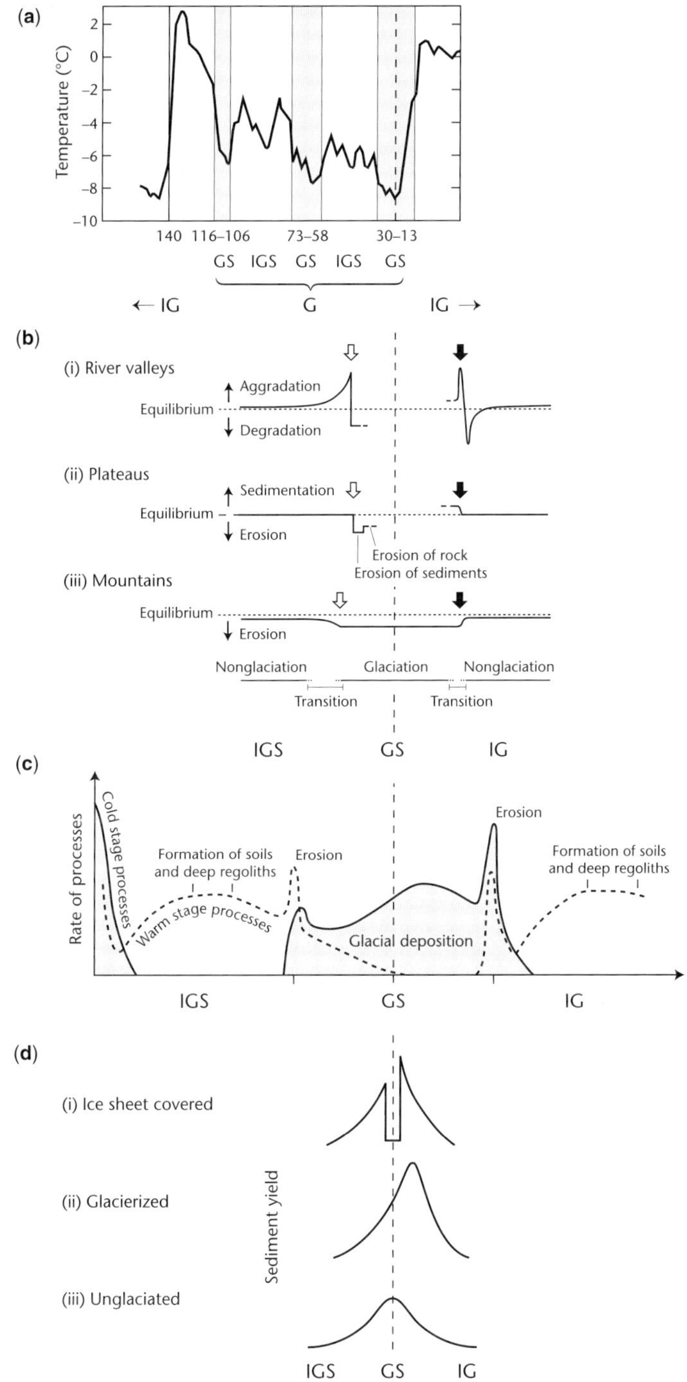

Fig. 11. Rhythm of glacial–interglacial cycles and geomorphic work. (**a**) The Vostok ice-core record of temperature fluctuation over the last glacial cycle (G) including three glacial stades (GS) and two interglacial stades (IGS) (Petit *et al.* 1990). (**b**) The alternation of periods of geomorphic activity with periods of comparative quiescence, plotted separately for river valleys, plateaus and mountains in British Columbia (Clague 1986). Focus is on the transition periods between interglacial and glacial stades. (**c**) Changing kinds of geomorphic work occurring during a full glacial cycle (Starkel 1987). (**d**) Hypothetical rhythms of sediment yield over the past 30 000 years as a function of: (i) continental ice-sheet glaciation; (ii) valley glaciation; and (iii) beyond the limits of glaciation (Owens & Slaymaker 2004).

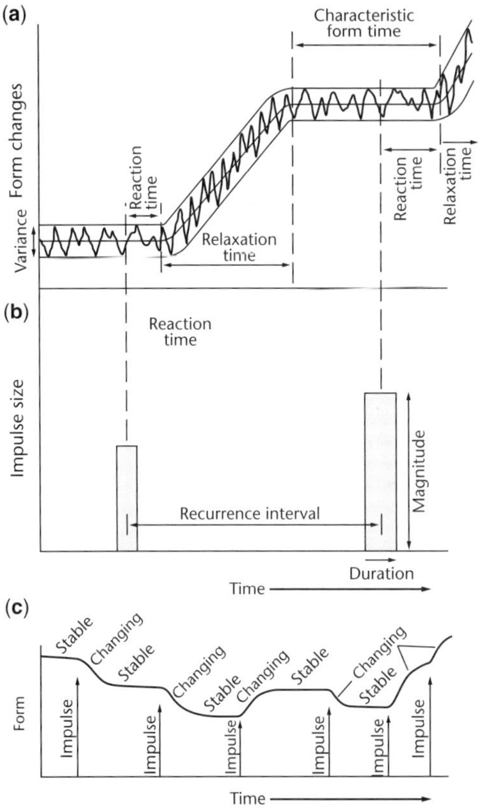

Fig. 12. General concepts of geomorphological time. (**a**) Landform changes over time, including reaction time (time taken for a disturbance, or impulse, to be recorded in the landscape), relaxation time (time during which the impact of the disturbance is being absorbed by the landscape) and characteristic form time (time during which the landscape is comparatively quiescent). (**b**) Dimensions of the disturbance: magnitude, duration and recurrence interval. (**c**) Distribution of disturbances over time and resultant form changes (after Brunsden 1980, 1993).

landforms or transitional landscapes. Paraglacial environments are thus one large example of a disturbance regime landscape. This conclusion opens up the question of how to measure and assess the extent to which glaciated landforms and landscapes at various spatial scales have been modified during the Holocene. The rate and trajectory of that change are what we define as paraglaciation.

References

ASHMORE, P. 1993. Contemporary erosion of the Canadian landscape. *Progress in Physical Geography*, **17**, 190–204.

BALLANTYNE, C. K. 2002*a*. Paraglacial geomorphology. *Quaternary Science Reviews*, **21**, 1935–2017.

BALLANTYNE, C. K. 2002*b*. A general model of paraglacial landscape response. *The Holocene*, **12**, 371–376.

BALLANTYNE, C. K. & WHITTINGTON, G. 1999. Late Holocene floodplain incision and alluvial fan formation in the central Grampian Highlands, Scotland: chronology, environment and implications. *Journal of Quaternary Science*, **14**, 651–671.

BEAUDOIN, A. B. & KING, R. H. 1994. Holocene paleoenvironmental record preserved in a paraglacial alluvial fan, Sunwapta Pass, Jasper National Park, Alberta. *Catena*, **22**, 227–248.

BENN, D. I. & EVANS, D. J. A. 1998. *Glaciers and Glaciation*. Arnold, London.

BOVIS, M. J. 1990. Rock slope deformation at Affliction Creek, southern Coast Mountains, BC. *Canadian Journal of Earth Sciences*, **27**, 243–254.

BOWEN, D. Q. 1978. *Quaternary Geology*. Pergamon, Oxford.

BRARDINONI, F. & HASSAN, M. A. 2006. Glacial erosion, evolution of river long profiles and the organization of process domains in mountain drainage basins of coastal BC. *Journal of Geophysical Research*, **111**, F01013, doi: 10.1029/2005JF000358.

BRUNSDEN, D. 1980. Applicable models of long term landform evolution. *Zeitschrift für Geomorphologie, Supplement Bund*, **36**, 16–26.

BRUNSDEN, D. 1993. The persistence of landforms. *Zeitschrift für Geomorphologie, Supplement Bund*, **93**, 13–28.

CHURCH, M. 1972. *Baffin Island Sandurs: A Study of Arctic Fluvial Processes*. Geological Survey of Canada Bulletin, **216**.

CHURCH, M. 1996. Space, time and the mountain: how do we order what we see? *In*: RHOADS, B. L. & THORN, C. E. (eds) *The Scientific Nature of Geomorphology*. Wiley, Chichester, 147–170.

CHURCH, M. & GILBERT, R. 1975. Proglacial fluvial and lacustrine sediments. *In*: JOPLING, A. V. & MCDONALD, B. C. (eds) *Glaciofluvial and Glaciolacustrine Sedimentation*. SEPM, Special Publications, **23**, 22–100.

CHURCH, M. & RYDER, J. M. 1972. Paraglacial sedimentation: a consideration of fluvial processes conditioned by glaciation. *Bulletin Geological Society of America*, **83**, 3059–3072.

CHURCH, M. & SLAYMAKER, O. 1989. Disequilibrium of Holocene sediment yield in glaciated British Columbia. *Nature*, **337**, 452–454.

CHURCH, M., HAM, D., HASSAN, M. & SLAYMAKER, O. 1999. Fluvial clastic sediment yield in Canada: scaled analysis. *Canadian Journal of Earth Sciences*, **36**, 1267–1280.

CLAGUE, J. J. 1986. The Quaternary stratigraphic record of British Columbia – evidence for episodic sedimentation and erosion controlled by glaciation. *Canadian Journal of Earth Sciences*, **23**, 885–894.

CRUDEN, D. M. & HU, X. Q. 1993. Exhaustion and steady state models for predicting landslide hazards in the Canadian Rocky Mountains. *Geomorphology*, **8**, 279–285.

CURRY, A. M. 2000. Observations on the distribution of paraglacial reworking of glacigenic drift in western Norway. *Norsk Geografisk Tidsskrift*, **54**, 139–147.

DeWolf, Y. 1988. Stratified slope deposits. *In*: Clark, M. J. (ed.) *Advances in Periglacial Geomorphology*. Wiley, Chichester, 91–110.

Eldredge, N. & Gould, S. J. 1972. Punctuated equilibria: an alternative to phyletic gradualism. *In*: Schopf, T. J. M. (ed.) *Models in Paleobiology*. Freeman Cooper, San Francisco, CA, 82–115.

Embleton-Hamann, C. 2004. Proglacial landforms. *In*: Goudie, A. S. (ed.) *Encyclopedia of Geomorphology*, Volume 2. Routledge, London, 810–813.

EPICA 2004. Eight glacial cycles from an Antarctic ice core. *Nature*, **429**, 623–628.

Eyles, N. & Kocsis, S. 1988. Sedimentology and clast fabric of subaerial debris flow facies in a glacially influenced alluvial fan. *Sedimentary Geology*, **59**, 15–28.

Forbes, D. L. & Syvitski, J. P. M. 1994. Paraglacial coasts. *In*: Carter, R. W. G. & Woodroffe, C. D. (eds) *Coastal Evolution: Late Quaternary Shoreline Morphodynamics*. Cambridge University Press, Cambridge, 373–424.

Forbes, D. L. & Taylor, R. B. 1987. Coarse-grained beach sedimentation under paraglacial conditions, Canadian Atlantic coast. *In*: Fitzgerald, D. M. & Rosen, P. S. (eds) *Glaciated Coasts*. Academic Press, San Diego, CA, 51–86.

French, H. M. 2007. *The Periglacial Environment*. Wiley, New Jersey.

Friele, P. A., Clague, J. J., Simpson, K. & Stasiuk, M. 2005. Impact of a Quaternary volcano on Holocene sedimentation in Lillooet River valley, B.C. *Sedimentary Geology*, **176**, 305–322.

Harbor, J. & Warburton, J. 1993. Relative rates of glacial and non-glacial erosion in alpine environments. *Arctic and Alpine Research*, **25**, 1–7.

Hasholt, B., Walling, D. E. & Owens, P. N. 2000. Sedimentation in arctic proglacial lakes: Mittivakkat Glacier, south-east Greenland. *Hydrological Processes*, **14**, 679–699.

Hewitt, K. 2002. Introduction. *In*: Hewitt, K., Byrne, M.-L., English, M. & Young, G. (eds) *Landscapes of Transition*. Kluwer Academic, Dordrecht, 1–8.

Hewitt, K. 2006. Disturbance regime landscapes: mountain drainage systems interrupted by large rockslides. *Progress in Physical Geography*, **30**, 365–393.

Hicks, D. M., McSaveney, M. J. & Chinn, T. J. H. 1990. Sedimentation in proglacial Ivory Lake, Southern Alps, New Zealand. *Arctic and Alpine Research*, **22**, 26–42.

Iturrizaga, L. 1999. Typical debris accumulation forms and formations in High Asia. *GeoJournal*, **47**, 277–339.

Johnson, P. G. 1984. Paraglacial conditions of instability and mass movement: a discussion. *Zeitschrift für Geomorphologie*, **28**, 235–250.

Jordan, P. & Slaymaker, O. 1991. Holocene sediment production in Lillooet River, BC: a sediment budget approach. *Géographie physique et Quaternaire*, **45**, 45–57.

Knox, J. C. 1984. Responses of river systems to Holocene climates. *In*: Wright, H. E. (ed.) *Late Quaternary Environments of the United States. Volume 2: The Holocene*. Longman, London, 26–41.

Matthews, J. A., Shakesby, R. A., Berrisford, M. S. & McEwen, L. J. 1998. Periglacial patterned ground in the Styggedalsbreen glacier foreland, Jotunheimen, southern Norway: microtopographical, paraglacial and geochronological controls. *Permafrost and Periglacial Processes*, **9**, 147–166.

Miall, A. D. 1978. Lithofacies types and vertical profile models in braided river deposits: a summary. *In*: Miall, A. D. (ed.) *Fluvial Sedimentology*. Canadian Society of Petroleum Geologists, Memoirs, **5**, 597–604.

Milliman, J. D. & Syvitski, J. P. M. 1992. Geomorphic/tectonic control of sediment transport to the ocean: the importance of small, mountainous rivers. *Journal of Geology*, **100**, 525–544.

Owens, P. A. & Slaymaker, O. (eds) 2004. *Mountain Geomorphology*. Edward Arnold, London.

Penck, A. & Bruckner, E. 1909. *Die Alpen im Eiszeitalter* (3 vols.) Tauchnitz, Leipzig.

Petit, J. R., Mounier, L., Jouzel, J., Korotkevitch, Y. S., Kotlyakov, V. I. & Lorius, C. 1990. Paleoclimatological and chronological implications of the Vostok core dust record. *Nature*, **343**, 56–58.

Ryder, J. M. 1971*a*. The stratigraphy and morphology of paraglacial alluvial fans in south-central British Columbia. *Canadian Journal of Earth Sciences*, **8**, 279–298.

Ryder, J. M. 1971*b*. Some aspects of the morphometry of paraglacial alluvial fans in south-central BC. *Canadian Journal of Earth Sciences*, **8**, 1252–1264.

Schumm, S. A. 1973. Geomorphic thresholds and complex response of drainage systems. *In*: Morisawa, M. (ed.) *Fluvial Geomorphology*. Publications in Geomorphology, **3**, 299–310.

Schumm, S. A. & Rea, D. K. 1995. Sediment yield from disturbed earth systems. *Geology*, **23**, 391–394.

Shaw, J. & Archer, J. 1979. Deglaciation and glaciolacustrine sedimentation conditions, Okanagan valley, BC. *In*: Schluchter, C. (ed.) *Moraines and Varves*. Balkema, Rotterdam, 347–355.

Siegert, M. J. 2001. *Ice Sheets and Late Quaternary Environmental Change*. Wiley, Chichester.

Simpson, G. G. 1963. Historical science. *In*: Albritton, C. C. (ed.) *The Fabric of Geology*. Addison-Wesley, Reading, MA, 24–48.

Slaymaker, O. 1987. Sediment and solute yields in British Columbia and Yukon: their geomorphic significance re-examined. *In*: Gardiner, V. (ed.) *International Geomorphology 1986*, Volume 1. Wiley, Chichester, 925–945.

Slaymaker, O. 2004. Paraglacial. *In*: Goudie, A. S. (ed.) *Encyclopedia of Geomorphology*, Volume 2. Routledge, London, 759–762.

Slaymaker, O. & Kelly, R. E. J. 2007. *The Cryosphere and Global Environmental Change*. Blackwell, Oxford.

Starkel, L. A. 1987. Long-term and short-term rhythmicity in terrestrial landforms and deposits. *In*: Rampiro, M. R., Sanders, J. E., Newman, W. S. & Konigsson, L. K. (eds) *Climate History, Periodicity and Predictability*. Von Nostrand Reinhold, New York, 323–332.

Thorn, C. E. & Loewenherz, D. S. 1987. Spatial and temporal trends in alpine periglacial studies: implications for paleo-reconstruction. *In*: BOARDMAN, J. (ed.) *Periglacial Processes and Landforms in Britain and Ireland*. Cambridge University Press, Cambridge, 57–65.

WASHBURN, A. L. 1979. *Geocryology: A Survey of Periglacial Processes and Environments*. Edward Arnold, London, 237–243.

WORSLEY, P. 2004. Periglacial geomorphology. *In*: GOUDIE, A. S. (ed.) *Encyclopedia of Geomorphology*, Volume 2. Routledge, London, 772–776.

WYRWOLL, K.-H. 1977. Causes of rock slope failure in a cold area: Labrador–Ungava. *Geological Society of America, Reviews in Engineering Geology*, **3**, 59–67.

On the interpretation of discrete debris accumulations associated with glaciers with special reference to the British Isles

W. BRIAN WHALLEY

School of Geography, Archaeology and Palaeoecology, Queen's University, Belfast BT7 1NN, UK (e-mail: b.whalley@qub.ac.uk)

Abstract: A selection of glacial deposits with distinct morphological or stratigraphic forms associated with glaciers is considered with respect to climatic signals and debris inputs at the time of formation. The relationships are by no means simple and consideration is given, in general terms, to the range of conditions that might apply to a range of depositional features now found 'relict' in the British Isles. These factors are spatial (including continentality) and altitudinal, as well as climate and climatic variability. Examples, mainly from present-day marginally glacierized environments, are given to illustrate the complexity of these interrelationships. Features included are: plateau glaciers and their outlets, where moraines of the outlet glaciers may not be representative of the overall behaviour; plateau glaciers and remnant blockfields related to time of formation; the formation of moraines and rock glaciers; and protalus ramparts and protalus lobes as functions of ice and debris input, as well as thermal regime. It is suggested that the relative amounts of ice and rock debris are important in the formation of certain features. Understanding these relationships is an on-going process and is required for effectively mapping an interpretation of past local- and medium-scale environmental conditions.

Since the days of Agassiz, Charpentier and Venetz in the Alps at the start of studies in glacial geology and geomorphology, uniformitarianism has been the significant guide to interpretations of the Quaternary (and older) landscape features (Chorley *et al.* 1973). In the mid 19th Century there were still objections to the significance of glaciers in the interpretation of features in the British Isles as being due to glaciation. Even Darwin was a proponent of the iceberg theory rather than terrestrial glaciers (Mills 1983) and the late conversion of William Buckland after the acceptance of the 'glacial theory' by the Geological Surveys in Edinburgh and Dublin is well known (Chorley *et al.* 1973). To this conceptual base of glacial observations, evidence has been added worldwide in both present-day (glacierized) and past (glaciated) regions. Since the 1950s a wide variety of dating, observational and analytical techniques have been added. Alongside the recognition of glacial landforms has been a growth in knowledge of other geomorphological processes, glaciology and sedimentology.

Despite this wide armoury of techniques, direct field observation is still important and process recognition remains a problem. In this paper I suggest that care must be used when interpreting landforms, especially when related to their past climatic history. This is especially important where rates of process are assumed and where, in particular, similar landforms might be produced by different processes (resulting in so-called 'equifinality' or 'form-convergence').

Accumulations of debris, whether directly deposited from a glacier or by some creep or flow mechanism in the permafrost and paraglacial domains, frequently have distinct forms, to which we give names and over which we perhaps dispute origins. These forms, here called *discrete debris accumulations*, provide the basis for mapping glacial, periglacial and paraglacial features. However, when we turn to interpretation, especially when making inferences about environmental conditions, we need to place observations within the context of a rather imperfect knowledge of behaviour and response to past environmental conditions in general or events in particular. This paper examines some of these constraints, and shows that both caution and more precise glacio-geomorphological investigations are still required. Examples are used to highlight some of these situations and problems, noting that 'glacial' does not always equate to glaciers. The situations are taken from present-day examples to highlight critical conditions and where knowledge from ice-free areas is poor. Answers, it should be noted, are not necessarily provided.

One purpose of this paper is to attempt to tie in some aspects of process-based geomorphology to Quaternary studies. It does not attempt to be a definitive guide, 'how to do' it or even how it might be done, but, rather, 'of what do we have to be more careful?'. This follows from the need to obtain good evidence to provide better interpretations of past events. Whether mapping is done on

From: KNIGHT, J. & HARRISON, S. (eds) *Periglacial and Paraglacial Processes and Environments*.
The Geological Society, London, Special Publications, **320**, 85–102.
DOI: 10.1144/SP320.7 0305-8719/09/$15.00

foot, by satellite or aerial photographs or a combination of these, there is almost always a pattern-recognition problem. Can we recognize features typical of something we wish to see, or are there things that we see but should be interpreted in a non-traditional way? As knowledge changes so do interpretations. This applies to specific geomorphological features as well as to the implications about their interpretation.

The geological literature is full of examples. For example, Peter Wilson, to his credit, has changed his mind about the interpretation of certain rock glaciers in Donegal (Wilson 2004). A simple and linear sequence of change in climate, leading to glacier mass balance change and a glacier leaving an interpretable and dated trace (such as a moraine), is useful in a regional, as well as temporal, manner. For example, Shakesby and co-authors (Shakesby *et al.* 1987; Shakesby & Matthews 1993; Shakesby 1997) have discussed problems related to protalus ramparts. There are different responses according to glacier size, as well as to the mode of precipitation input (winter storms, summer monsoon) and the effects of continentality. For example, Harrison *et al.* (1998) have suggested that a small glacier existed in the prevailing wind leeside of the Exmoor plateau based on their interpretation of a small moraine or protalus rampart at the foot of a small combe (corrie). Further, the mode of debris input, which affects the visible trace, also needs to be considered. Mapping some of these discrete features, which might nevertheless be somewhat indistinct topographically, may be a problem in interpretation for the mapper. Recording such features in the literature presents another difficulty, especially if no photographic images are used. For example, interpreting the debris accumulations in the English Lake District (Sissons 1980*b*) presents problems of climatic interpretation, when it is not clear what the features actually are, or represent, in a genetic sense. Similarly, Harrison *et al.* (2008) have discussed the range of features that have generally been called 'rock glaciers' and how they may be interpreted environmentally. More precise matching of formation processes and mechanisms to environmental conditions will allow improved modelling of ice mass extents and volumes, as well as associated climate and environments (Hubbard 1999; Gollege & Hubbard 2005).

Here I examine some problems associated with linking climatic signals to deposits in the landscape. This does not aim to be a complete review but merely to point out the difficulties involved – both for present-day interpretation and for those who follow with new interpretations of the complex relationships of climate, ice and deposited features.

Major environmental controls on glacier extents

A basic interpretation of morainic deposits relates to glacier mass balance as a surrogate for precipitation (usually winter) and temperature (usually summer), and follows the sequence shown embedded in Figure 1 (Meier 1965). This is a starting point for many discussions about geological traces in the landscape produced by climate changes. However, this simple scheme needs to sit in a wider framework of climatic variables with spatial and altitudinal controls. Furthermore, there are feedbacks within this simple system. The area or extent of a glacier can be stimulated as the glacierized area itself increases and, especially, gains in altitude. Conversely, as a glacier decreases in size the reduction is exacerbated by increased long-wave radiation from the low albedo surrounding rock surfaces. The lasting indication can be a moraine (as one form of discrete debris accumulation), but also might be a trimline. Sometimes the trimline might be part of a moraine that can be dated, thus allowing an ice-volume estimate to be made. In some cases total ice volumes may be rather different than regional Little Ice Age (LIA) or Younger Dryas limits might suggest (e.g. Evans *et al.* 2002). It depends when and where we take the measurements and the size of the sampled area. Because of the effects of continentality (e.g. Chorlton & Lister 1971) a mass balance input signal may produce a moraine, but one which represents rather different water-equivalent volumetric conditions according to location, altitude and time. Backward interpretation from a moraine may thus produce misleading results if we try to predict a past environmental gradient. This may also be true with variation in longitude. We are still unsure of the timing of LIA maxima in Norway or, indeed, of the reasons for this advance. Vincent *et al.* (2005) have suggested precipitation changes for the termination of the LIA in the Alps, but despite various investigations (Nesje & Dahl 1993; Winkler 1996, 2003; Nesje *et al.* 2007*b*, 2008) the patterns for all of Norway are still unclear. Scandinavia is also a good example of the operation of the altitude/continentality differences, which are compounded by long-term variations of onshore storm tracks and the position of the Polar Front (for further discussion see Whalley 2004 and Bakke *et al.* 2008). Where present-day glaciers are concerned, techniques such as 'glaciation limit' (Østrem 1966, 1972) – perhaps better termed glaciation altitude – have been used for regional mapping and shown on a regional basis (Østrem *et al.* 1973), although considerable variation occurs within this spatial domain. Glacier size variations can be seen; for

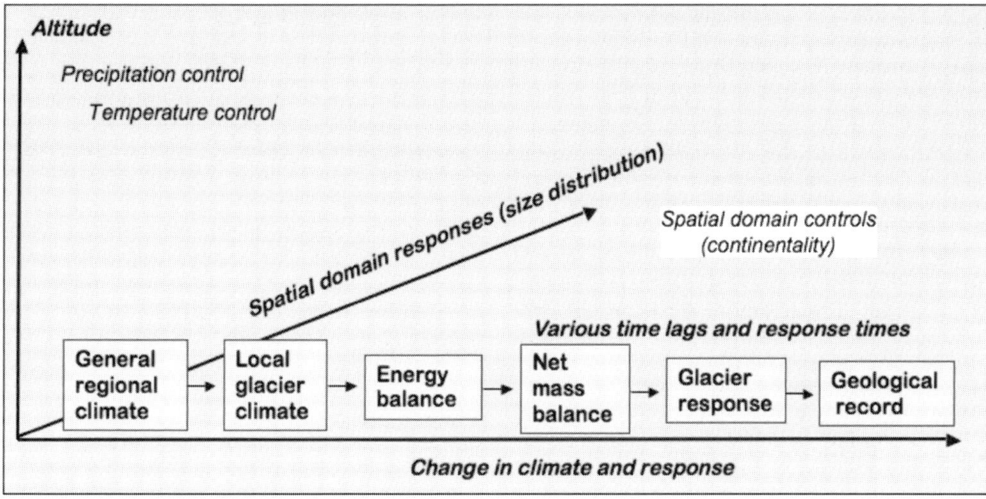

Fig. 1. The basic controls on glaciers and their responses in the geological record. The boxed sequence, following Meier (1965), is embedded within domains that affect the geological record. No feedbacks are shown in the sequence, although some are mentioned in the text.

example, across the Pyrenees (Fig. 2). Evans (1990) has considered similar variations in British Columbia as Østrem had done earlier (Østrem 1972). Now estimates of 'climate' can be made using data for Equilibrium Line Altitudes (ELAs) and sampled local meteorology (Ohmura *et al.* 1992; Nesje 2007), but this generalization, although commonly used (Leonard 1989; Torsnes *et al.* 1993), may be difficult for retrodiction because of regional and temporal variability. Techniques such as Accumulation Area Ratio (AAR) (Osmaston 2002) have been used to help investigate these past

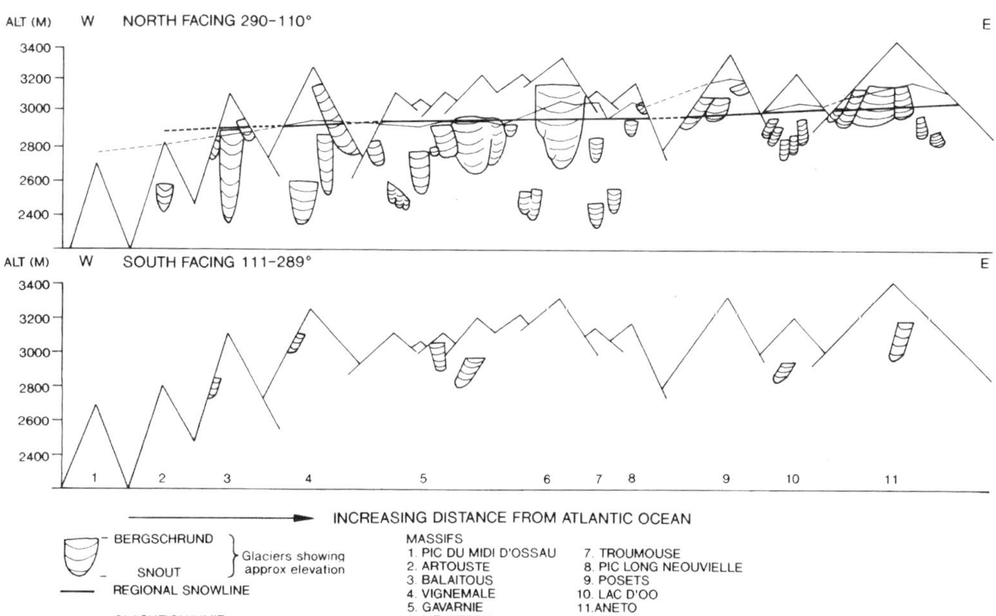

Fig. 2. Variation of regional snowline and glaciation limit in a transect across the Pyrenees for both north- and south-facing glaciers (A. Gellatly pers. commun.).

major controls. For example, the AAR method and its variations have been used in the British Isles to help reconstruct Younger Dryas glacier climates (Sissons & Sutherland 1976; Sissons 1980*a*; Sutherland 1984). However, as mentioned later, this approach has to fit within domains of changing environmental conditions, as indicated in Figure 1; altitude and continentality both affect temperature and precipitation to varying degrees.

Aside from the glacier size problem (mentioned later) Figure 1 can be used as the basis for the investigation of past mass balances by the reconstruction of glacier volumes as indicated by moraines and trimlines. Dating these relicts is, of course, important in modelling budget histories (Nye 1965; Oerlemans 1986) and to see how glaciers respond to climate change (Oerlemans 1989). Davis (1985) discussed a number of problems associated with glaciation (especially Neoglacial) synchronicity and moraine chronologies, aside from the difficulties and reliability of dating methods. More recently, various authors have examined trends over large areas (Porter 2000); and for a more restricted area, Evans *et al.* (2002) have examined radiating glaciers from Øksfjordjøkelen in north Norway.

Glaciers are normally considered to have distinct accumulation and ablation areas, with an ELA dividing them at the end of the ablation year. For temperate glaciers of 1 km or more in length, the snowline altitude at the end of the summer gives a good idea of the ELA. However, smaller glaciers have accumulation and ablation years (or periods of years) rather than spatially delimited areas, and thus no distinct snowline is visible. Hence, AARs for small, present-day, glaciers are not meaningful. This suggests that small glaciers in deglaciated areas cannot supply an AAR on which to map regional ELAs. Other devices, such as Terminus–Head Altitude Ratio (THAR) and glaciation limit calculations, need to be used with great care (Porter 2000). This applies to plateau glaciers (Rea *et al.* 1999), but is especially important where the glaciers are small, i.e. corrie glaciers, and we can see this in marginally glacierized areas today. The altitude of the terminus itself may reflect the covering of debris at some stage as much as the extent of glacier ice. It follows that there is a transition from valley glaciers, which do have a defined equilibrium line and where palaeoenvironmental calculations (via AARs) may be applicable, to small corries where this does not apply. Several authors (Kuhle 1988; Nesje 1992) have investigated topographic elements in glacial systems that bear upon these relationships. Despite work by, for example Bakke *et al.* (2005), we still lack an understanding of this size-related effect for an area or transect. All the glaciers in Figure 2 would be problem cases for an AAR calculation; the variability with respect to a spatially mapped glaciation limit is clear. Snow input directions have changed over time, as well as in year-to-year variations. In mountainous regions a cloudy summer can have great effect on a small glacier in reducing melt, so that in some years no glacier ice may be exposed. We should assume that such effects were present for small glaciers in the past. Hence, care must be taken in using glaciers to interpret past glacial history, and especially the associated climate, unless glacier size is considered. It is possible that a multiproxy approach is needed, and that tree ring and, especially, lacustrine data need to be added to moraine–glacier deposits (Matthews & Karlén 1992; Bakke *et al.* 2005).

Plateau glaciers

The ideas of Gordon Manley (1955) on the development of glaciers on plateaus and rounded summits of northern England prompted investigation of the plateau glaciers of north Norway (Gellatly *et al.* 1986, 1989; Whalley *et al.* 1989, 1994*a*, 1995*a*; Gordon *et al.* 1995). The findings provided a plateau-based approach to looking at deglaciation as much as a valley-orientated viewpoint (Gordon *et al.* 1987). Subsequently, McDougall used some of these ideas and examined the way in which traces of moraines near High Raise in the Lake District could be identified as a consequence of summit ice fields with little debris input (McDougall 2001). An interpretative paper of the Øksfjordjøkelen area (north Norway) by Evans *et al.* (2002) shows the importance of wide-coverage mapping and of looking at the whole glacierized area rather than just individual outlet glaciers. The interpretation of features on a fully deglaciated area is important but difficult to implement climatically without a good understanding of the reasons for the variability of glaciers' responses to climate.

In general, examination of moraines in an area cannot be defined precisely enough to provide more than a basic volume–chronology relationship. This may be done in terms of moraine size, as well as distance down valley. Useful though this is, it does not help refine the timing of glacier advances. Size–position relationships are generally considered as coeval (Ballantyne 1990). However, we know from the landforms and detailed studies of Jostedalsbreen (Norway) that the advances and retreats of the outlet glaciers are often non-synchronous (Winkler 1996; Nesje 1989*a*; Bickerton & Matthews 1993). Typically, north-facing glaciers are out of phase with south-flowing tongues. This has been attributed to winter precipitation changes (Nesje *et al.* 2001; Nesje 2005). Although there is work still needed to provide a more precise explanation, evidence suggests that changes over the whole of the ice cap might have a part to play. Nigardsbreen, a glacier

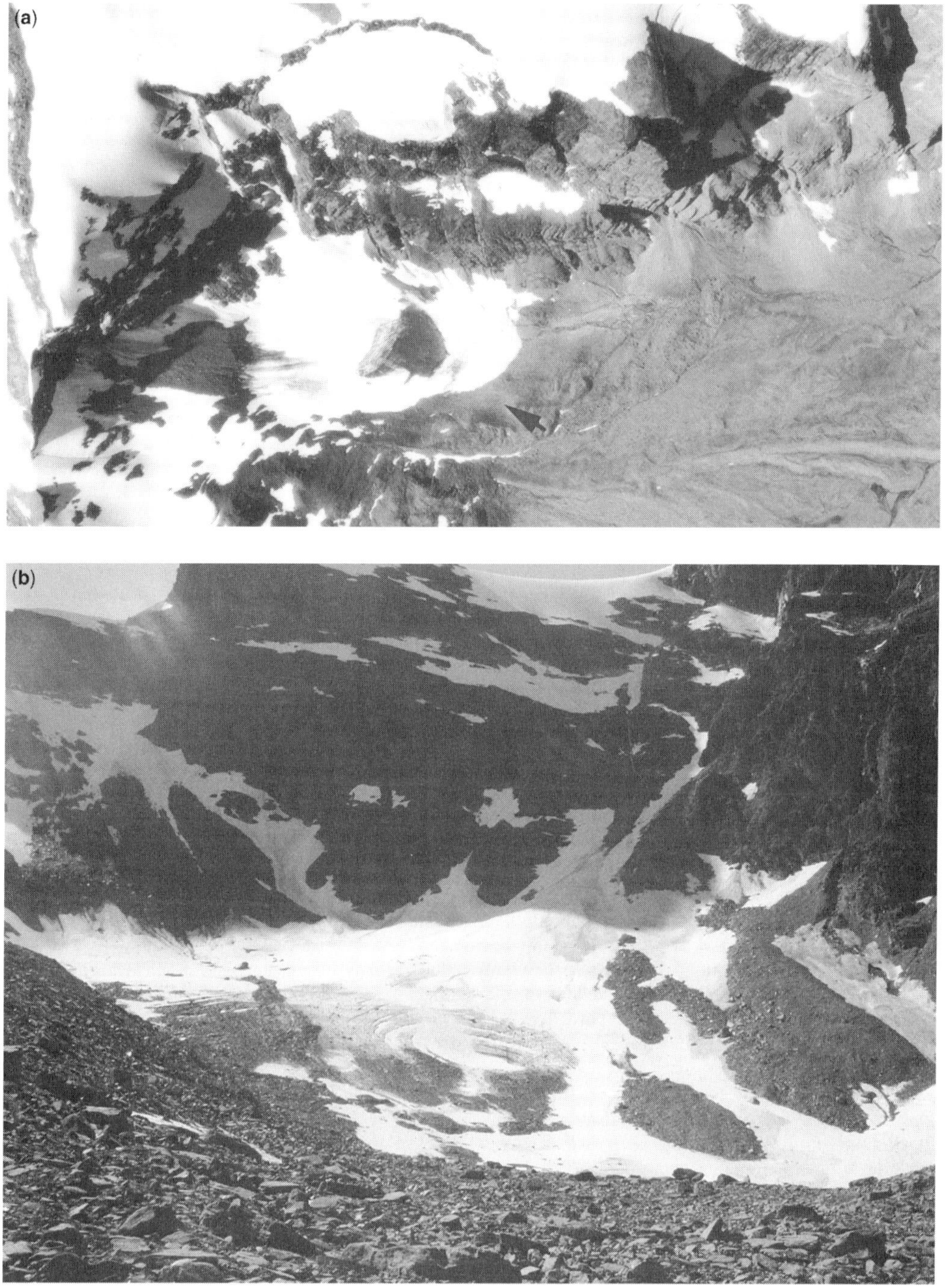

Fig. 3. (**a**) A small glacier (Galbmarieheppi) at about 600 m asl below the plateau ice of Vage'gevarri, Troms, north Norway. Distance left to right, south to north on the image is 1200 m; vertical distance from the centre of image to the plateau area (at top of the image) is 900 m; the arrow shows the direction and origin for (b). (Aerial photograph 1986, courtesy of Terratec, Oslo). (**b**) Shows the same glacier 20 July 2005. The history of the moraine complex is difficult to determine on site, the more so without knowledge of the behaviour of the plateau glacier. The notion of an equilibrium line (and hence AAR) for the corrie glacier is meaningless on its own, not only because of its size but because it is a portion of the ablation area of the plateau above. The dirty ice patch in the lower centre of this image is the same as that just left of centre in (a).

with a long record of substantial frontal movements (Østrem *et al.* 1976) has been used to model climate changes (Oerlemans 1986, 1992, 1997). However, it may be significant that the summit plateau area of this outlet glacier is relatively thin. Thus, accumulation here will be very significant in inducing the response of the snout. We can also add a less well-known factor, that of a changing mass balance to the whole of a plateau. Østrem & Tvede (1986) in mapping the Folgefonna ice plateau in southern Norway between 1959 and 1981 noted not only a decreasing volume but that the 'centre of mass' of the plateau had shifted position. An analysis of wind directions suggested an increase in the westerly component of precipitation-bearing winds from 29% in 1951–1959 to 43% between 1960 and 1979. A consequence of such subtle changes of area–mass balance disposition may be an incorrect association of moraines and trimlines with direct climatic conditions. At the very least the manifestations of volumetric change across a large area should be explored. Unfortunately, we have too little data to evaluate such effects, even for present-day conditions, although Hannah *et al.* (1999) and Fealy & Sweeney (2005), amongst others, have pointed out some synoptic-scale climatological relationships to mass balances.

Moraines (for example, in front of outlet glaciers from extant glacierized plateaus such as Jostedalsbreen) reflect accumulation area conditions, perhaps modulated by regional snow input directions, so very small glaciers although disconnected from the plateau reflect activity on the plateau. However, in this case it may well be switching the ice supply on or off. Figure 3 shows a small glacier below a plateau where the ice input is restricted as the glacier above wanes. The moraine sequence in this case relates to the edge topography and the (lagged) supply of ice from above. This particular glacier is supplied by avalanches from above, but is rapidly decaying. The moraine system associated with this small glacier is quite complex for its size as it reflects the variation of location on the plateau edge and temporal variation of the ice input. This glacier also shows the difficulty of defining an equilibrium line and, hence, using a concept such as AAR for glacier–climate reconstruction.

Debris input to glacial systems

To the glacial parameters of Figure 1 we also need to add debris into the system. This is a complication that is rarely considered; not as a morainic marker of glacier extent, but to consider where and when the debris addition may have an effect on the system as a whole. Not only may the total amount be important, for example for preserving dead ice at the snout of a glacier long after the debris-free glacier has melted, but the debris flux at any one time may have an effect on the ice extent. For example, a glacier in equilibrium that receives a debris input near the snout (as from a large rockfall) would produce a glacier advance; in effect, the ablation area is reduced. Indeed, the timing of debris input (at the start or end of a glacier advance phase, for example) may be significant. Figure 4a

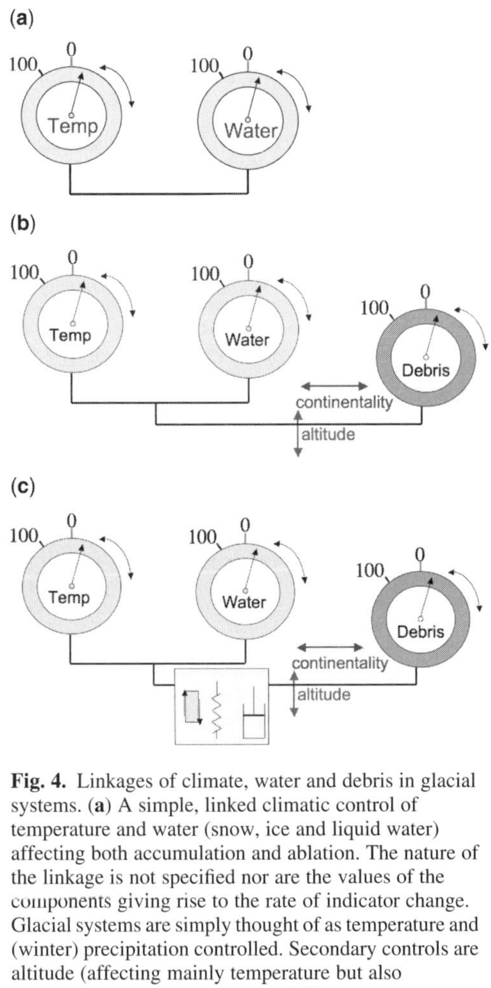

Fig. 4. Linkages of climate, water and debris in glacial systems. (**a**) A simple, linked climatic control of temperature and water (snow, ice and liquid water) affecting both accumulation and ablation. The nature of the linkage is not specified nor are the values of the components giving rise to the rate of indicator change. Glacial systems are simply thought of as temperature and (winter) precipitation controlled. Secondary controls are altitude (affecting mainly temperature but also precipitation) and continentality (affecting mainly precipitation but also temperature). (**b**) The 'simple' components are now linked to debris supply (but still considering altitude and continentality) that may affect the basic glacier system; for example, a debris-laden snout descending much further down the valley (and thus affecting the mapped glacier extent). (**c**) This diagram considers the dynamics, the mechanical responses in the system. The symbols in the (black) box represent shear stresses, elastic/plastic (rheological) and dashpot (damping and resisting motion). These have to be taken into account as they themselves change over time as, for example, as glacier thickness declines so does flow rate.

illustrates the potential complexity here, again related to altitude continentality and temporal input variations as mentioned above. This is modified slightly in Figure 4b, where the addition of debris may affect the glacier response. In Figure 4c consideration is given to the dynamics of the situation where debris has an input into a glacial, permafrost or periglacial system. Here we need to know rather more about shear and resistance in the system(s) as well as lags. The next consideration will be to the way in which basal ice of an ice sheet may or may not deposit till or deform underlying sediment, and of a debris deposit that is discrete, namely blockfield.

Figure 5 is a composite showing the main discrete debris accumulations mentioned here. As each of these depositional features represents some complex relationship with climate over time (Fig. 4c) then not only do we need to continue to investigate the mechanisms of formation of each and their relationship to climate, but also the way in which site variability can be minimized by spatial integration/differentiation over time.

Plateau summit detritus and blockfield

There has been long-standing discussion about refugia on mountain and plateau summits (Dahl 1966; Nordal 1987); see also Fickert *et al.* (2007) for a discussion about the relevance of debris-covered glaciers. However, as seen earlier, there is no doubt that in north Norway glaciers have covered high (1600 m asl (m above sea level)) plateaus. After ice had departed from summit plateaus, blockfields can be found. Ballantyne (1998) has reviewed work in this area, augmented by his own fieldwork in Scotland. After considering previous work, especially of Nesje (1989*b*), he considered three basic interpretations, that: mountain-top detritus predates the Last Glacial Maximum (LGM) and survived as nunataks; or predates the LGM but survived under cold-based ice; or post-dates the LGM and has developed under periglacial conditions.

However, it is not always clear how any one blockfield should necessarily be interpreted in accord to one or other of these possibilities (or, indeed, if they are linked). Where, in north Norway, high blockfields have not had any vegetation cover, more specific evidence for an allochthonous origin can be found. Investigations in the Lyngen Alps show banding of the gabbroic bedrock revealed in aerial photographs (Fig. 6). Field examination shows coarse surface blockfield on top of bedrock, so the assumption is that neither local ice cap nor continental glacial conditions (with the ice assumed to be some 1 km above the plateau tops) produced any distortion of the blockfield. It is not clear whether the basal shear stresses developed were insufficient to move unfrozen blockfield (i.e. wet base glacier cap) or the blockfield itself had a high shear strength being frozen (dry base assumption). The presence of clay minerals in the blockfield and the resistance of the gabbros to weathering suggest that it is Neogene in origin (Whalley *et al.* 2004). The thermal regime of the overriding LGM (or earlier) ice is not known. Surface exposure cosmogenic ratios are increasingly helpful to constrain such age estimates (Linge *et al.* 2006; Nesje *et al.* 2007*a*). Where plateau gradients are steep and Holocene ice has moved off the plateau tops then blockfield may be removed and striations are visible together with localized moraines. The distance between blockfield and glacially scoured blockfield areas may be only a few metres in some cases, so assumptions about blockfield origin need to be guided by such observations.

Debris supply to small glaciers and rock glaciers

Most mountain glaciers have their origin in corries. If the glacier extent is (or was) small then, as mentioned earlier, using an AAR approach for estimating regional snowlines (or ELAs) is difficult. Similarly, corries that have ice contributed by a plateau may also need to be considered carefully. Furthermore, the relative quantities of ice and debris input to the glacier–moraine system needs to be appreciated. This is important because of the protecting effect of glacial debris (when over a few centimetres thick) or the possible formation of rock glaciers. Figure 7 is a schematic suggesting that the relative proportions (or fluxes) of debris and ice in a system have different consequences in terms of morphology. The end members are 100% glacier ice and 100% debris (= scree). The following questions are pertinent, even if not definable:

- Where is the debris now? (what form does the deposit take?)
- Where did it come from? (weathering origin?)
- How much debris (and its flux)? (compared to ice input?)
- When did it join the system? (early or late in the glacier growth 'cycle'?)
- How much ice (and its flux)? (compared to debris input?)
- What temperature regime operates (now/past)? (related also to flow rates)
- What is the ice preservation potential over (long) time periods?

Further, one should not assume that debris supply is constant in time and that present-day observational

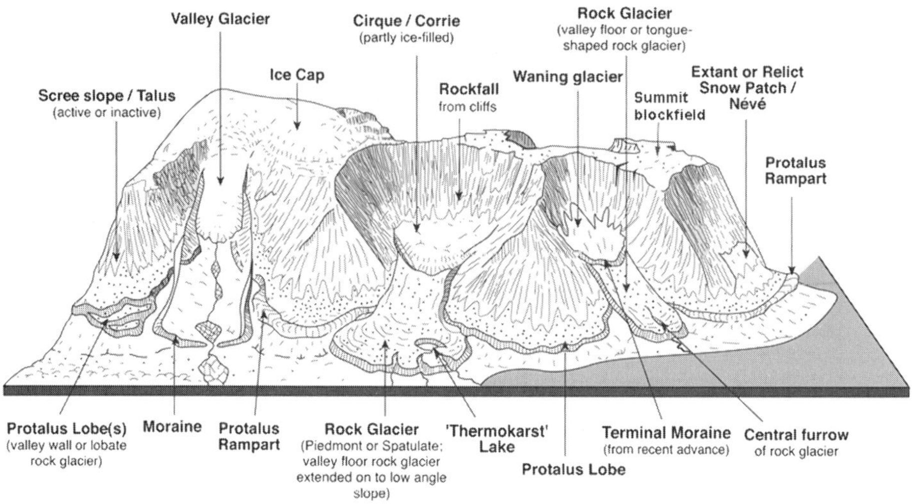

Fig. 5. Representation of the main discrete debris accumulations in a mountainous area, after Whalley & Azizi (2003).

Fig. 6. A portion of the Balgesvarri plateau glacier (Central Lyngen Alps, Norway, *c.* 1600 m asl; distance left–right, east–west is *c.* 400 m). The plateau glacier flows from bottom left towards the image centre. The banded gabbro bedrock can be clearly seen beyond the ice margin in the image centre. This is undeformed, autochthonous, blockfield, over 1 m deep, and is thought to be at least Neogene in age. Notice that there was virtually no plateau snow remaining at this altitude towards the end of a fine, hot summer (1998 image, courtesy of Terratec, Oslo).

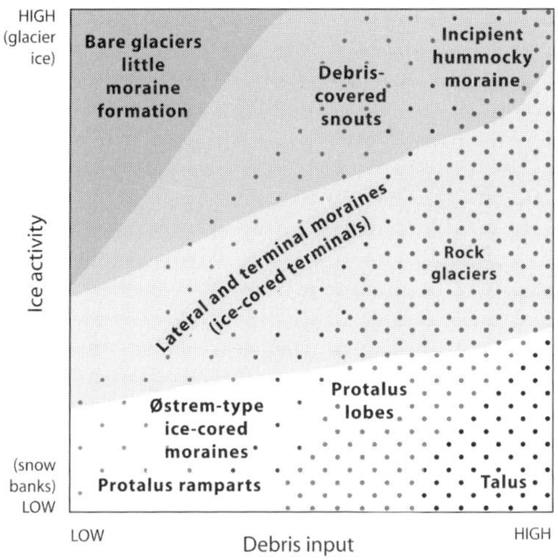

Fig. 7. This schematic illustrates the relative flux of debris and ice into a system. It is implicit that changes are involved as well as absolute quantities. In effect, it is possible to have a trajectory of ice and rock debris inputs to determine the end point and the feature so formed. Working backwards, from the deposit to the debris and ice origins, might help identify formative mechanisms and their relation to climatic controls.

time slices are restricted. This list seems to add more questions than it solves! However, it does help guide investigations about formative processes and how features should be interpreted. Some short illustrative examples are now presented to illustrate the ideas developed here.

In Figure 7 'Østrem-type ice-cored moraines' refers to the type of featured studied by Østrem where the ice origin is thought to be related to snow-bank ice at the glacier snout (Østrem 1964). Although Barsch (1971) considered these moraines to be rock glaciers, there seems to be enough evidence in terms of both morphology and ice origin to differentiate these two forms, although here is a possible example of form convergence. Some features interpreted as lateral push moraines or as decaying ice under debris (Haeberli 1979; Whalley 1979) near Grubengletscher (Switzerland) may be related to Østrem-type moraines, but it would appear that such features are ripe for more detailed investigation and mapping.

Buried glacier ice has long been noted and particularly its significance with respect to glacier mapping ice extents (Haeberli & Epifani 1986; Whalley *et al.* 1986), especially with respect to ice-dammed lakes. However, although lateral moraines with masked or buried glacier ice are probably not uncommon, they are rarely seen because slumping moraine debris rapidly recovers any exposed ice (as in Fig. 8c, for a rock glacier snout). One example was recorded at the glacier Tsijiore Nouve in Switzerland and was associated with a subsequent glacier advance (Whalley 1973). Generally, substantial lateral moraines have sufficient strength to keep the ice in check, but an advance may deform a moraine wall as in this case. Occasionally, a complete breach of a moraine wall by glacier ice may take debris with it and produce a feature that looks like (or can be defined as) a rock glacier (Messerli & Zurbuchen 1968).

The debate continues about the origin of 'rock glaciers' and it will not be continued here. Suffice it to say that the term is used morphologically only and does not imply a mode of origin (Hamilton & Whalley 1995). However, it is explicit that at least some rock glaciers do contain glacier ice (Whalley *et al.* 1995*b*; Potter *et al.* 1998; Whalley & Azizi 2003) and that this is reflected in discussion about the presence (or absence) of permafrost as a requirement for rock glacier formation. It has been argued (Blagborough & Farkas 1968) that if rock glaciers are seen but no other debris deposit (such as a moraine) occurs then that area had permafrost but no glaciers. This reasoning must be doubted as rock glaciers can be found outside permafrost limits and with glacier-ice cores.

In this example critical points are taken in both ice and debris supply. Figure 8a shows an aerial photograph of two corries in Skoldalur, north Iceland. The right (corrie B, western) contains a small rock glacier (Nautardalur), which is glacier-ice cored (Martin *et al.* 1994; Whalley & Martin 1994; Whalley *et al.* 1994*b*, 1995*b*). Several other corries in the vicinity contain rock glaciers, but corrie A to the east does not. In fact, it has barely a moraine in the valley bottom, just a spread of boulders (Fig. 8b). Corrie A is somewhat smaller than B, but is at a similar altitude. It appears to have developed only a very small glacier, however. (In Fig. 8b, much of what is seen at the corrie headwall is snow rather than ice, this being a very cold summer.) The aerial photograph gives a rather better idea of the ice conditions in the two glacier systems. The simplest interpretation here is that corrie A only ever had a small glacier (during the LIA), but with a sparse supply of surficial debris. The thickness of this cover was evidently insufficient to protect the small ice body which has now disappeared and no moraine-like ridges can be seen. The rock glacier in corrie B (Fig. 8c) shows the snout of the feature (and the ice core exposed) and the bulk provided by the glacier ice that will (ultimately) decay (Whalley *et al.* 1994*b*). This illustrates the fine dividing line between glacier and no glacier, rock glacier and no rock glacier. Harrison *et al.* (2008) illustrated other interpretational problem examples from the British Isles. This debate is also related to the

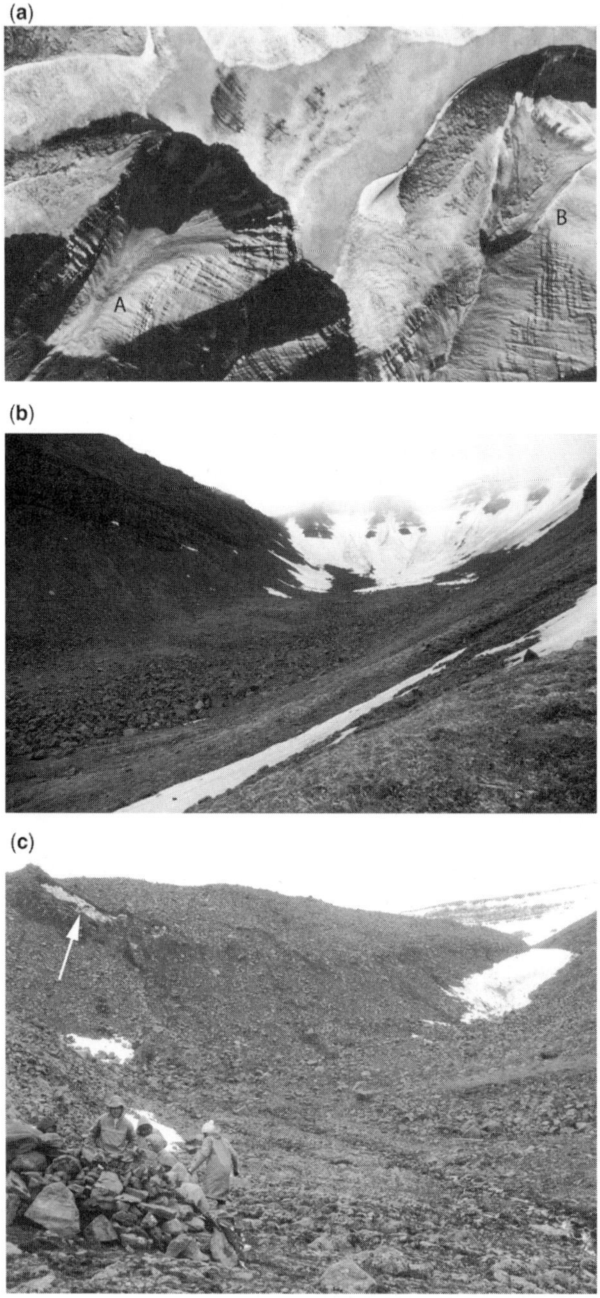

Fig. 8. (**a**) Two neighbouring corries compared; Skóldalur, Tröllaskagi, north-central Iceland. The left (easterly) corrie has debris input but only ice present on the backwall, Nautardalur glacier and rock glacier (right) has small glacier and rock glacier. See text for discussion. Distance left–right (approximately east–west) is 2000 m; Google Earth Pointer: 65°25′36.33″N, 18°17′ 56.21″W, elevation 900 m, (photograph courtesy of Landmaelingar Islands). (**b**) View of the left-hand corrie A (1993) with the thin boulder spread visible. It is probably related to a supraglacial moraine, by analogy with nearby corries, but no distinct rock glacier or moraine is evident. The snow at the head of the corrie is the result of a cold, cloudy summer. In other years very little snow (no glacier ice) is visible. The photograph was taken from a location shown by the lower-left leg of the A in (a). (**c**) Front of the rock glacier in Nautardalur (corrie B) in 1975. The arrow shows the exposure of glacier ice, which, together with others on the feature (but out of sight), show that the whole feature is glacier ice-cored. This area is in shadow in (a).

interpretation of substantial rockfalls or rockslides and rock glacier formation, for example that of Beinn Alligin in Wester Ross (Sissons 1975; Whalley 1976; Gordon 1993) or in Cwm Bochlwyd in North Wales (Harrison 1992).

The interpretation of any rock glacier system thus depends on the proportion of debris to ice and the way in which debris is combined with ice. For the glacier in corrie A of Figure 7b this would have an effect (depending on when the observations were taken) on the interpretation of the glaciation level, if indeed a glacier was considered to have existed here at all. The rock glacier in Figure 9a, National Creek rock glacier in the Wrangell Mountains, Alaska, contrasts markedly with that in Figure 8c. No visible glacier exists, yet the scree input from the headwall is carried along surface flowlines. Away from the headwall, scree fans stop abruptly on the rock glacier surface and, despite being in a permafrost area, do not form independent rock glaciers or protalus lobes. The feature is interpreted as being a function of high input of debris on a declining ice input. This rock glacier lies in the southern Wrangell Mountain zone of the altitude–distance diagram of Figure 9b. Although this diagram perhaps gives an indication of feature-related trends (see also Fig. 2), its interpretation for glacier–climate interactions is rather poor until we know exactly what is being mapped.

Protalus ramparts

The literature on protalus ramparts is voluminous (Ballantyne & Harris 1994) and will not be reviewed here. In summary, however, we may consider them as being perennial snow patches or small glacier ice bodies over which debris accumulates in an

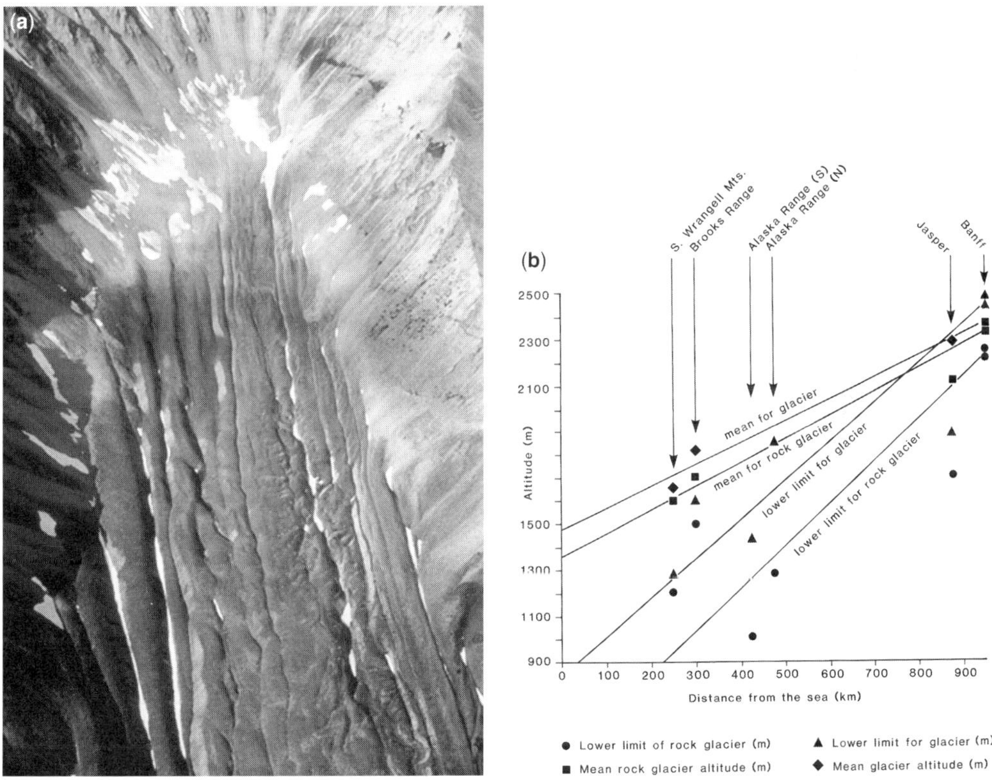

Fig. 9. (**a**) National Creek rock glacier, Wrangell Mountains, Alaska. Paraglacial scree movement appears to be taking over from the, now relict, glacial phase of sediment movement by the rock glacier. Notice that the longitudinal debris strings emanate from the corrie headwall; debris supply from the sidewalls, left and right, only incorporate debris in a very limited fashion to the rock glacier tongue. Although this is a permafrost area the rock glacier itself is considered to be of glacial origin. Width across image is approximately 400 m. (**b**) Diagram showing the relationships (for superimposed transects from the Pacific margin inland) between altitude and distance from the sea for mapped-glacier and rock-glacier distributions in Alaska. (Data from H. E. Martin and W. B. Whalley, unpublished.)

Table 1. *Discrete debris accumulation features and related landforms, some of which may present problems of interpretation. Factors are suggested that may need to be taken into account for future investigations of formation and dynamics, together with possible difficulties of their use for environmental interpretation. Not all the features listed are discussed or mentioned in the text but may be useful as diagnostic criteria. Note that gradations of form (Fig. 7) and debris content (Fig. 11) occur naturally. The features are listed alphabetically; the interested reader or investigator might care to rank these according to their own criteria of significance*

Feature name	Comments on formation, etc.	Environmental interpretation use or caution
Blockfield	If autochthonous: (i) Was it deformed by ice sheets? (ii) How old is it?	If undeformed or not removed how is this interpreted? Possible cosmogenic ratio exposure data Use of tors related to blockfield might be helpful
Blockstream*	(i) Periglacial movement of blockfield (ii) Periglacial/post-glacial feature?	Still uncertainty in interpretation
Central furrow (of rock glacier)	Reveal buried glacier ice core? See also 'thermokarst lake'	Glacial rather than permafrost origin; no evidence remaining in relict rock glacier feature
Debris covered snout	(i) Identifiable today, may have changed dynamics in the past (ii) Relationship to rock glacier?	May not have produce a distinctive moraine Relationship to amount and time of debris input
Hummocky moraine	Passive formation (ablation) In some cases might be related to push moraines	Various interpretations, related to moraines, debris transport location, ice deformation
Lateral moraine ridge	Lateral deposition from glacier, often continuous with terminal moraine; location may be re-occupied by different ice advances	Constrain ice volume (if dated)
Østrem-type moraine	Originally, frontal debris deposition over 'old' snowbank; Possible confusion with: (i) Push moraine (ii) Rock glacier (glacier ice or permafrost) (iii) Protalus lobe	Relict feature difficult to interpret due to lack of ice and (as far as known) a significant relict feature. May look like a rock glacier – which then provides possible interpretation problems.
Protalus lobe†	(i) Involvement with glacier/snowbank ice + debris input flux (ii) Involvement with permafrost-derived + ice debris input flux	Glacial, nival or permafrost maintenance, length of time of preservation; dating problems possible
Protalus rampart‡	(i) Debris passively over snowbank (ii) Construction by small glacier (iii) Might develop into rock glacier (permafrost or glacier related?)	Size may indicate origin of ice; assumption that snow-derived relates to regional snowline rather than possible glacierization altitude
Push moraine	Topographic forms may have various formative processes	Interpretation as glacier margin movement or permafrost-related dynamics?
Rock glacier	(i) Glacier origin (ii) Permafrost origin (iii) Rockslide relict (iv) Composite origin (v) Breach of lateral moraine wall	Permafrost formative conditions or glacier; use in constructing regional trends for glacier ice (below regional limit) or assumption that all rock glaciers are of permafrost origin; Difficult to trace if rockfall-related
Talus (scree slope)	Usually unambiguous; length of time of formation may be considerable	Paraglacial reactivation of old feature possible; may grade into other features down-slope (protalus lobe, protalus rampart, rock glacier)

(*Continued*)

Table 1. *Continued*

Feature name	Comments on formation, etc.	Environmental interpretation use or caution
Terminal moraine	Usually distinct, 'absolute' chronology desirable	Multiple advance positions may be reached at different times; help determine ice volume at a given time; linked to lateral moraine can give debris input volume over (a given) time
Thermokarst lake	May indicate glacier ice presence in rock glacier	No evidence (?) in relict feature
Trimline	Links to lateral moraine ridge and thence terminal moraine	Constrain ice volume, dating if not related to datable lateral moraine

*English/US usage rather than German, Blockström = rock glacier.
†Equivalent to lobate rock glaciers or valley wall rock glaciers of some workers.
‡Also known as winter nival ridge, pronival ridge or snow-bed feature.

ice–rock mixture models used (Whalley & Azizi 1994). Just as cosmologists invoke different models of the universe according to the trend in 'dark matter/dark energy', glacial geomorphologists need to keep exploring models of how these fit together (Fig. 5). Table 1 is an attempt to identify and summarize some of these problems.

As hinted in several places earlier, better areal differentiation and use of geostatistical models related to features found and spatial mapping of climatological parameters is promising. Not just because some variables may be better identified, but also because interpretations of past glacial–permafrost–climate conditions have varied in space as well as time.

I thank A. Nesje, A. Finlayson and J. Knight for comments and suggesting improvements of this paper. I also thank colleagues with whom I have discussed some of the issues raised here.

References

AZIZI, F. & WHALLEY, W. B. 1995. Finite element analysis of the creep of debris containing thin ice bodies. *In*: *Proceedings of the 5th International Offshore and Polar Engineering Conference*, Volume 2. International Society of Offshore and Polar Engineers, The Hague, 336–341.

BAKKE, J., DAHL, S. O., PAASCHE, Ø., LØVLIE, R. & NESJE, A. 2005. Glacier fluctuations, equilibrium line altitudes and palaeoclimate in Lyngen, Northen Norway, during the Lateglacial and Holocene. *The Holocene*, **15**, 518–540.

BAKKE, J., LIE, Ø., DAHL, S. O., NESJE, A. & BJUNE, A. E. 2008. Strength and spatial patterns of the Holocene wintertime westerlies in the NE Atlantic region. *Global and Planetary Change*, **60**, 28–41.

BALLANTYNE, C. K. 1990. The Holocene glacial history of Lyngshalvøya, northern Norway: chronology and climatic implications. *Boreas*, **19**, 93–117.

BALLANTYNE, C. K. 1994. Glaciological constraints on protalus rampart development. *Permafrost and Periglacial Processes*, **5**, 145–153.

BALLANTYNE, C. K. 1998. Age and significance of mountain-top detritus. *Permafrost and Periglacial Processes*, **9**, 327–345.

BALLANTYNE, C. K. & HARRIS, C. 1994. *The Periglaciation of Great Britain*. Cambridge University Press, Cambridge.

BARSCH, D. 1971. Rock glaciers and ice-cored moraines. *Geografiska Annaler*, **53A**, 203–206.

BICKERTON, R. W. & MATTHEWS, J. A. 1993. 'Little Ice Age' variations of outlet glaciers from the Jostedalsbreen ice cap, southern Norway; a regional lichenometric-dating study of ice-marginal moraine sequences and their climatic significance. *Journal of Quaternary Science*, **8**, 45–66.

BLAGBOROUGH, J. W. & FARKAS, S. E. 1968. Rock glaciers in the San Mateo Mountains, southcentral New Mexico. *American Journal of Science*, **266**, 812–823.

CAPPS, S. R. J. 1910. Rock glaciers in Alaska. *Journal of Geology*, **18**, 359–375.

CHORLEY, R. J., DUNN, A. J. & BECKINSALE, R. P. 1973. *The History of the Study of Landforms: or the Development of Geomorphology: Volume 1: Geomorphology Before Davis*. Methuen, London.

CHORLTON, J. C. & LISTER, H. 1971. Geographical controls of glacier budget gradients in Norway. *Norsk Geografisk Tidsskrift*, **25**, 159–164.

DAHL, R. 1966. Block fields, weathering pits and tor-like forms in the Narvik Mountains, Nordland, Norway. *Geografiska Annaler*, **48A**, 55–85.

DAVIS, P. T. 1985. Neoglacial moraines on Baffin Island. *In*: ANDREWS, J. T. (ed.) *Quaternary Environments*. Allen & Unwin, London, 682–718B.

EVANS, D. J. A., REA, B. R., HANSOM, J. D. & WHALLEY, W. B. 2002. Geomorphology and style of plateau icefield deglaciation in fjord terrains: the example of Troms–Finnmark, north Norway. *Journal of Quaternary Science*, **17**, 221–239.

EVANS, I. S. 1990. Climatic effects on glacier distribution across the southern coast mountains, B.C. Canada. *Annals of Glaciology*, **14**, 58–64.

FEALY, R. & SWEENEY, J. 2005. Detection of a possible change point in atmospheric variability in the North Atlantic and its effect on Scandinavian glacier mass balance. *International Journal of Climatology*, **25**, 1819–1833.

FICKERT, T., FRIEND, D., GRÜNINGER, F., MOLNIA, B. & RICHTER, M. 2007. Did debris-covered glaciers serve as Pleistocene refugia for plants? A new hypothesis derived from observations of recent plant growth on glacier surfaces. *Arctic, Antarctic and Alpine Research*, **39**, 245–257.

GELLATLY, A. F., WHALLEY, W. B. & GORDON, J. E. 1986. Topographic control over recent glacier changes in Southern Lyngen Peninsula, North Norway. *Norsk Geografisk Tidsskrift*, **41**, 211–218.

GELLATLY, A. F., WHALLEY, W. B., GORDON, J. E. & HANSON, J. D. 1989. Thermal regime and debris supply on plateau ice caps of north Norway: preliminary observations and implications. *Geology*, **16**, 983–986.

GOLLEGE, N. R. & HUBBARD, A. 2005. Evaluating Younger Dryas glacier reconstructions in part of the western Scottish Highlands: a combined empirical and theoretical approach. *Boreas*, **34**, 274–286.

GORDON, J. E. 1993. Beinn Alligin. *In:* GORDON, J. E. & SUTHERLAND, D. G. (eds) *Quaternary of Scotland*. Joint Nature Conservation Committee. Chapman & Hall, London, 118–122.

GORDON, J. E., WHALLEY, W. B. & GELLATLY, A. F. 1995. Fluctuations of glaciers in Lyngsdalen, Troms, Norway, during the 20th Century. *Zeitschrift für Gletscherkunde und Glazialgeologie*, **31**, 125–134.

GORDON, J. E., WHALLEY, W. B., GELLATLY, A. F. & FERGUSON, R. I. 1987. Glaciers of the southern Lyngen Peninsula, Norway: a possible model for mountain deglaciation. *In:* GARDINER, V. (ed.) *International Geomorphology*. John Wiley, Chichester, 743–758.

HAEBERLI, W. 1979. Holocene push-moraines in alpine permafrost. *Geografiska Annaler*, **61A**, 43–48.

HAEBERLI, W. & EPIFANI, F. 1986. Mapping the distribution of buried glacier ice. An example from Lago delle Locce, Monte Rosa, Italian Alps. *Annals of Glaciology*, **8**, 78–81.

HAMILTON, S. J. & WHALLEY, W. B. 1995. Rock glacier nomenclature: a re-assessment. *Geomorphology*, **14**, 73–80.

HANNAH, D. M., GURNELL, A. M. & MCGREGOR, G. R. 1999. Identifying links between large-scale atmospheric circulation and local glacier ablation climates in the French Pyrenees. *In*: *Interactions Between the Cryosphere, Climate and Greenhouse Gases*. International Association of Hydrological Sciences, Publications, **256**, 155–163.

HARRISON, S. 1992. Sturzstrom or fossil rock glacier? A reinterpretation of the relict boulder deposit, Cwm Bochlwyd, North Wales. *Quaternary Newsletter*, **6**, 22–25.

HARRISON, S., ANDERSON, E. & PASSMORE, D. G. 1998. A small glacial cirque basin on Exmoor, Somerset. *Proceedings of the Geologists' Association*, **109**, 149–158.

HARRISON, S., WHALLEY, W. B. & ANDERSON, E. 2008. Relict rock glaciers and protalus lobes in the British Isles: implications for Late Pleistocene mountain geomorphology and palaeoclimate. *Journal of Quaternary Science*, **23**, 287–304.

HUBBARD, A. 1999. High-resolution modelling of the advance of the Younger Dryas ice sheet and its climate in Scotland. *Quaternary Research*, **52**, 27–43.

ISAKSEN, K., ØDEGÅRD, R. S., EIKEN, T. & SOLLID, J. L. 2000. Composition, flow and development of two tongue-shaped rock glaciers in the permafrost of Svalbard. *Permafrost and Periglacial Processes*, **11**, 241–257.

KUHLE, M. 1988. Topography as a fundamental element of glacial systems. *GeoJournal*, **17**, 545–568.

LEONARD, F. M. 1989. Climate change in the Colorado Rocky mountains: Estimates modern climate at Late Pleistocene equilibrium lines. *Arctic and Alpine Research*, **21**, 245–255.

LINGE, H., BROOK, E. J., NESJE, A., RAISBECK, G. M., YIOU, F. & CLARK, H. 2006. In situ ^{10}Be exposure ages from southeastern Norway: implications for the geometry of the Weichselian Scandinavian ice sheet. *Quaternary Science Reviews*, **25**, 1097–1109.

MANLEY, G. 1955. On the occurrence of ice-domes and permanently snow-covered summits. *Journal of Glaciology*, **2**, 453–456.

MARTIN, H. E., WHALLEY, W. B., ORR, J. & CASELDINE, C. 1994. Dating and interpretation of rock glaciers using lichenometry, south Tröllaskagi, North Iceland. *Münchener Geographische Arbeiten*, **12**, 205–224.

MATTHEWS, J. A. & KARLÉN, W. 1992. Asynchronous neoglaciation and Holocene climate change reconstructed from Norwegian glaciolacustrine sedimentary sequences. *Geology*, **20**, 991–994.

MCDOUGALL, D. A. 2001. The geomorphological impact of Loch Lomond (Younger Dryas) Stadial plateau icefields in the central Lake District, northwest England. *Journal of Quaternary Science*, **16**, 531–543.

MEIER, M. F. 1965. Glaciers and Climate. *In:* WRIGHT, H. E. & FREY, D. G. (eds) *The Quaternary of the United States*. Princeton University Press, Princeton, NJ, 795–805.

MESSERLI, B. & ZURBUCHEN, M. 1968. Blockgletscher im Weissmies und Aletsch und ihre photogrammetrische Kartierung. *Die Alpen*, **3**, 139–152.

MILLS, W. 1983. Darwin and the Iceberg Theory. *Notes and Records of the Royal Society of London*, **38**, 109–127.

NESJE, A. 1989*a*. Glacier-front variations of outlet glaciers from Jostedalsbreen and climate in the Jostedalsbre region of western Norway in the period 1901–80. *Norsk Geografisk Tidsskrift*, **43**, 3–17.

NESJE, A. 1989*b*. The geographical and altitudinal distribution of Block Fields in southern Norway and its significance to the Pleistocene Ice sheets. *Zeitschrift für Geomorphologie*, Supplement band, **72**, 41–53.

NESJE, A. 1992. Topographical effects on the equilibrium-line altitude on glaciers. *GeoJournal*, **27**, 383–391.

NESJE, A. 2005. Briksdalsbreen in western Norway: AD 1900–2004 frontal variations as a combined effect of variations in winter precipitation and summer temperature. *The Holocene*, **15**, 1245–1252.

NESJE, A. 2007. Palaeo ELAs. *In:* ELIAS, S. A. (ed.). *Encyclopedia of Quaternary Science*, Volume 2. Elsevier, Amsterdam, 882–892.

NESJE, A., BAKKE, J., DAHL, S. O., LIE, Ø. & MATTHEWS, J. A. 2008. Norwegian mountain glaciers in the past, present and future. *Global and Planetary Change*, **60**, 10–27.

NESJE, A. & DAHL, S. O. 1993. Late Glacial and Holocene glacier fluctuations and climate variations in western Norway: a review. *Quaternary Science Reviews*, **12**, 255–261.

NESJE, A., DAHL, S. O. *ET AL*. 2007*a*. The surface geometry of the Last Glacial Maximum ice sheet in the Andoya-Skanland region, northern Norway, constrained by surface exposure dating and clay mineralogy. *Boreas*, **36**, 227–239.

NESJE, A., DAHL, S. O., THUN, T. & NORDLI, Ø. 2007*b*. The 'Little Ice Age' glacial expansion in western Scandinavia: summer temperature or winter precipitation? *Climate Dynamics*, **30**, 789–801.

NESJE, A., MATTHEWS, J. A., DAHL, S. O., BERRISFORD, M. S. & ANDERSSON, C. 2001. Holocene glacier fluctuations and winter-precipitation changes in the Jostedalsbreen region, western Norway, based on glaciolacustrine sediment records. *The Holocene*, **11**, 267–280.

NORDAL, I. 1987. Tabula rasa after all? Botanical evidence for ice-free refugia in Scandinavia reviewed. *Journal of Biogeography*, **14**, 377–388.

NYE, J. F. 1965. A numerical model of inferring the budget history of a glacier from its advance and retreat. *Journal of Glaciology*, **5**, 589–607.

OERLEMANS, J. 1986. An attempt to simulate historic front variations of Nigardsbreen, Norway. *Theoretical and Applied Climatology*, **37**, 126–135.

OERLEMANS, J. 1989. On the response of valley glaciers to climatic change. *In:* OERLEMANS, J. (ed.). *Glacier Fluctuations and Climatic Change*. Kluwer, Dordrecht, 353–371.

OERLEMANS, J. 1992. Climate sensitivity of glaciers in southern Norway: application of an energy-balance model to Nigardsbreen, Hellstugubreen and Alfotbreen. *Journal of Glaciology*, **38**, 223–232.

OERLEMANS, J. 1997. A flowline model for Nigardsbreen: projection of the future glacier length based on dynamic calibration with the historic record. *Annals of Glaciology*, **24**, 382–389.

OHMURA, A., KASSER, P. & FUNK, M. 1992. Climate at the equilibrium line of glaciers. *Journal of Glaciology*, **38**, 397–411.

OSMASTON, H. A. 2002. Estimates of glacier equilibrium-line altitudes by the Area × Altitude, the Area × Altitude Balance Ratio and the Area × Altitude Balance Index methods and their validation. *Quaternary International*, **138–139**, 22–31.

ØSTREM, G. 1964. Ice-cored moraines in Scandinavia. *Geografiska Annaler*, **46A**, 282–237.

ØSTREM, G. 1966. The height of the glaciation limit in southern British Columbia and Alberta. *Geografiska Annaler*, **48A**, 126–138.

ØSTREM, G. 1972. Height of the glaciation level in Northern British Columbia and Southeastern Alaska. *Geografiska Annaler*, **54A**, 76–84.

ØSTREM, G. & TVEDE, A. 1986. Comparison of glacier maps – a source of climatological information. *Geografiska Annaler*, **68A**, 225–231.

ØSTREM, G., HAAKENSEN, N. & MELANDER, O. 1973. *Atlas over breer i Nord-Skandinavia*. [Atlas of Glaciers in Northern Scandinavia.] Norges vassdrags- og energiverk, Vassdragsdirektoratet, Meddelelse frä Hydrologisk avdeling, Oslo, 315.

ØSTREM, G., LIESTØL, O. & WOLD, B. 1976. Glaciological investigations at Nigardsbreen, Norway. *Norsk Geografisk Tidsskrift*, **30**, 187–209.

PORTER, S. C. 2000. Snowline depression in the tropics during the Last Glaciation. *Quaternary Science Reviews*, **20**, 1067–1091.

POTTER, N. J., STEIG, E. J., CLARK, D. H., SPEECE, M. A., CLARK, G. M. & UPDIKE, A. B. 1998. Galena Creek rock glacier revisited - new observations on an old controversy. *Geografiska Annaler*, **80A**, 251–265.

REA, B. R., WHALLEY, W. B., DIXON, T. S. & GORDON, J. E. 1999. Plateau icefields as contributing areas to valley glaciers and the potential impact on reconstructing ELAs: a case study from the Lyngen Alps, North Norway. *Annals of Glaciology*, **28**, 97–102.

RICHMOND, G. M. 1952. Comparison of rock glaciers and block streams in the La Sal Mountains, Utah. *Geological Society of America Bulletin*, **83**, 1292–1293.

SHAKESBY, R. A. 1997. Pronival (protalus) ramparts: a review of forms, processes, diagnostic criteria and palaeoenvironmental implications. *Progress in Physical Geography*, **21**, 394–418.

SHAKESBY, R. A. & MATTHEWS, J. A. 1993. Loch Lomond Stadial glacier at Fan Hir, Mynydd Du (Brecon Beacons), South Wales: critical evidence and palaeoclimatic implications. *Geological Journal*, **28**, 69–779.

SHAKESBY, R. A., DAWSON, A. G. & MATTHEWS, J. A. 1987. Rock glaciers, protalus ramparts and related phenomena, Rondane, Norway: a continuum of large-scale talus-derived landforms. *Boreas*, **16**, 305–317.

SISSONS, J. B. 1975. A fossil rock glacier in Wester Ross. *Scottish Journal of Geology*, **92**, 182–190.

SISSONS, J. B. 1980*a*. Paleoclimatic influences from Loch Lomond advance glaciers. *In:* LOWE, J. J., GRAY, J. M. & ROBINSON, J. E. (eds) *Studies in the Lateglacial of Northwestern Europe*. Pergamon, Oxford, 31–43.

SISSONS, J. B. 1980*b*. The Loch Lomond Advance in the Lake District, northern England. *Transactions of the Royal Society of Edinburgh: Earth Sciences*, **71**, 13–27.

SISSONS, J. B. & SUTHERLAND, D. G. 1976. Climatic inferences form former glaciers in the South East Grampian Highlands, Scotland. *Journal of Glaciology*, **17**, 325–346.

SUTHERLAND, D. G. 1984. Modern glacier characteristics as a basis for inferring former climates with particular reference to the Loch Lomond Stadial. *Quaternary Science Reviews*, **3**, 291–309.

TORSNES, I., RYE, N. & NESJE, A. 1993. Modern and Little Ice Age equilibrium line altitudes on outlet glaciers from Jostedalsbreen, western Norway, an evaluation of different approaches to their calculation. *Arctic and Alpine Research*, **25**, 106–116.

VINCENT, C., LE MEUR, E., SIX, D. & FUNK, M. 2005. Solving the paradox of the end of the Little Ice Age in the Alps. *Geophysical Research Letters*, **32**, L090706, doi: 10.1029/2005GL022552.

WHALLEY, W. B. 1973. An exposure of ice on the distal side of a lateral moraine. *Journal of Glaciology*, **12**, 327–329.

WHALLEY, W. B. 1976. A fossil rock glacier in Wester Ross. *Scottish Journal of Geology*, **12**, 175–179.

WHALLEY, W. B. 1979. The relationship of glacier ice and rock glacier at Grubengletscher, Kanton Wallis, Switzerland. *Geografiska Annaler*, **61A**, 49–61.

WHALLEY, W. B. 2004. Glacier research in mainland Scandinavia. *In:* CECIL, L. D., GREEN, J. R. & THOMPSON, L. G. (eds) *Earth Paleoenvironments: Records Preserved in Mid- and Low-latitude Glaciers*. Kluwer, Dordrecht, 121–143.

WHALLEY, W. B. & AZIZI, F. 1994. Models of flow of rock glaciers: analysis, critique and a possible test. *Permafrost and Periglacial Processes*, **5**, 37–51.

WHALLEY, W. B. & AZIZI, F. 2003. Rock glaciers and protalus landforms: analogous forms and ice sources on Earth and Mars. *Journal of Geophysical Research, Planets*, **108**, 8032, doi: 8010.1029/2002JE001864.

WHALLEY, W. B. & MARTIN, H. E. 1994. Rock glaciers in Tröllaskagi, their origin and climatic significance. *Münchener Geographische Arbeiten*, **B12**, 289–308.

WHALLEY, W. B., GELLATLY, A. F. & GORDON, J. E. 1989. Effects of topographic and climatic controls on 19th and 20th Century glacier changes in the Lyngen and Oksfjord areas, North Norway. *In:* OERLEMANS, H. (ed.) *Symposium on Glacier Fluctuations and Climatic Change*. Kluwer, Amsterdam, 153–172.

WHALLEY, W. B., GELLATLY, A. F. & MARTIN, H. E. 1986. The problem of hidden ice in glacier mapping. *Annals of Glaciology*, **8**, 181–183.

WHALLEY, W. B., GORDON, J. E. & GELLATLY, A. F. 1994*a*. Plateau and valley glaciers in North Norway: Relationships to climate change. *Annales Geophysicae*, **12,** Supplement II, 387.

WHALLEY, W. B., GORDON, J. E., GELLATLY, A. F. & HANSOM, J. G. 1995*a*. Plateau and valley glaciers in north Norway: responses to climate over the last 100 years. *Zeitschrift für Gletscherkunde und Glazialgeologie*, **31**, 115–124.

WHALLEY, W. B., HAMILTON, S. J., PALMER, C. F., GORDON, J. E. & MARTIN, H. E. 1995*b*. The dynamics of rock glaciers: data from Tröllaskagi, north Iceland. *In:* SLAYMAKER, O. (ed.) *Steepland Geomorphology*. Wiley, Chichester, 129–145.

WHALLEY, W. B., PALMER, C., HAMILTON, S. & GORDON, J. E. 1994*b*. Ice exposures in rock glaciers. *Journal of Glaciology*, **40**, 427–429.

WHALLEY, W. B., REA, B. R. & RAINEY, M. M. 2004. Weathering, blockfields, and fracture systems and the implications for long-term landscape formation: some evidence from Lyngen and Øksfjordjøkelen areas in North Norway. *Polar Geography*, **28**, 93–119.

WILSON, P. 2004. Relict rock glaciers, slope failure deposits, or polygenetic features? A reassessment of some Donegal debris landforms. *Irish Geography*, **37**, 77–87.

WINKLER, S. 1996. Front variations of outlet glaciers from Jostedalsbreen, western Norway, during the twentieth century. *Norges geologiske undersøkelse, Bulletin*, **431**, 33–47.

WINKLER, S. 2003. A new interpretation of the date of the 'Little Ice Age' glacier maximum at Svartisen and Okstindan, northern Norway. *The Holocene*, **13**, 83–95.

Paraglacial rock slope failure as an agent of glacial trough widening

DAVID JARMAN

Mountain Landform Research, Findon Cottage, Ross-shire IV7 8JJ UK
(e-mail: david.jarman914@virgin.net)

Abstract: Rock slope failure (RSF) generates the largest single erosional events in the glacial–paraglacial land system, leaving numerous obvious cavities and less obviously weakened valley walls. Its contribution to trough widening in a mountain range has not previously been systematically quantified. Map-based measures of RSF 'depth of bite' are applied to five sample areas in the Scottish Highlands, and a comparator area in north Norway, all in metasediments structurally conducive to mass deformation and block sliding. Problems in applying map-based measures include bedrock cavities remaining partially occupied by failed debris or subsequent infill, and multiple planes of reference. The most practical measure is of maximum recess depth on any single contour (D_{MAX}). This is a standardizable single-point indicator of visible impact, not a measure of actual cavity depth, nor an average applying to the whole RSF. In four of the five areas, average D_{MAX} is consistent at 40–45 m. RSF breadth averages 270–600 m over the five areas. RSF affects 9% and 14% of total valley wall length in the two densest RSF areas, rising to 47% and 52% on two specific valley sides. The depth:breadth ratio in areas dominated by slope deformation can be twice that in areas of translational sliding. An evolutionary model of glacial–paraglacial cycling proposes a 'zone of paraglacial relaxation' in which RSF is intense in early cycles as fluvial profiles adjust to ice discharge, diminishing with maturity as trough walls become stress-hardened, and reviving in response to neotectonic and glaciological perturbations, notably ice piracy via transfluent breaching. However, a major unknown is the efficacy of glacial exploitation of RSFs: if it takes several cycles to evacuate debris and pare back cavity angles, cumulative RSF impact is lessened. Glacial–paraglacial cycling is a classic positive feedback loop, promoting valley widening beyond the parabolic norm. Preferential exploitation of structure by RSF promotes asymmetrical trough profiles. RSF acts both as a scarp retreat process, and as a slope reduction counterpoint to glacial slope steepening. In landscape evolution, it is a powerful agent in destruction of paleic relief, notably around watersheds that are undergoing breaching by transfluent ice, where trough development and widening is still vigorous.

Paraglacial rock slope failure (RSF) is the most dramatic mode of paraglacial activity, as scoped by Ballantyne (2002). It is widespread, but not endemic, in glaciated mountain ranges. It is an episodic (high-magnitude–low-frequency) process, occurring in susceptible locations where deglaciation stresses are sufficient to rupture bedrock slopes over extents from tens of metres to kilometres. It tends to relax glacially steepened and deepened slopes towards lower angles and conditional stability. It occurs in diverse modes, with landshaping effects including trough widening and scarp retreat (Jarman 2003*a*, 2006).

RSF is now becoming recognized as a major contributor to erosion in young mountain ranges, which are often glaciated (Hewitt 1988; Korup *et al.* 2007). In these areas the paraglacial signal is enmeshed with tectonic, fluvial incision and gravitational effects. In older glaciated ranges, such as Britain and Scandinavia, the paraglacial RSF contribution to their gross landscape evolution is clearer but barely acknowledged (Evans 1997; Jarman 2002). Glacial geomorphology has focused on the classical parabolic cross-section (Harbor 1992; Augustinus 1995), and on processes and rates of *deepening* of the floor (Glasser & Hall 1997; Hebdon *et al.* 1997), rather than on trough *widening*. Sugden & John (1976, p. 209) noted that 'the role of sub-aerial slope processes in modifying and widening troughs has not been examined in any depth'. Even with recognition that 'slopes are likely to be unstable and prone to collapse, thus modifying the form of the trough by paraglacial reworking' (Benn & Evans 1998, p. 352), there has been little systematic attempt to quantify it over wider areas, least of all with respect to RSF.

Yet the cavities left by RSFs are often obvious (Fig. 1) and represent the highest-impact single erosional events in the glacial–paraglacial land system. Debris may fall onto the valley glacier (e.g. Gordon *et al.* 1978; Sigurđsson & Williams 1991; Evans *et al.* 2006), or be emplaced at or near the slope foot for export by the next glacier, e.g. at Coire Gabhail, Glencoe (Ballantyne 2007*b*), or remain partly or wholly within the failure cavity. Since many RSFs are not fully evacuated, the term 'bite'

From: KNIGHT, J. & HARRISON, S. (eds) *Periglacial and Paraglacial Processes and Environments*.
The Geological Society, London, Special Publications, **320**, 103–131.
DOI: 10.1144/SP320.8 0305-8719/09/$15.00

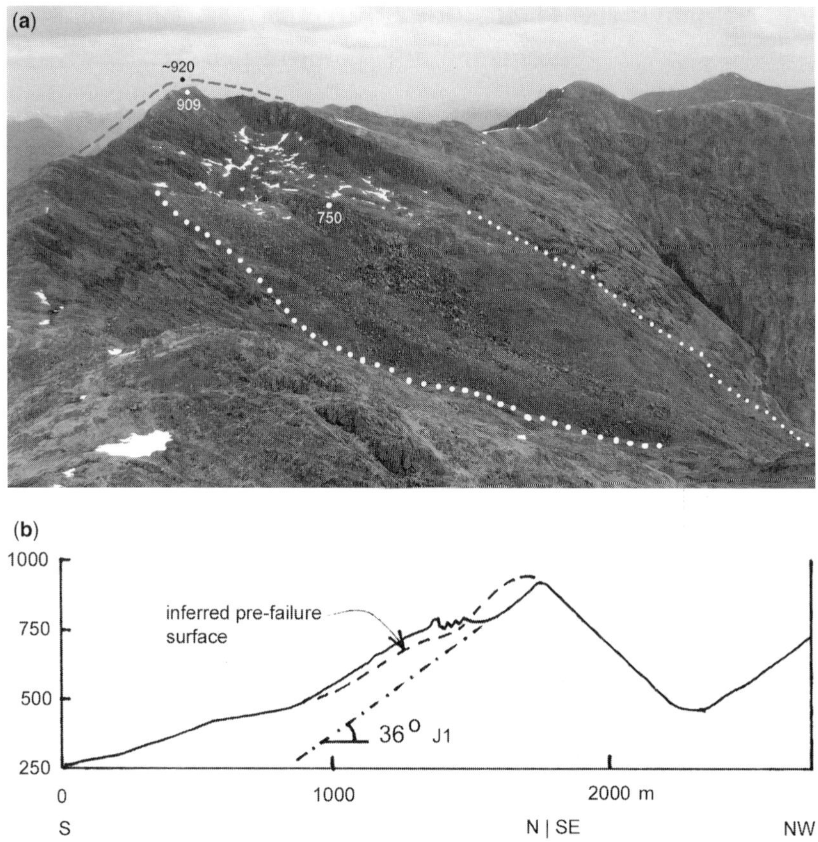

Fig. 1. Streap RSF, Glenfinnan, Western Highlands [NM 944860], a 0.30 km^2 arrested translational slide with progression to subcataclasmic failure (located in Cluster 2, Fig. 6). (**a**) Sliding is on the angled joint-set exposed in the left source scarp. The slide cavity has created bounding arêtes and a horn summit. The pre-failure summit is inferred (from Watters 1972) to have been some 10 m higher and a more rounded dome. The arrested failed mass impounds a lochan, and has a prominent extruded pinnacle (point 750). D_{MAX} (see Fig. 5) is 155 m, one of the largest recorded in the Highlands (see Table 2). View from SW, photograph by Hamish Johnston. (**b**) Long section through the RSF derived partly from Watters (1972).

is adopted to embrace both visible and occupied cavities. It also expresses the erosional dimension to RSF, which is too often considered only as a process or deposit. The practical and conceptual challenges of measuring it are explored in a Scottish Highlands context:

- Can the '*depth of bite*' of extant RSFs be defined and consistently measured?
- Can typical/maximal rates of paraglacial scarp retreat by the RSF process be obtained? And how do they vary across a range of geomorphic contexts?
- How might RSF incidence have varied and cumulated over the Quaternary? And can a model of RSF contribution to landscape evolution over many glacial–paraglacial cycles be inferred from its present incidence?
- How does the glacial–paraglacial process actually work? And can it really facilitate wholesale trough widening, beyond the contribution visibly achieved by extant RSFs?

Paraglacial RSF types, ages and geotechnics in the Scottish Highlands

A simple RSF typology (Jarman 2006) distinguishes: (a) cataclasmic and (b) subcataclasmic failures, where the debris has reached the slopefoot or lower slope, thus fully evacuating the cavity; (c) arrested translational slides, where the slipped mass still partly occupies a reasonably discernible cavity; and (d) extensional and (e) compressional slope deformations characterized by creep features and antiscarp arrays, where margins are diffuse and no readily definable cavity may exist. Individual cases may display compound character, with evolution of mode across a slope or by nested subsequent events. Cavities may be rectilinear (armchair), acute or obtuse wedges (often multiple), or planar slices. RSF locations range from fjord flanks at sea level to high summits at 1200 m asl (m above sea level). Many weak or modified candidate sites are not confirmable without geotechnical investigation, and remain probable/possible.

Very few RSFs in the Scottish Highlands have been dated, and only the small minority of (sub)cataclasmic cases with fragmented debris lend themselves to it (Ballantyne 2007*a*, *b*). It is assumed for this study that all extant Highland RSFs are paraglacial, with many occurring at or soon after deglaciation, and a diminishing tail (Cruden & Hu 1993) by progressive failure or delayed reaction. A handful of cases appear to be triggered by (glaci-)fluvial incision; no substantial cases are known within the last millennium. RSF interpretation is complicated by the minimally erosive Loch Lomond (Younger Dryas) Stadial. Although the majority of extant RSFs are within Loch Lomond Stadial limits, all are arguably responding primarily to slope stresses engendered by the Last Glacial Maximum and its deglaciation. Present RSF incidence thus excludes a lost population of immediately post-maximum RSFs exported by the Loch Lomond glaciers. Some of their cavities may survive as 'debris-free scarps' inventorized by Holmes (1984) but these are excluded from this study.

Most Highland RSFs are in the prevalent Neoproterozoic metasediments, with varying degrees of tectonization yielding extensive through-going discontinuities; RSF is rare on the granite intrusions. This paper does not consider the geotechnics of the sample areas or case studies (see Jarman 2007*a* and associated site reports, with photographs of the full range of RSF types; see also Watters 1972; Holmes 1984). Generally, structure and lithology in the sample areas are conducive to RSF, but actual incidence and absence respond to differences in joints, faults, foliation surfaces, dip, aspect, incidence of fallible rock units, etc. Likewise, whether failure is triggered by periglacial, hydrological or seismic events is less important than slope preconditioning by glacial erosion.

Methodology for measuring RSF 'bite'

The key parameters required to quantify an RSF cavity are its *depth* and *breadth* (Fig. 2). This is straightforward in homogenous terrain such as Icelandic basalt-plateau rims (Bentley & Dugmore 1998), but is more difficult in complex Scottish geology. Breadth is usually evident, although slope deformations may have diffuse margins. Depth of bedrock bite is unclear in the majority of Highland RSFs, which are not fully evacuated and/or have subsequent fill (talus, solifluction). Cataclasmic cases with rockwall cavities such as Beinn Alligin (Ballantyne 2007*a*) are exceptionally rare. Further complications arise with irregular source geometries such as multiple wedges (e.g. Fig. 3), or where source scarps are degraded and indefinite, or with compound sites of different modes or ages of failure.

This occupied-cavity problem vitiates any attempts at accurate measurement of topographic depth parameters, whether by field survey,

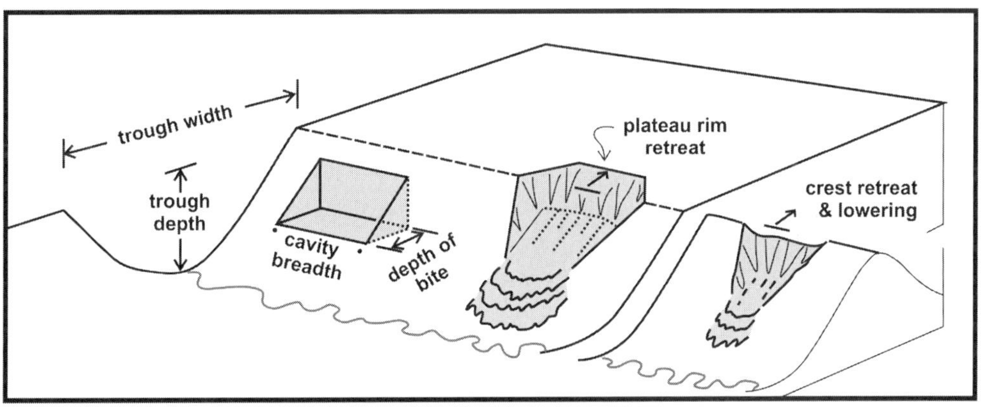

Fig. 2. Terminology for scarp retreat by paraglacial RSF. 'Cavity breadth' is used rather than 'width' since it is orthogonal to 'trough width' and must not be confused with valley widening. 'Depth of bite' can also be confused with 'trough depth', but no suitable alternative term exists. The depth of bite depends on the planes of reference adopted (see Fig. 4). Further complications arise where RSF daylights behind a ridge crest, since the bite is both widening the valley laterally and lowering the ridge vertically.

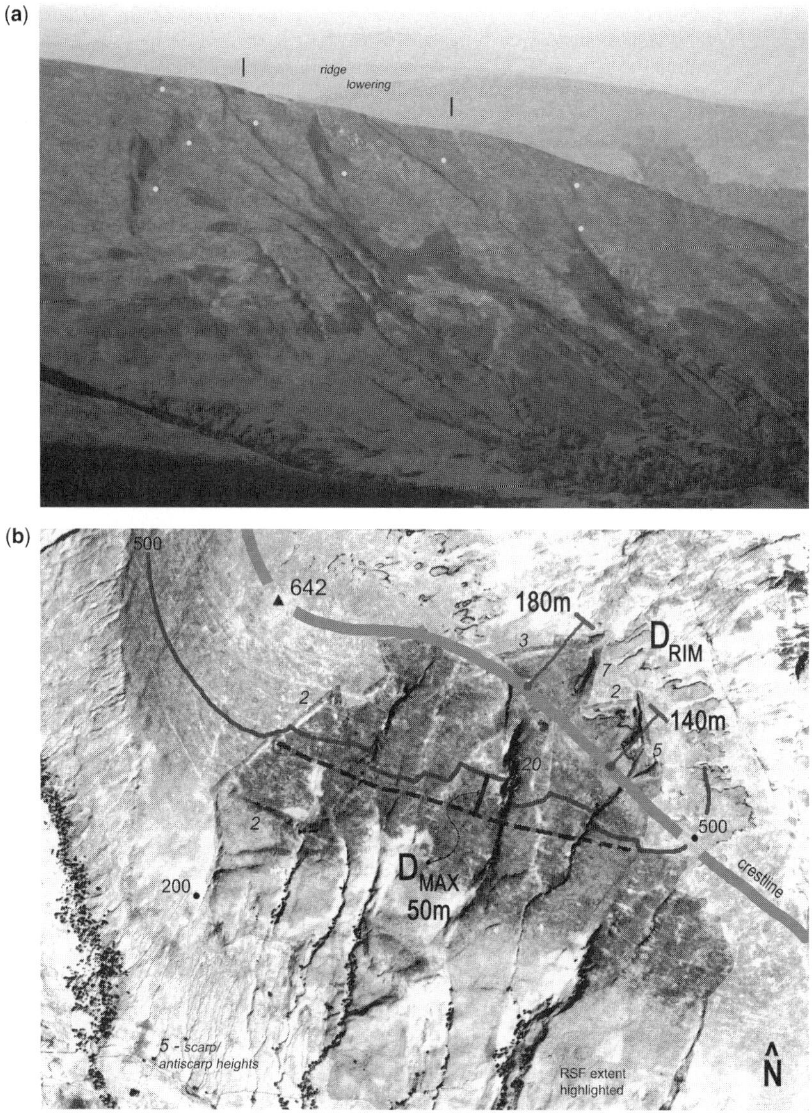

Fig. 3. Beinn Dubh [NS338948] in Glen Luss (for location see Figs 7 and 14). (**a**) View downvalley from the west. Three main translational slide wedges have developed within a large (0.5 km^2) deformation. The several bedrock cavities (marked with dots) remain partly obscured by non-evacuated debris and subsequent infill, so that a representative depth measure is difficult to derive. The ridge crest has been lowered by approximately10 m for 350 m between the bars. (**b**) The depth measures defined in Figure 5 give D_{MAX} of 50 m in the main cavity, and D_{AVE} of 20 m across the site. Two of the wedges break the crestline, daylighting downslope towards Loch Lomond with D_{BAY} of 180 m and 140 m. This incipience creates a weakness likely to be exploited by glacial parafluence, breaching the ridge and isolating its lower end as a 'pap', a common form in the Highlands. Vertical air photograph © RCAHMS.

photogrammetry or DEM (digital elevation model). Calculating scarp retreat rates from debris volumes (e.g. Holmes 1984) only applies to the minority of (sub)cataclasmic RSFs. Geophysical surveys of RSF subsurface extent have not been attempted in British mountains.

There are at least five possible pre-failure planes of reference from which to measure RSF depth (Fig. 4). The map-based measure D_{PLAN} is selected for this study as most relevant for landscape evolution, but measures into the rockmass are also reported since deep-seated failures will

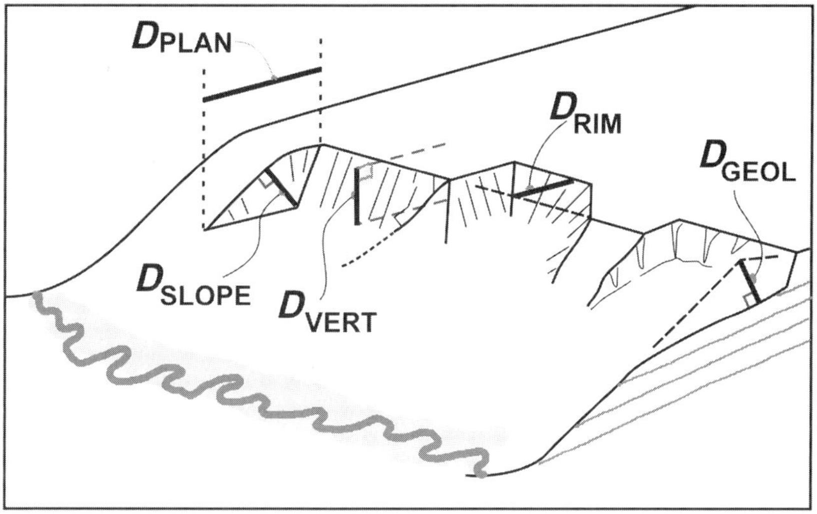

Fig. 4. Five possible planes of reference from which RSF depth of bite could be measured. D_{PLAN}, orthogonal to the valley axis (as a plan measure of lateral valley widening); D_{RIM}, a variant of D_{PLAN}, from plateau rim (as a measure of scarp retreat); D_{SLOPE}, parallel to the valley-side slope; D_{VERT}, vertical depth to base of failure; D_{GEOL}, parallel to the geological failure surface.

be more influential in trough widening than shallow ones.

Map-based measures of depth of bite

Other than in the simplest rectangular evacuated cavity, no single measure of D_{PLAN} can fully represent RSF depth of bite. Ideally, a composite measure would be generated from the average bite along each contour. However, with many RSFs having irregular contours such sophistication is unwarranted. A more robust method simply identifies the deepest contour recess; this recess may well occur where retained debris or infill is shallowest and where it may most closely approximate true cavity depth. Four D_{PLAN} variants employed here bracket trough widening with a range of values (Fig. 5).

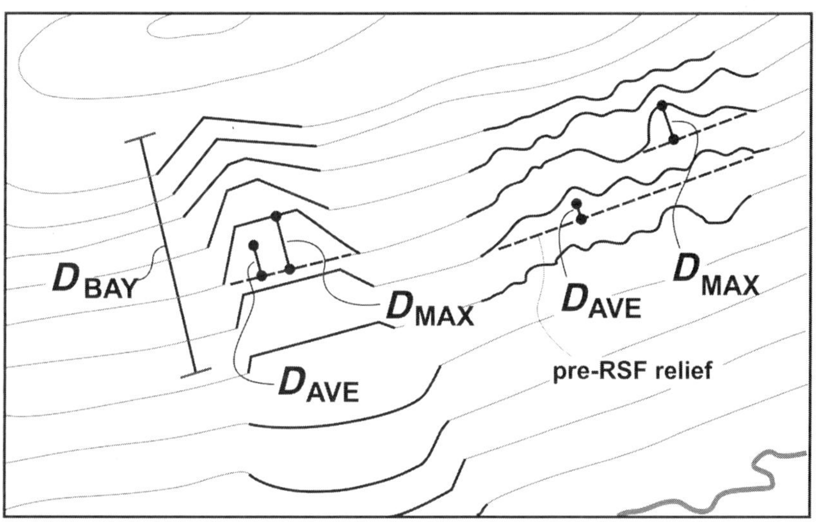

Fig. 5. Three measures of D_{PLAN} obtained in this study (D_{RIM} is shown on Fig. 4). The difficulties of applying them meaningfully to irregular RSF slips and slope deformations are illustrated on the right.

- *Maximum contour recess* (D_{MAX}). This measure best represents the visual impact of the failure. The deepest contour recess from a projection of the pre-failure contour across the site is taken. At Ordnance Survey 1:25 000 scale, contours are at 10 m vertical interval, but the bold 50 m contours are most reliable for reconstruction. The D_{MAX} contour can be at the top of the failure, but is more usually in the upper third.
- *Average contour recess* (D_{AVE}). A prevailing figure (not a calculated mean) is taken along the same contour. In a 90° wedge, it would be half D_{MAX}. It is a corrective for broad deformations where D_{MAX} can be a local aberration, but is difficult to apply consistently.
- *Bay depth* (D_{BAY}). This is the generalized plan depth from the mouth of the failure bay where it interrupts the valley wall to the source scarp or fractures (the term 'bay' is preferred to 'cavity' as that refers to the bedrock shape). D_{BAY} can reach an order of magnitude greater than D_{MAX}, especially on long or gentle slopes (and thus does not usually represent actual scarp retreat).
- *Rim bite* (D_{RIM}) (Fig. 4). This applies in a minority of cases where RSFs cut into a plateau, and to incipient failure behind a rim or brow; it can coincide with D_{MAX}.

Where no measurable contour recess or bay exists, as with failures on convexities and some slope deformations, the site is excluded from the depth analysis, but its breadth is still included in the total. Note that breadth and depth data cannot be multiplied to obtain failure volumes. Statistical analysis of the data is inappropriate given the diversity of RSF shapes and settings.

Sample areas in Scotland and Norway

The mainland Scottish Highlands displays the most extensive array of primarily paraglacial RSFs in the British Isles. There are at least 550 definite and probable sites (>0.01 km^2), their average area being 0.21 km^2 (Jarman unpubl. data). Five sample areas have been selected (Fig. 6 and Table 1). Three are within the main clusters of large RSFs (>0.25 km^2) identified by Jarman (2006), and two are in areas of sporadic or smaller-scale RSF. They span the Linton zonal typology of increasing glacial trough dissection from east to west (in Clayton 1974). All are in the prevailing Neoproterozoic metasediments, traversing their full diversity of lithology and structure. Slope angle and aspect have not been measured: while RSF occurs on lower- and higher-angle slopes alike, it is more frequent on valley sides with geological dip conducive to failure, or with open S-to-W aspects (cirque dissection reduces scope for RSF on N to E faces).

In the main RSF clusters, the *Cowal–Arrochar–Luss* sample area (Fig. 7) is intensely dissected by glaciated troughs and breaches. It is one of the densest clusters in Scotland, with RSF affecting 7–8% of the terrain in core areas (Jarman 2007*a*). The structural dip and micaceous schist interbedding are often conducive to translational sliding (Fig. 3). RSF is found at all levels and in all topographic contexts, but is least common on the flanks of mature valleys of preglacial origin. In this area, a mountain core is flanked by lower hills, both with available relief of 500–800 m. By contrast, in the *Ericht–Gaick* plateau sample area (Hall & Jarman 2004) RSF is clustered along the rims of two transectional breaches; one follows a major Caledonide fault, the other being sinuous and of uncertain origins (Fig. 8) (Jarman 2004*a*). Low-angle structures and arenaceous lithology are not conducive to sliding here, and failure tends to be by slope deformation or in deeply-weakened material. Available relief is only about 400 m. At the other extreme, the *Kintail–Affric* sample area has some of the highest available relief (<1000 m) and steepest terrain in Scotland. Long, well-defined ridges are separated by major transectional breaches, but only locally interrupted by cross-breaches, where RSF is most intense, affecting 6% of the terrain (Jarman 2003*c*). The geology of high-angle indurated metasediments dissected by faults promotes failure in both deformational and slide modes, but RSF can occur without structural assistance; it is sparse on the mature east-flowing valley sides.

Away from the main clusters, the *Monar–Strathfarrar* sample area is in high mountains with transectional breaches and similar geology to Kintail–Affric, but displays sparse albeit bold RSF, such as Sgurr na Conbhaire (Fig. 9) and Sgurr na Ruaidhe (Fig. 10). The *Dearg–Wyvis* sample area (Fig. 11) has an intermediate plateau book-ended by higher massifs, with smaller-scale RSF along a narrow internal valley rim and in minor breaches, which are probably fault-directed.

Inventories of RSF for all these sample areas have been compiled from an air photography search (Holmes 1984); from those British Geological Survey 1:50 000 maps and unpublished field slips that show landslips; and from field explorations. The 193 sites identified cover about a third of the Highland RSF population. Of these 146 had measurable breadths, but only 96 yielded map-measurable depths, including 36 cases of more than 0.25 km^2, which comprise 24% of the better-verified large-RSF database (Jarman 2006).

A comparator area in *north Norway* offers an initial check on the possible wider relevance of the Scottish results. NE Troms (east of Tromsø) has affinities with the Highlands in its 1000 m

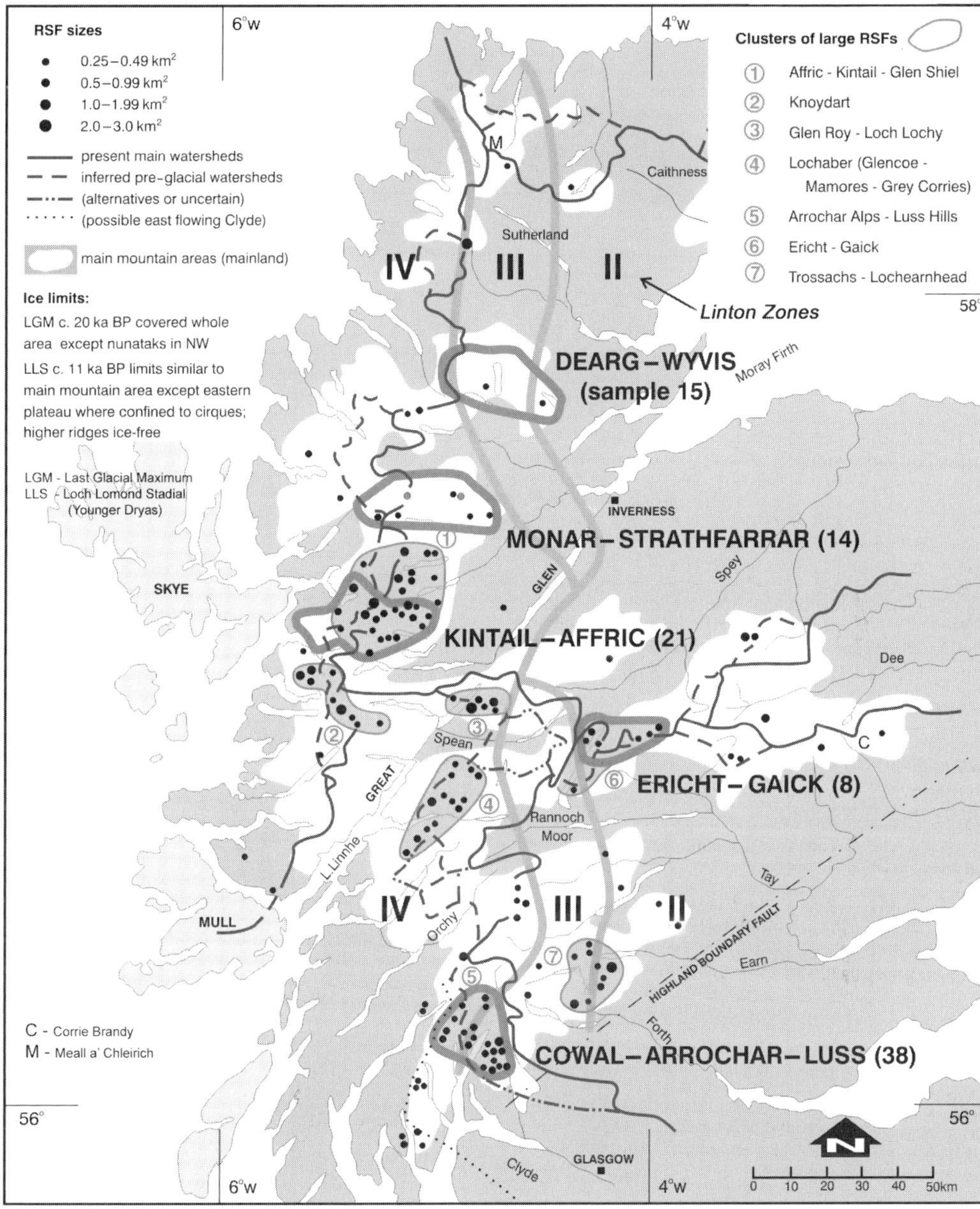

Fig. 6. The five sample areas in the Scottish Highlands, in relation to clusters of larger RSFs and the Linton zones of increasing glacial dissection from east to west (from Clayton 1974). The near-complete population of larger RSFs (>0.25 km^2) is a fair proxy for the population of all RSFs, which are widely distributed across all three zones but with great variation in clustering and sparsity. Adapted from Jarman (2006).

available relief, its subhorizontal Caledonide metasedimentary geology and its dissected paleic relief. It possesses one of the densest reported RSF clusters in the Scandes (Kverndal & Sollid 1993). An area east of Lyngen fjord (Fig. 12) has been geomorphologically mapped at 1:50 000 (Tolgensbakk & Sollid 1988), identifying 50 mass movement sites that appear to span the five Highland categories of paraglacial RSF. Most are subcataclasmic debris masses sufficiently evolved to resemble rock glaciers (Kverndal & Sollid 1993); only seven are cataclasmic events filling the cirque or trough floor.

Table 1. *Sample areas, populations and results*

Sample areas *Sub-areas*	Linton zone	Total identified RSFs*	Average RSF depth of bite (D): Contour recess n	Contour recess D_{AVE} (m)	Contour recess D_{MAX} (m)	Bay depth D_{BAY} (m)	RSF breadth (B): n	Average breadth (m)	Total breadth (km)	% valley wall†	Ratio Average D_{MAX}: Average B
Cowal–Arrochar–Luss	IV	72	38	25	**45**	260	60	**385**	24	14	***8.5***
Cowal–Ardgoil S			*14*	*30*	***50***	*270*	*16*	***380***	*6*		***7.5***
Ardgoil N–Arrochar			*10*	*10*	***25***	*175*	*22*	***300***	*8*		***12.0***
Luss Hills			*14*	*35*	***55***	*300*	*22*	***475***	*5*		***8.5***
Ericht–Gaick	II	18	8	–	**30**	235	14	**480**	6	3.5	**16.0**
Kintail–Affric	IV	52	21	15	**45**	400	38	**600**	23	9	**12.0**
Monar–Strathfarrar	IV	22	14	30	**40**	290	18	**370**	5	3	**9.0**
Dearg–Wyvis	III	29	15	–	**40**	170	16	**270**	4	1.5	**7.0**
Total sample areas (Scotland)		193	96				146				
Norway–Kåfjord	[III]	50	17	–	**60**‡	–	24	**400**	10	7	**7.0**

*Identified RSFs may not be quantifiable either because they have not been verified in the field and map evidence is inadequate, or because the site is a slope deformation or minor crag collapse not yielding a measurable cavity at 1:25 000.

†Proportion of total valley-wall length in sample area affected by RSF – coarse estimate based on main glaciated valley walls legible at 1:250 000, excludes corries and lower relief.

‡Excluding one exceptional site (Table 2), which inflates the figure to 70 m.

[Full datasets for each area available from the author.]

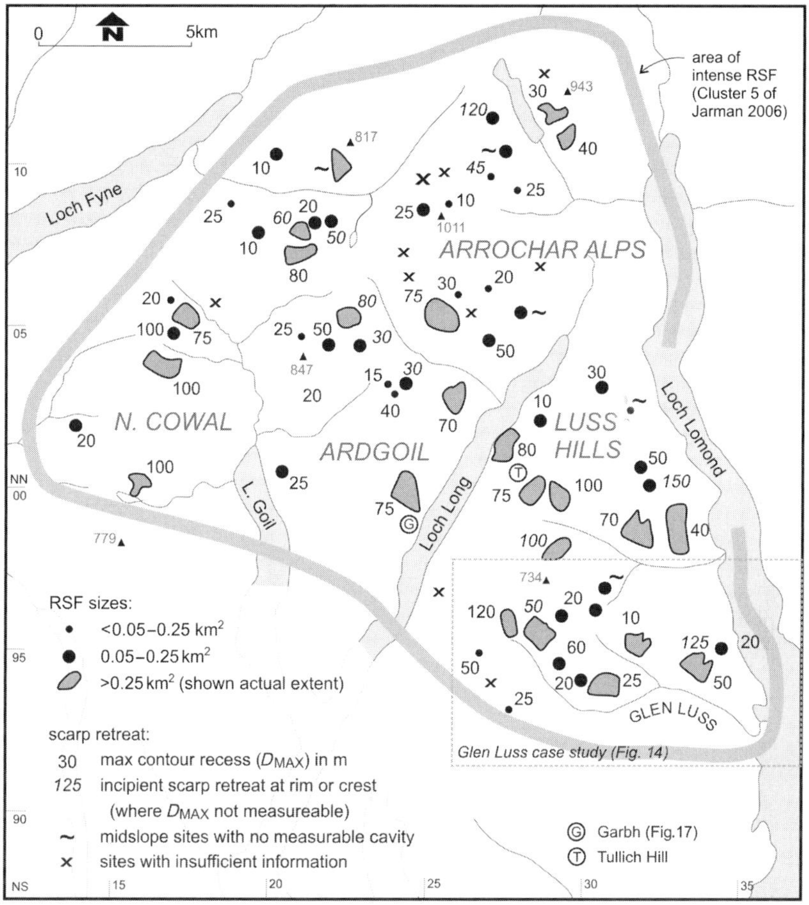

Fig. 7. The Cowal–Arrochar–Luss sample area, SW Highlands, is one of the densest RSF clusters in Britain. Smaller failures predominate in the northern (Arrochar Alps) part in higher, more glacially scoured terrain. Many RSFs are contributing to widening the complex network of glacial breaches of regional and local watersheds. Updated from Jarman (2003*a*).

Fig. 8. The Gaick Pass, a narrow glacial breach of the main Grampian watershed east of Drumochter, looking south. The plateau rim is 825–850 m asl and Loch an Dùin [NN723800] is at 490 m. Despite the limited available relief of approximately 350 m, the left (east) side displays slope deformation along 1 km of trough wall with remarkably deep incipient extension (D_{RIM} of 175 m to white dashes) into the paleic surface of A' Chaoirnich. On the right (west), the nose of An Dùn is sharpened by slippage with a D_{MAX} of 40 m (inferred former crestline white dotted).

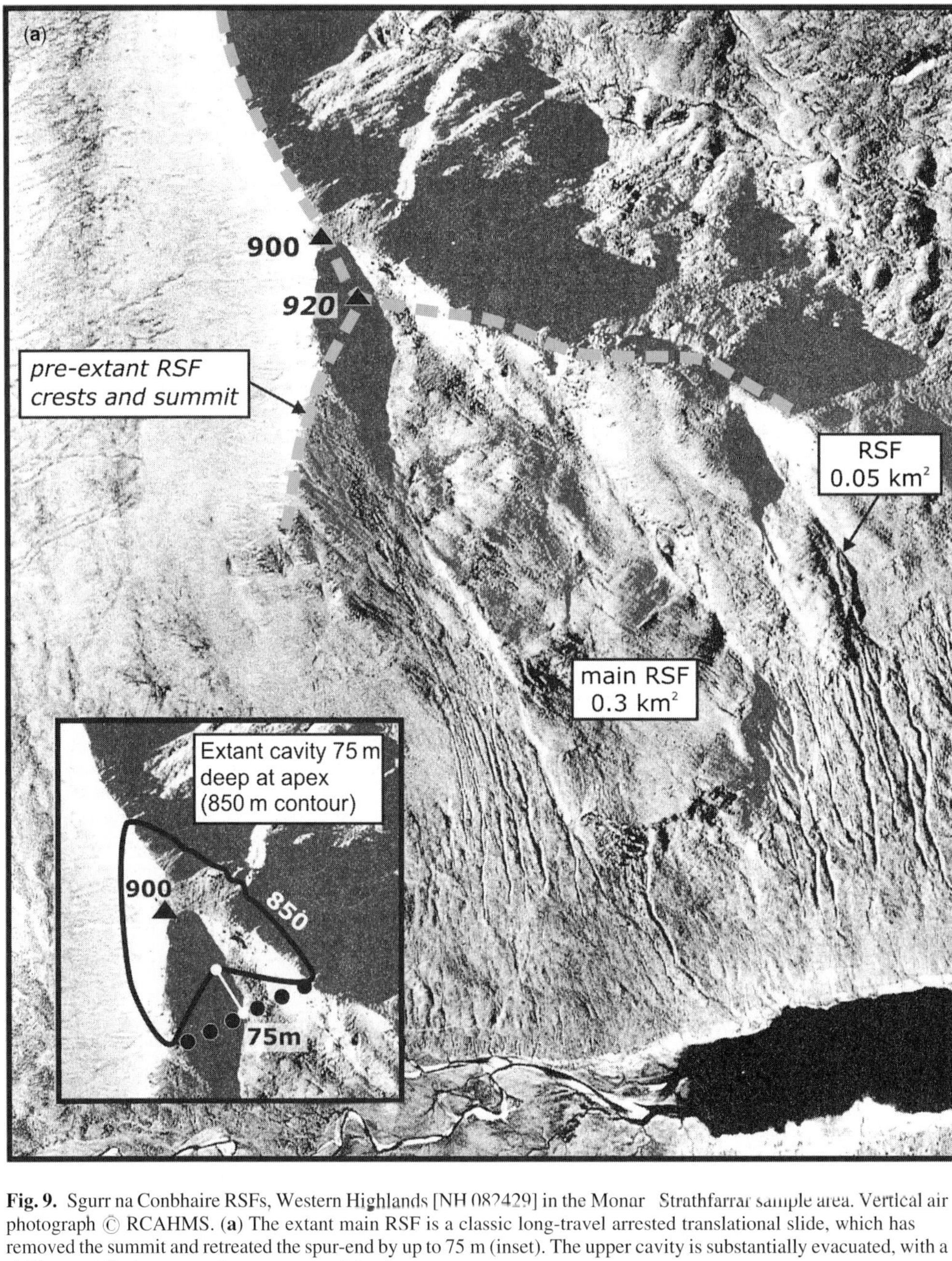

Fig. 9. Sgurr na Conbhaire RSFs, Western Highlands [NH 082429] in the Monar–Strathfarrar sample area. Vertical air photograph © RCAHMS. (**a**) The extant main RSF is a classic long-travel arrested translational slide, which has removed the summit and retreated the spur-end by up to 75 m (inset). The upper cavity is substantially evacuated, with a slabby west flank scarp acting as a main slide plane.

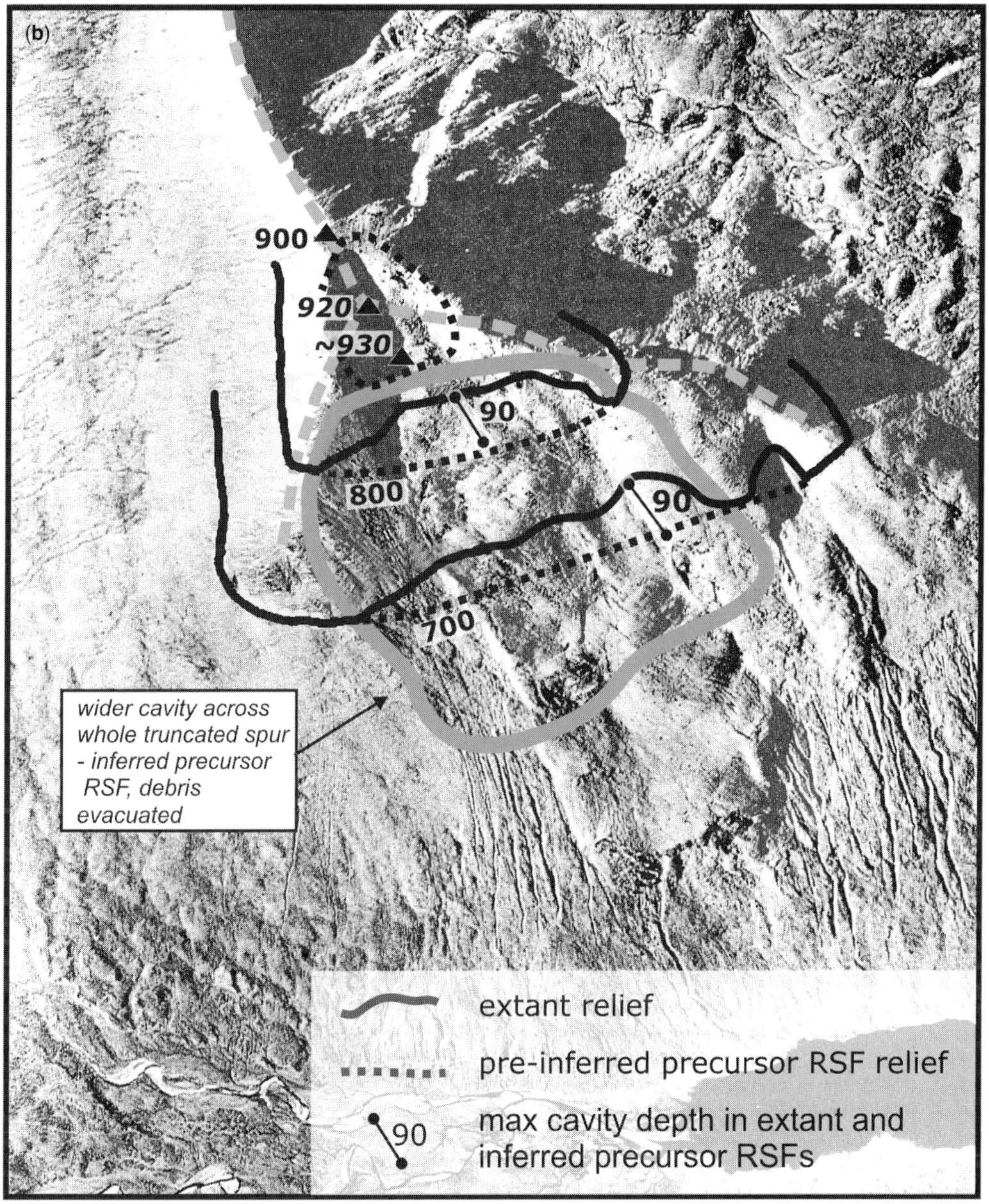

Fig. 9. (*Continued*) (**b**) It is inferred from the cavity-within-a-bay form that an earlier, broader RSF occurred here. The process of wholesale glacial truncation of this spur is thus being assisted by extensive RSF with D_{MAX} coincidentally of *c.* 90 m in both generations of cavity. Fenton (1991) estimated D_{GEOL} at 150 m, reflecting the thickness of the main arrested mass. The earlier cavity may originally have been deeper, as its mid-slope arms appear to have been pared back by the valley glacier during an intervening stadial.

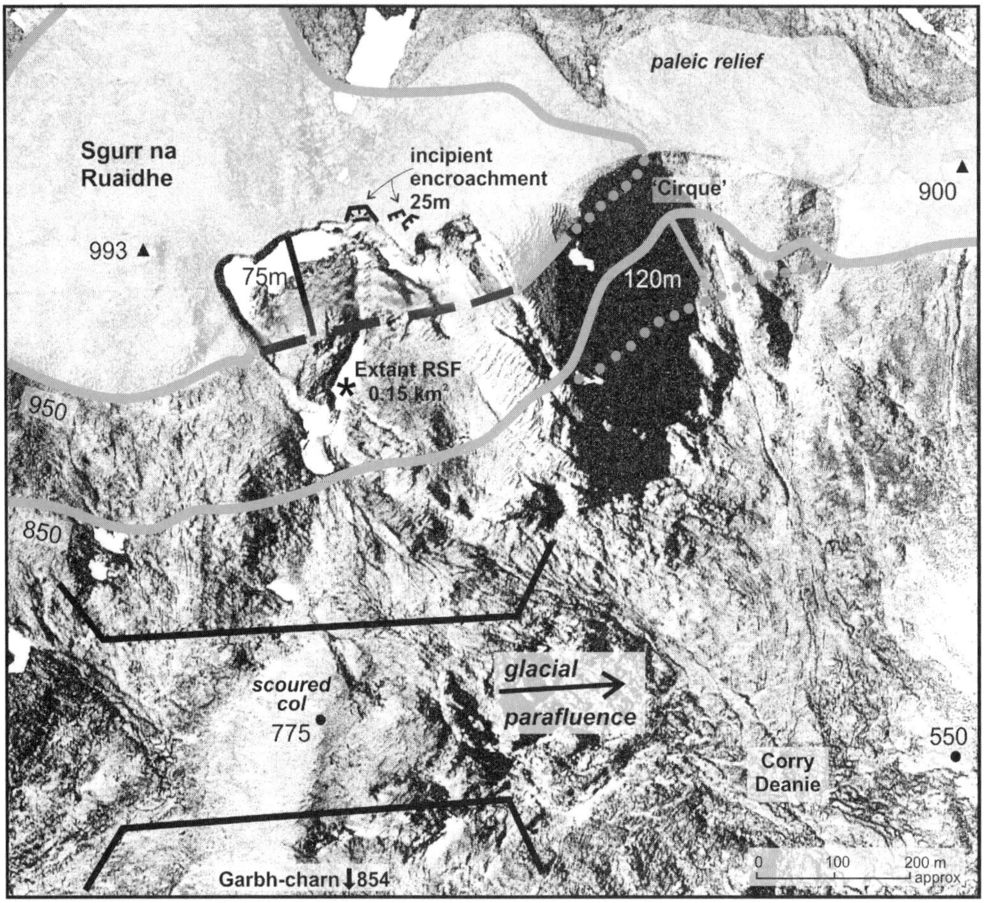

Fig. 10. Sgurr na Ruaidhe RSF [NH292424] in the Monar–Strathfarrar sample area. The deep, fresh-looking short-travel arrested slide bites by 25 m (D_{MAX}) at * and 75 m (D_{RIM}) into paleic relief (pale tone). The adjacent cavity to the east may be the scar of an earlier RSF with all debris since evacuated, with an inferred D_{MAX} of 120 m. It is at present a poorly developed SE-facing corrie. The extant RSF locus is separated from the main immature Strathfarrar trough by the promontory of Garbh-charn, and failure is thus inferred to have been provoked by glacial 'parafluence' through the col parallel to the Farrar ice flow, with an available relief of only 200 m. Note incipient encroachments 25 m into the plateau above the main RSF headscarp, with ground lowering of *c.* 2 m. Vertical air photograph © RCAHMS.

None face E or NE. Seventeen deposits have measurable cavities above them, which are probable RSF sources.

Results showing significant and consistent depth of bite

The results (Table 1) confirm that the typical paraglacial RSF makes a very substantial 'bite' into the trough wall. There is also consistency between the diverse sample areas, suggesting that RSF processes respond to slope scales, structures and stresses of widespread applicability. Averaged D_{MAX} is consistently in the range 40–45 m across four of the five areas; it is coincidental with a 40 m figure reported for 30 RSFs in the modal 0.5–1.0 km breadth class in northern Iceland (Bentley & Dugmore 1998). A higher figure in north Norway (60 m) might reflect a greater degree of cavity evacuation in somewhat larger-scale relief, but may simply be sampling bias to map-recognizable cases. The Ericht–Gaick D_{MAX} result is lower because the small sample is mainly of deformations and sags that tend not to yield pronounced contour recesses. There is greater localized variation within Cowal–Arrochar–Luss where the steep, laterally-convex Arrochar Alps favour smaller RSFs, and the open slopes of the Cowal and Luss hills accommodate broader RSFs. The mapped results for D_{MAX} in two contrasting

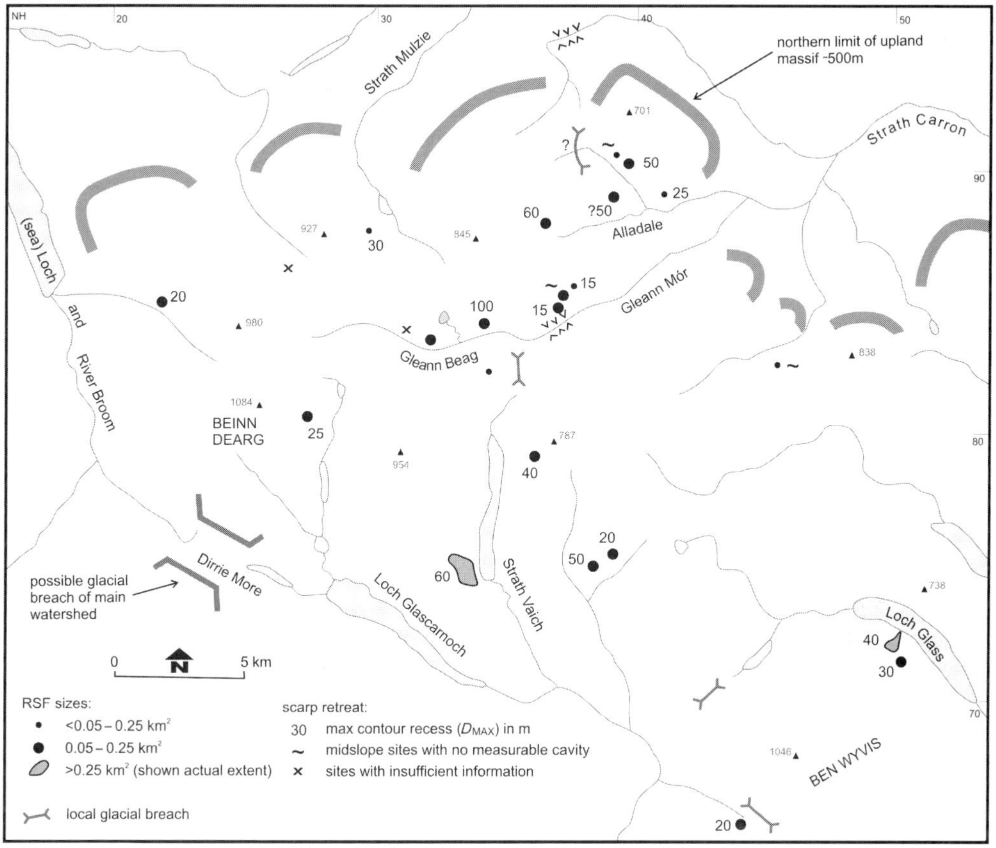

Fig. 11. The Dearg–Wyvis sample area, Northern Highlands. By contrast with Figure 7, RSF is sparse and small scale, with only two larger cases. As the geological factors are not dissimilar, this may reflect its lightly dissected character at a centre of ice dispersal where the ice cap was thinning northwards. RSF incidence may here respond to local deepenings at a relatively early stage of glacial adaptation. Gleann Mór–Beag is a narrow trough incising the intermediate plateau by only 300–400 m. Alladale is a short but perhaps aggressively enlarging trough-corrie, a branch of which has transected the external plateau margin. Ben Wyvis is flanked by deep narrow glacial breaches across the NE–SW mountain barrier. Other RSF loci range from summit crests (where they are enlarging corries) to low valley sides. Nevertheless, D_{MAX} is not significantly different from the intense clusters.

sample areas (Figs 7 and 11) demonstrate the consistency of paraglacial 'bite' for individual RSFs.

The $\boldsymbol{D_{AVE}}$ measure is about half D_{MAX} in the most typical area (Cowal–Arrochar–Luss); more than half in Monar–Strathfarrar with its bolder bites; and one-third in Kintail–Affric with its significantly broader failures. The numerical value of D_{AVE} is of little intrinsic value, but confirms that D_{MAX} is not aberrant, and that regional variations are reasonably explicable. Average bay depth ($\boldsymbol{D_{BAY}}$) is typically six times D_{MAX} and ranges from 170 to 400 m; in Monar–Strathfarrar (290 m) it is about 30% of prevailing half-valley-width, suggesting substantial unexhausted potential for long-term trough widening. Greater bay depth goes with longer trough walls.

Average RSF breadth (**B**) ranges from 270 to 600 m. RSF affects 14% of total valley wall length in Cowal–Arrochar–Luss and 9% in Kintail–Affric, but such significant cumulative impact is limited to the dense clusters. Although a direct relationship between depth and breadth is reported by Bentley & Dugmore (1998), the results presented here suggest considerable variability. The depth : breadth ratio is lowest where structures are conducive to sliding, and where dissection limits RSF width. It is markedly higher where slope deformation is prevalent.

These averaged results meld the full range of failure sizes, types and topographic contexts; they do not consider variations in lithology, gradient, aspect or valley-side roughness. Some individual

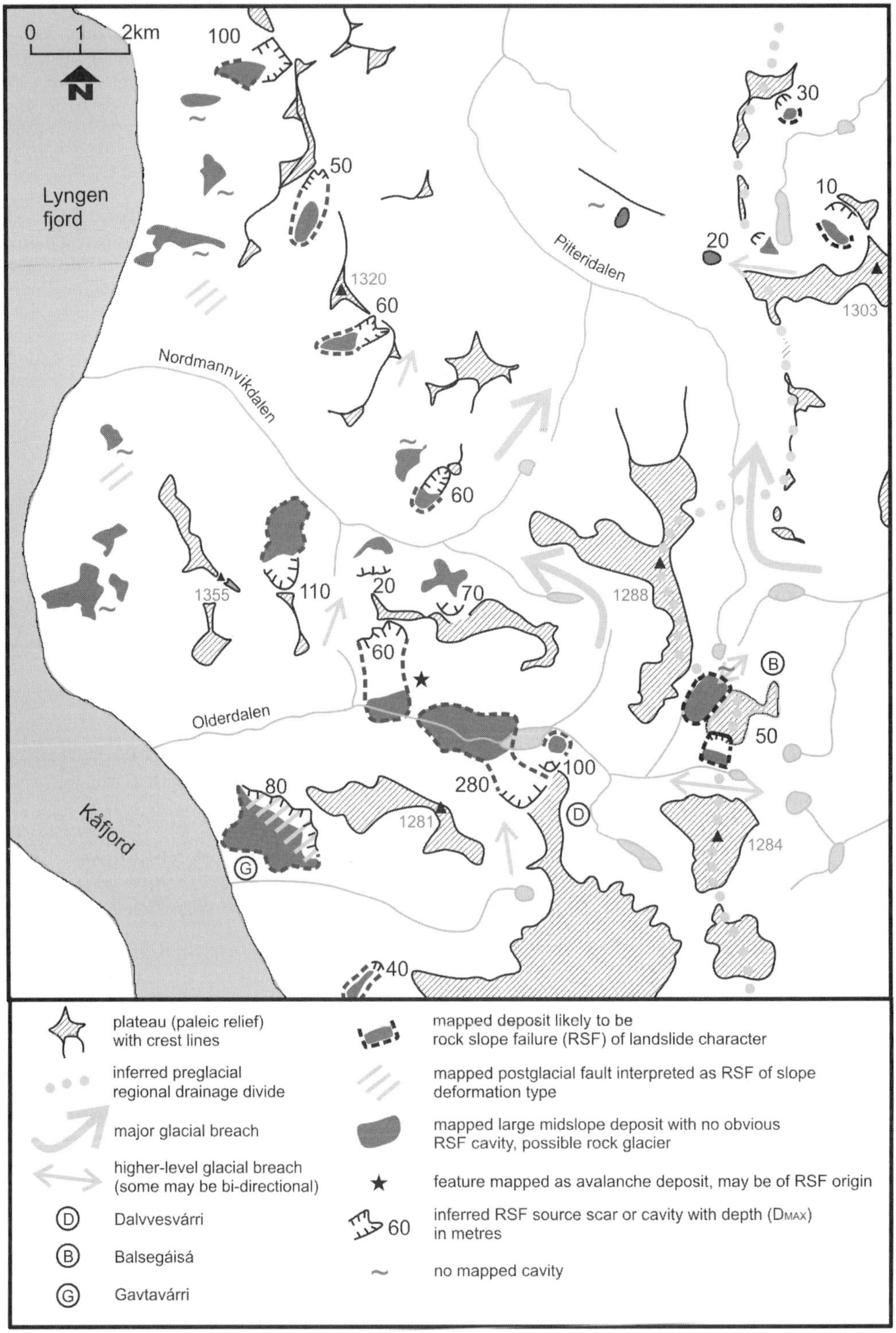

Fig. 12. The north Norway sample area, at Kåfjord east of Tromsø [69°40′N, 21°45′E]. Probable RSFs are interpreted from the 1:50 000 geomorphological map (peripheral areas omitted) with limited field verification in Olderdalen and

Table 2. *Significant individual cases of RSF trough-widening*

Sample area RSF site	Ref.	Depth of bite		Breadth (m) *Combined adjacent sites
		D_{MAX} (m)	D_{BAY} (m)	
Cowal–Arrochar–Luss				
Mullach Coire a' Chuir	[1]	100	650	1200
Garbh (L Long)	Fig. 17	75	800	700
The Cobbler S	[2]	*none – convex slope*		1000
Tullich Hill W + E	[3]	75/100	300/250	1250*
Cruach an t-Sithein	Fig. 14	120	500	450
Ericht–Gaick				
Meall Cruaidh/The Fara		25/70	100/600	2550*
Kintail–Affric				
Beinn Fhada	[4]	*slope deformation*		3000
Sgurr na Ciste Duibhe	[5]	100	1000	750
Carn na Con Dhu	[6]	120	300	1250
Monar–Strathfarrar				
Sgurr na Conbhaire	Fig. 9	90	400	575*
Sgurr na Feartaig A/B		90/50	240/140	1250*
Sgurr na Ruaidhe	Fig. 10	80 [D_{RIM}]	350	400
Dearg–Wyvis				
Cail Mòr	Fig. 11	120 [D_{RIM}]	200	120
Other Highlands				
Streap (Glenfinnan)	Fig. 1	155	670	450
Corrie Brandy (cirque)	Fig. 15	130	–	400
North Norway				
Olderdalen	[7]	280	–	

Sources: [1] Jarman 2003*a*; [2] Jarman 2004*b*; [3] Jarman 2003*b*; [4] Jarman & Ballantyne 2002; [5] Jarman 2007*b*; [6] Holmes 1984; [7] Jarman 2002.

sites are very large (Table 2), in areas of both dense and sparse RSF. Six such RSFs bite into wall or rim by 100 m or more, notably in Cowal–Arrochar–Luss (Fig. 7), with Streap (Fig. 1) attaining 155 m. Twelve sites or adjacent groups exceed 1 km in breadth, with Beinn Fhada at 3 km being the largest RSF in the British mountains.

Concentrated RSF trough widening effects

Extant paraglacial RSF is clustered both regionally and locally. Its quantitative geomorphic impact over a whole mountain range is relatively small: the interest here lies in where and why it becomes intense. This may be in specific valleys, in over-enlarging corries (cirques) or on particular plateau rims.

Whole-valley impacts: two case studies

While overall valley-wall RSF incidence does not exceed 14% even in dense clusters, locally it attains 18% in the Gaick Pass (Fig. 8) and 25% in the north Loch Ericht breach.

In *Knoydart* (Fig. 13), RSF affects 28% of the valley sides of Gleann an Dubh-Lochain and Gleann Meadail, while it is almost absent in adjacent Gleann na Guiserein which is of similar character. One possible difference lies in the inputs of transfluent ice over breaching cols to the first two valleys. The north flank of Gleann an

Fig. 12. (*Continued*) Nordmannvikdalen. Paleic relief and D_{MAX} contour recesses are plotted from the 1:50 000 topographical map. Balsegáisá is a paleic relief remnant undergoing demonstrable shrinkage by RSF. Dalvvesvárri is a cataclasmic slide damming a lake; its extreme cavity depth relies on reconstruction of a lost promontory (Jarman 2002) and is discounted from the average D_{MAX}. Gavtavárri is the largest slope deformation mapped, with progression to slide lobes.

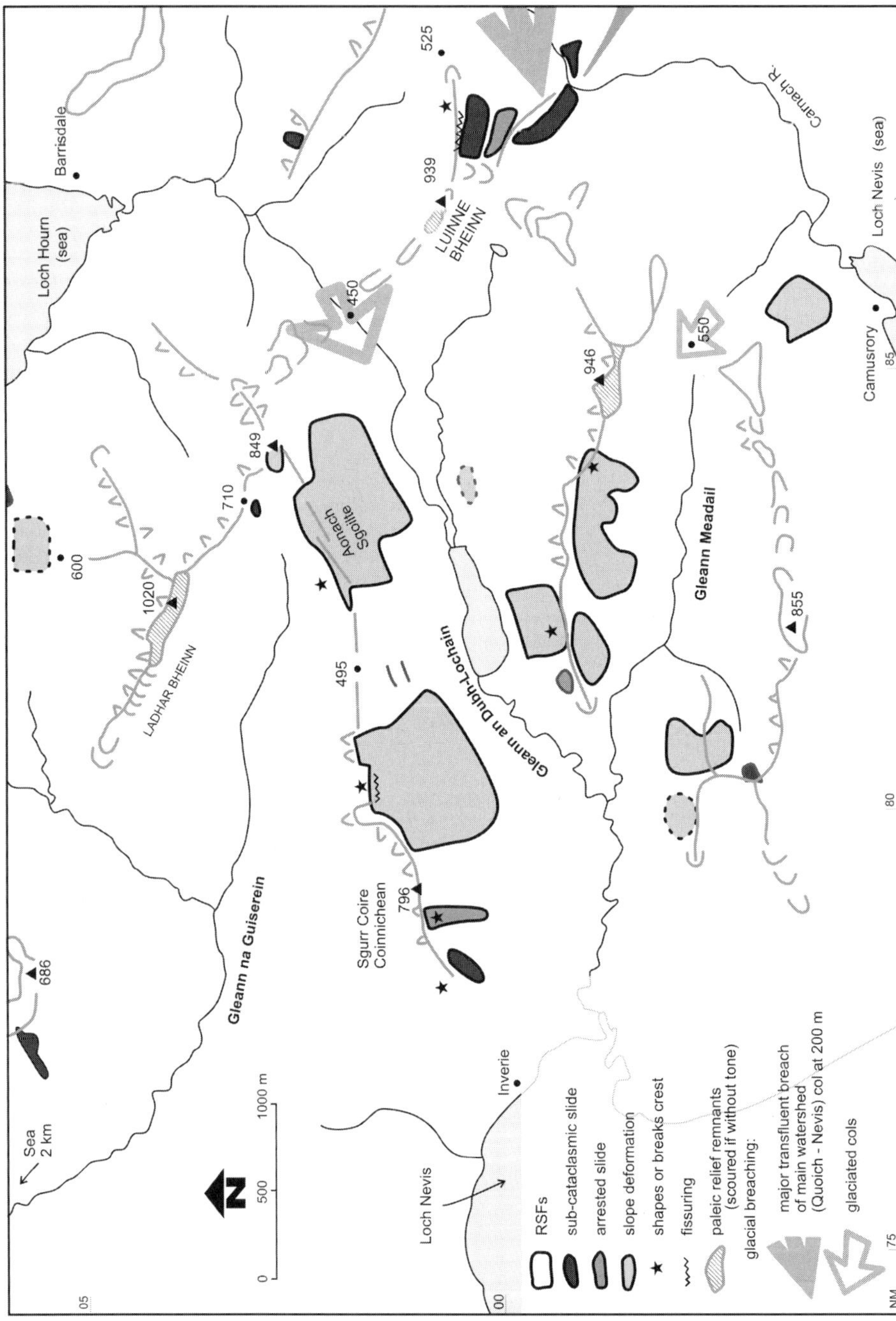

Fig. 13. High RSF incidence in central Knoydart valleys (for location see Fig. 6; NW part of Cluster 2). Available relief ranges from 600 to 1000 m. Deformational RSF is intense in twc valleys with glaciated cols at their heads, but sparse elsewhere except locally on Luinne Bheinn below a breach of the main watershed.

Dubh–Lochain is affected by RSF for at least 47% of its length, with Aonach Sgoilte displaying the most dramatic split ridge in Britain, comparable to an alpine 'doppelgrat' (see fig. 2.12 in Jarman 2007*a*). Since most of the RSFs are broad slope deformations, depth measures have not been taken.

At *Glen Luss* (Fig. 14), a remarkable sequence of RSFs ranging from near-*in situ* slope deformations to long-travel slides occupies 52% of the north valley side (whole-valley impact *c.* 26%). Taking D_{AVE} and weighting it by RSF breadth, extant RSF cavities alone are widening about half of one side of the trough by 25 m, along the site-by-site contour of greatest indentation. There is no obvious explanation for such concentrated activity here, nor similarly on the north side of adjacent Glen Douglas. Failure is predominantly on south aspects partly in response to a favourable structural dip (e.g. Fig. 3); north aspects seem generally less susceptible to RSF. Asymmetric trough profiles are common, and may evolve by such preferential widening.

Other individual valleys with high RSF incidence include Glen Ample, which follows the major Caledonide Loch Tay Fault and is one of the better candidates for a neotectonic association (located in Cluster 7 of Fig. 6) (see Jarman 2007*d*). Glen Roy (located in Cluster 3) is noted for its proglacial lake jökulhlaups, but the RSFs are of varied dates and characters (Fenton 1991), and the valley is a breach of the main pre-glacial watershed (Jarman 2008). RSF mini-clusters are also associated with breach systems such as Tyndrum–Orchy–Lyon, Tay–Almond–Amulree and Cluanie–Affric.

Corrie widening

While RSF commonly contributes to widening main valleys and troughs (in about 70% of cases in the two largest clusters; Jarman 2003*a*, *c*), it also promotes side bay and corrie enlargement (15% of cases). RSF is uncommon in classic cirque bowls, most cases being in elongated side-troughs and open embayments. It occurs mainly on corrie flanks rather than headwalls: even at Ben Hee, where the head of Gorm-choire appears to have failed, this is inferred to be a former spur within a compound corrie (Jarman & Lukas 2007). This suggests that headwalls retreat by incremental attrition, with stresses continually relieved by rockfalls that rarely reach the RSF-scale seen in the Garbh Choire Mòr (Fannaich) slabslide (Holmes 1984). Headwalls may be buttressed against larger-scale failure by their arch-form in plan. Corrie flanks behave more like valley sides, with RSF promoting lateral rather than headward enlargement (e.g. Coire Gabhail; Ballantyne 2007*b*). This may help to account for the evolution of the broad compound cirques common in parts of the Highlands (Gordon 1977).

In Glen Clova, a sequence of disproportionately large cirques facing SW has attracted attention (Holmes 1984). Here, a cluster of RSFs has developed, including two on either flank of Corrie Brandy (Fig. 15), the larger biting into the flank rim by 130 m. This cirque is undergoing significant lateral widening by RSF; adjacent cirque flanks display angular shapes and scars suggesting post-late Devensian failures removed by the last corrie glaciers.

Incipient trough widening

Incipient RSF is identified where a rock mass has clearly failed along fissures or fractures but has only displaced by a few metres or decimetres; no contour cavity is revealed, but the potential bite into plateau or ridge measured by D_{RIM} can exceed 100 m (Table 3). It often develops above manifest RSFs (in 19 out of 72 RSFs in Cowal–Arrochar–Luss), where it is additional to valley-wall D_{PLAN} measures, and attests to upward migration of failure. Few of these incipient RSFs appear active, a rare case (evidenced by torn vegetation) being on Meall a' Chleirich (Fig. 16). More common are apparently dormant step-scarps (Fig. 17b) or false antiscarps (uphill-facing source scarps where the incipient failure daylights behind the crest, e.g. Fig. 3).

Beinn Fhada and A' Chaoirnich are the largest reported incipient RSFs, both affecting kilometric lengths of valley wall to hectometric depths, clearly bounded by a low (1–3 m) scarp or antiscarp fracture, and with extensive disturbance of the intervening preglacial land surface. Both are located above troughs transecting main watersheds, but in some of the greatest and least relief to bear RSF in Britain (850 and 330 m floor to rim, respectively).

The case at *Garbh*, Ardgoil [NN 241002] is exceptional and has not been discussed previously (Fig. 17). This RSF complex is on a broad shoulder breaking at the trough rim into overhanging crags with slipped and broken masses below. The trough is occupied by Loch Long, a fjord on the Caledonian trend. RSF extent is uncertain: Holmes (1984) identifies two sites totalling 0.79 km^2, but Geological Survey field slips by C. T. Clough (*c.* 1897) delineate 1.75 km^2. His failure boundary extends well up the heavily dislocated shoulder above the crag collapse rim to a linear feature (shown as C–C_1 on Fig. 17a) noted as 'small cracks and slips'. This is, in fact, a sharp step-break of headscarp-and-furrow character 1–2 m high (Fig. 17b), which is distinct for approximately 500 m on the Caledonoid NNE–SSW trend. It is

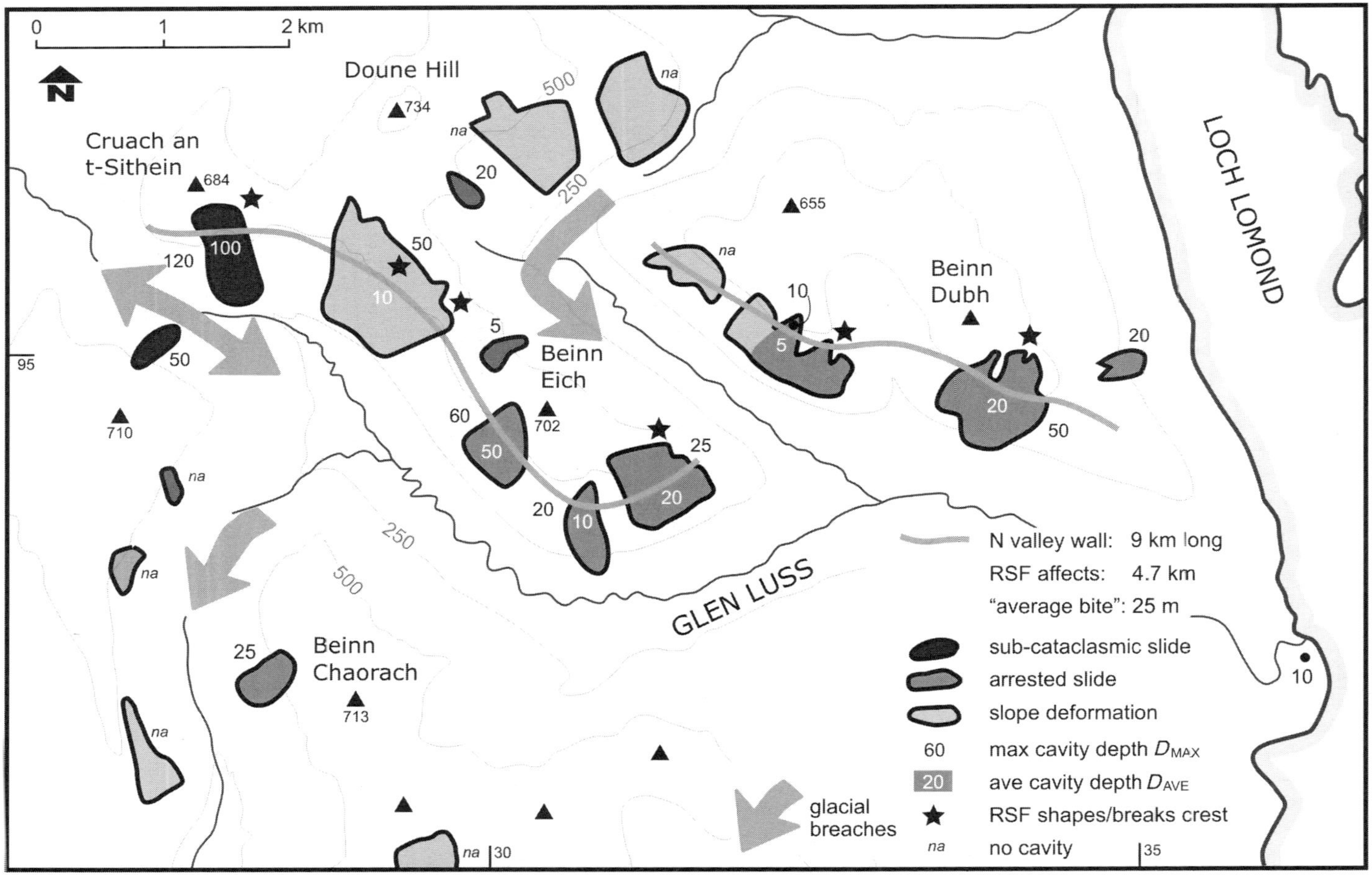

Fig. 14. High RSF incidence in Glen Luss (for location see Fig. 7). Despite available relief not exceeding 500 m, paraglacial trough widening is pronounced. RSF is favoured on the SW-facing valley side by a consistent regional dip in fallible lithologies with extensive through-going discontinuities. On other aspects in the Luss area, RSF tends to be limited to progressive deformation or small-scale collapses.

Fig. 15. Corrie Brandy [NO343756] in Glen Clova, a glacial trough incising an intermediate plateau in the SE Highlands (for location see Fig. 6). (**a**) Lateral widening of the atypically SW-facing corrie by the extant main RSF slippage (D_{RIM} of 130 m), by incipient extension above and adjacent to it (dotted line), and by a thin slice partially detached from the west rim; vertical air photograph © RCAHMS. (**b**) The main RSF on the east flank, with incipient scarplets top left; note the paleic relief in Linton Zone II of limited dissection (see Fig. 6).

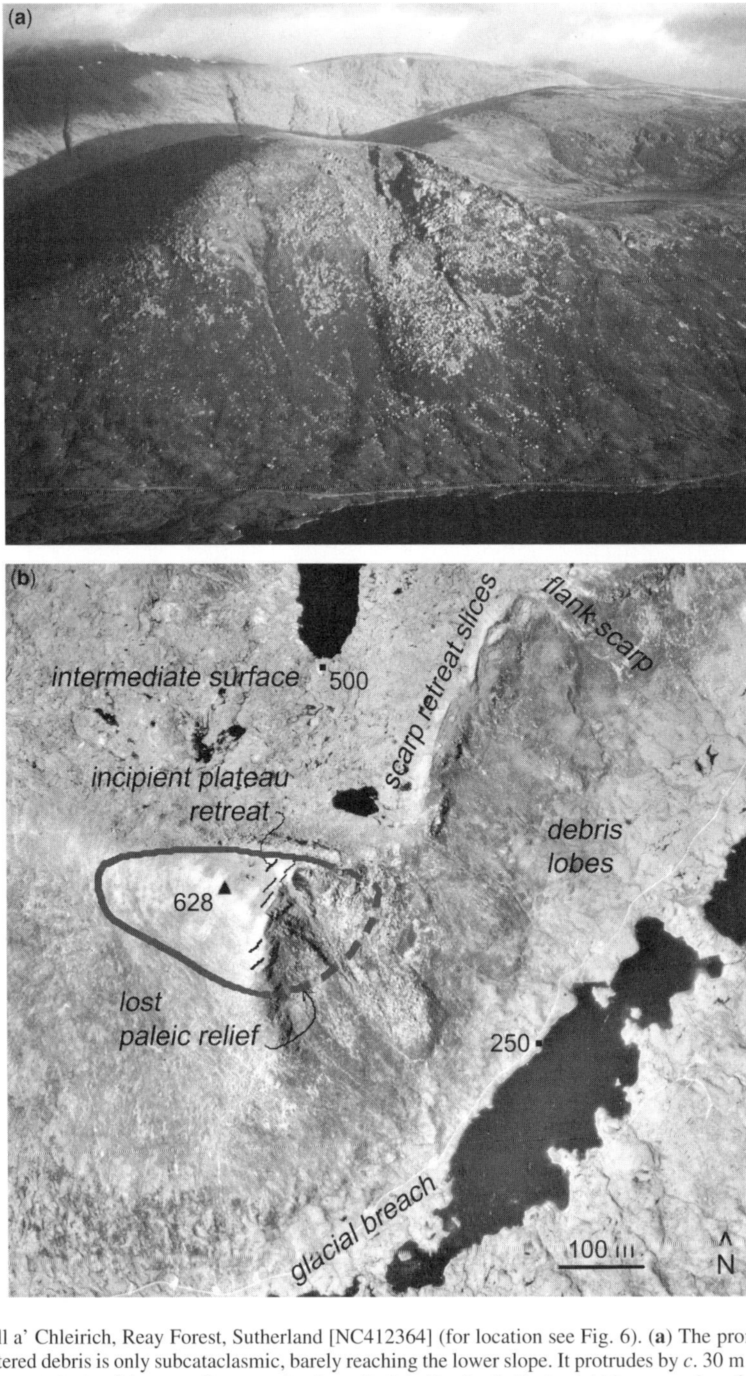

Fig. 16. Meall a' Chleirich, Reay Forest, Sutherland [NC412364] (for location see Fig. 6). (**a**) The prominent lobe of coarse shattered debris is only subcataclasmic, barely reaching the lower slope. It protrudes by *c*. 30 m and is ripe for removal by the next glacier. It leaves a D_{MAX} cavity above it of similar depth. On its right is a precarious failed mass that has only just parted company from the rim; there are large fresh rockfalls and extensive incipient fracturing behind the rim. On its left is a debris-free cavity with stepped source scarp inferred to date from earlier event(s). View NW across breach of main watershed. (**b**) The broad 0.5 km^2 RSF complex encroaches into the flat summit (a paleic relief residual) and a lower etch-surface for 1 km along the breach rim. The dashed line suggests the loss of paleic relief to RSF in recent cycles. Platy slippage NE of point 500 has a D_{RIM} of *c*. 20 m. Vertical air photograph © RCAHMS.

Table 3. *Large incipient rim retreats in the sample areas*

Sample area RSF site	Ref.	Indicators	Bite (m) D_{RIM}
Cowal–Arrochar–Luss			
Garbh (Ardgoil)	Fig. 17	Dislocations across broad shoulder, parallel to trough, normal scarp 1.5 m (D_{RIM} to D–D rupture is 675 m)	550 *(675)*
The Brack (Ardgoil)		Fissuring in summit plateau	30
Ben Donich (Ardgoil)		Fissure in ridge *c.* 200 m long, 10 m deep	70
Ben Vorlich (Arrochar)		Broad graben splitting summit ridge, 8 m uphill face, source crag fissuring	70–120
Beinn Bhreac (Luss)		Fissuring behind north ridge rim	150
Beinn Dubh (Luss)	Fig. 3	Double wedge bite across summit ridge, dropped <7 m (false antiscarps)	140/180
Ericht–Gaick			
Meall Cruaidhe (L. Ericht)		Wedge bite behind crest, dropped <2 m (false antiscarp)	90
An Dun E (Gaick)	[1] Fig. 8	Armchair bite into plateau island above slip, dropped *c.* 2 m	70
A' Chaoirnich (Gaick)	[1] Fig. 8	1–2 m dislocation across summit surface, parallel to trough rim	175
Kintail–Affric			
Beinn Fhada	[2]	0.5 m antiscarp across summit surface, parallel to trough rim	300
Sgurr na Ciste Duibhe	[3]	Rectilinear fault scarp 1–10 m high, behind summit ridge (false antiscarp)	80–250
South Cluanie Ridge		Fissures, false antiscarps behind crests – several tops	25–50
Aonach Meadhoin		Lineament across summit plateau above RSF	50
Monar–Strathfarrar			
Sgurr na Feartaig (Monar)	[4]	Wedge bites into ridge above slip scarps, dropped <3 m	15/25
Sgurr na Ruaidhe (Strathfarrar)	Fig. 10	Plateau incipience above armchair cavity, dropped <2 m	25

Sources: [1] Jarman 2004*a*; [2] Jarman & Ballantyne 2002; [3] Jarman 2007*b*; [4] Fenton 1991.

clearly post-glacial and not erosional. It resembles the Nordmannvikdalen 'neotectonic fault scarp' in north Norway (Dehls *et al.* 2000). Of two parallel features upslope on the air photograph, the farther E–E is simply erosional, but the nearer D–D is locally a submetric scarplet and furrow, and marks the extent of dryer-ground vegetation. The trough rim is degraded by scouring, but is interpolated along the truncated spurs at 500–550 m asl (A–A). Deep-seated incipience up the 1 km-broad shoulder could attain (D_{RIM}): 400 m to B–B, a bold 30 m high scarp at 660 m asl [NN 23850045]; 550 m to C–C, the sharp step at 670 m asl [NN 23750055]; and 675 m to D–D, the lineament at 690 m asl [NN 23630062].

Such large-scale sequential slicing back of this truncated spur along successive Caledonoid lineaments is similar to Mullach Coire a' Chuir nearby (Jarman 2003*a*). However, the long-section (Fig. 17c) shows that the basal discontinuity cannot be steeper than *c.* 18°, which is below the usual threshold permitting creep to progress to sliding in schists (Watters 1972). This gives a depth of incipient RSF of up to 75 m (D_{GEOL}), which would increase if an even gentler rupture zone existed, as with the 14° slope obtained for the very large Ben Our RSF (Jarman 2007*d*). The cause of such pervasive slope instability in relatively subdued relief on both sides of Loch Long (Fig. 7) may be glacial overdeepening along a Caledonian structural weakness transecting former interfluves. These may have been breached by transfluent ice, or a pre-glacial Clyde-river capture of the Forth headwaters may have been enlarged by diffluent Loch Lomond ice (Linton & Moisley 1960; Jarman 2003*a*).

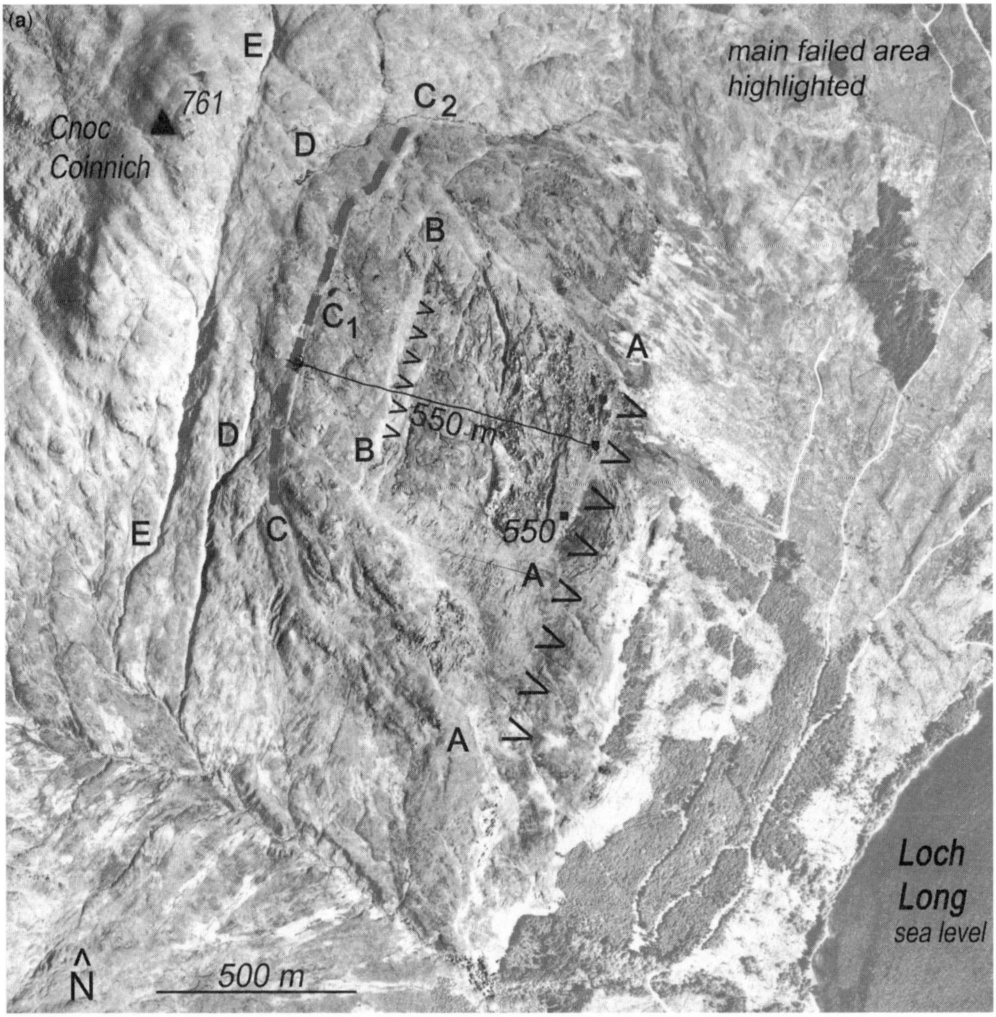

Fig. 17. Garbh RSF, Ardgoil [NS243998] (for location see Fig. 7). (**a**) The broad shoulder of Cnoc Coinnich (761 m asl) has Caledonoid lineaments exploited successively as A–A collapse scar along the trough rim; B–B source scarp to subsidence zone with extensive visible antiscarping; C–C_1 step scarp (Fig. 17c) continuing down the north flank to C_2 as head of the zone of antiscarped creep; D–D ground rupture of uncertain relationship to the main failed area; and E–E fluvial erosion gully. C–C marks by far the greatest known depth of incipient scarp retreat (D_{RIM}) in Britain at *c.* 550 m. The extent of failure is uncertain; C.T. Clough mapped failure down to the shore, now obscured by forestry. The photograph has the sun angle from SE, and shows distinctly darker vegetation within the main failed area up to D–D, indicative of freer-draining (fractured) terrain, endorsed by extensive springs below the northern boundary. Vertical air photograph © RCAHMS.

Comparative rockwall retreat rates

Previous studies in the Scottish Highlands have not taken map-based measures of RSF depth. Watters (1972) inferred long profiles through 13 large cases, from which D_{GEOL} can be scaled ranging 15–90 m and averaging 50 m. Fenton (1991) suggested that the bolder RSFs in the NW Highlands have D_{GEOL} typically of 30–100 m, with Sgurr na Conbhaire (Fig. 9) reaching 150 m. These figures suggest the moderate extent to which D_{MAX} contour bites may understate true cavity depths. But caution is needed with large slope deformations, where deep arcuate failure surfaces (Jarvis 1985) are less likely than shallower stepped surfaces or transitional creep zones as inferred at Beinn Fhada (Jarman 2006).

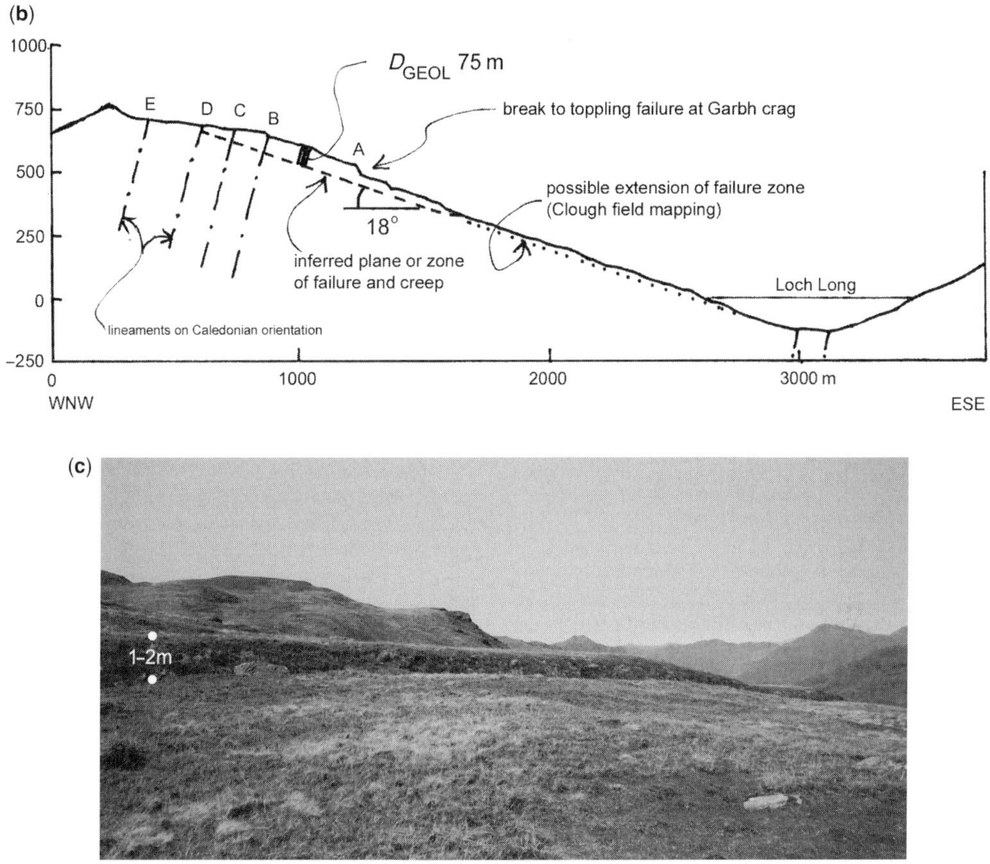

Fig. 17. (*Continued*) (**b**) Long section showing that only a very low angle of failure is possible; the effect of RSF is as much trough-deepening as trough-widening in such cases. (**c**) The fresh-looking (unglaciated) step-scarp at C–C, view at C_1 looking north.

Holmes (1984) obtained D_{SLOPE} values in six small (sub)cataclasmic cases in the 0.1 – 1 million m^3 range with quantifiable debris volumes. His retreat rates average 25 m, which is consistent with the D_{MAX} result for Arrochar/Ardgoil where such RSFs predominate. Scarp retreats of 6 m and 14.3 m have also been calculated from debris volumes at Baosbheinn, Gairloch (Sissons 1976) and Beinn Shiantaidh, Jura (Dawson 1977). These deposits had been assumed to be rock glaciers, but are now regarded as more likely to be RSFs (Ballantyne & Harris 1994; Ballantyne 1997). Ballantyne (2007*a*) found a D_{SLOPE} of 25 m by contour extrapolation across the exceptional evacuated cavity of Beinn Alligin. At Ben Hee, where on a convexity the RSF does not display contour recesses, Jarman & Lukas (2007) applied D_{VERT} to a terrain reconstruction with balanced failure volume and pre-failure relief. This gives a maximum vertical depth of surface lowering of 60 m, with 'scarp retreat' of the crestal position of 80–120 m.

It is difficult to compare singular RSF depths with incremental para/periglacial rates of rockwall retreat. Fifteen studies of talus slopes, debris flows and pronival ramparts yield retreat rates of 0.01–3.3 m ka^{-1} (Ballantyne & Harris 1994). Over an interglacial these are orders of magnitude less than RSF in Scotland, but upper-end rates could be comparable under prolonged periglacial conditions. In a rare study of contemporary RSF as a continuing process from rockslides onto an Alaskan glacier, Arsenault & Meigs (2005) calculated a mean valley-wall retreat rate of 6.7 m ka^{-1}, which scarcely increases if all rockfall debris is included. Even these modest process rates greatly exceed surface-dating results of less than 2 m of glacial erosion on Scandian trough walls in the entire last glacial cycle (Li *et al.* 2005).

Discussion

However vivid these illustrations of paraglacial RSF 'bite', they are only a response to one deglaciation amongst many, while the quantitative measures are of limited value in themselves. Their significance has to be assessed within the context of the glacial–paraglacial cycle as it has evolved over the Quaternary, and must bear in mind problems associated with the Loch Lomond Stadial. The efficacy of glacial exploitation of RSF bite must then be critically evaluated. A further problem arises with the sporadic incidence of RSF along trough walls, making the 'paraglacial relaxation' process stochastic rather than deterministic. Finally, the implications of RSF trough widening for landscape evolution merit discussion.

Glacial–paraglacial cycling: an evolutionary model

In a simple glacial–paraglacial sequence, glaciers erode trough walls, then retreat and debuttress them; the walls fail paraglacially; the next glaciers remove the failed material, and then repeat the cycle. The earliest known perception of the scale and cyclical significance of paraglacial RSF is by Holmes (1984, p. 232), who concluded: 'Millions of cubic metres of failed debris, mainly located in the Western Highlands, presently await glacial transportation. If similar volumes to those found today were created repeatedly in the past, then rock slope failure combined with glacial transportation must have been a major component of the denudation of the areas of the Highlands where RSF is important'.

Subsequent recognition of many glacial–paraglacial cycles over the Quaternary makes cumulative impact of RSF even more substantial, especially in the most geologically susceptible areas (Evans 1997). This impact may be greater than it might appear from the present sporadic incidence of RSF, if it was considerably more prevalent at early stages of fluvial valley adaptation to ice discharge (Jarman 2003*a*).

A schematic model (Fig. 18) demonstrating this evolving contribution over successive glacial–paraglacial cycles is necessarily speculative as past RSF incidence cannot be extrapolated from extant evidence, and there can be no comparator regions at immature 'early Quaternary' stages of development. An analogy might be found in young mountain ranges, where landsliding is now being recognized as the largest contributor to their erosion (Hovius & Stark 2006). There, RSF rates tail off as tectonic uplift wanes; here, they decline as glacial trough profiles mature and, conjecturally, slopes become 'stress-hardened' over repeated cycles, with exhaustion of the main fallible weaknesses. Rejuvenation will invigorate trough-floor incision and promote renewed RSF pulses, if system perturbations are vigorous enough. These might be regional, such as glacioisostatic rebound, or localized, such as glacial breaching. Present RSF clusters are often associated with glacial breaches (Jarman 2002, 2003*a*).

Interstadials, interglacials and the Loch Lomond Stadial

RSF efficacy over the Quaternary in this model is clearly dependent on the number and erosional severity of glacial–paraglacial cycles. In ranges such as northern Iceland (Bentley & Dugmore 1998) where the main troughs remain occupied by ice during interstadials, cycle frequency is much reduced. Conversely, shorter stadials with less glacial bedrock erosion will leave lower rock-mass stresses upon deglaciation, as with the Loch Lomond Stadial which, in the Highlands, merely reworked valley debris (Godard 1965). However, this takes a static view of slope stress fields: if rebound stresses (Hutchinson 1988) are dynamic, it might be that the abrupt loading and unloading of Loch Lomond ice exercised a 'trampoline effect', triggering a spate of RSFs. The Loch Lomond Stadial glacial–paraglacial cycle could either be a freak event overstating the typical mature-stage incidence of RSF (or understating it if its glaciers removed the evidence of some lateglacial RSFs) or it could suggest 'trampolining' as a significant factor during earlier rapid climatic fluctuations.

How effectively can glaciers exploit RSF cavities?

The glacial–paraglacial cycle model also implies that glaciers will substantially exploit the failed slopes during each stadial, in two ways: (1) by excavating the weakened material from the cavities; and (2) by gaining purchase on the cavity angles to increase erosion rates on intervening segments of trough wall. The latter process is inferred from Bentley & Dugmore (1998, p. 14) who noted that RSFs 'create irregularities in the trough side [which] could enhance trough widening as the glacier re-establishes a smoother channel form'. Figure 19 tests these critical assumptions. If all its obstacles or alternatives to cavity exploitation apply, it could take several glacial cycles to exhaust the deeper RSF bites.

The 'zone of paraglacial relaxation'

In Scotland and Scandinavia, as against higher ranges, extant RSF spacing is rarely close enough for intervening trough-wall segments to be

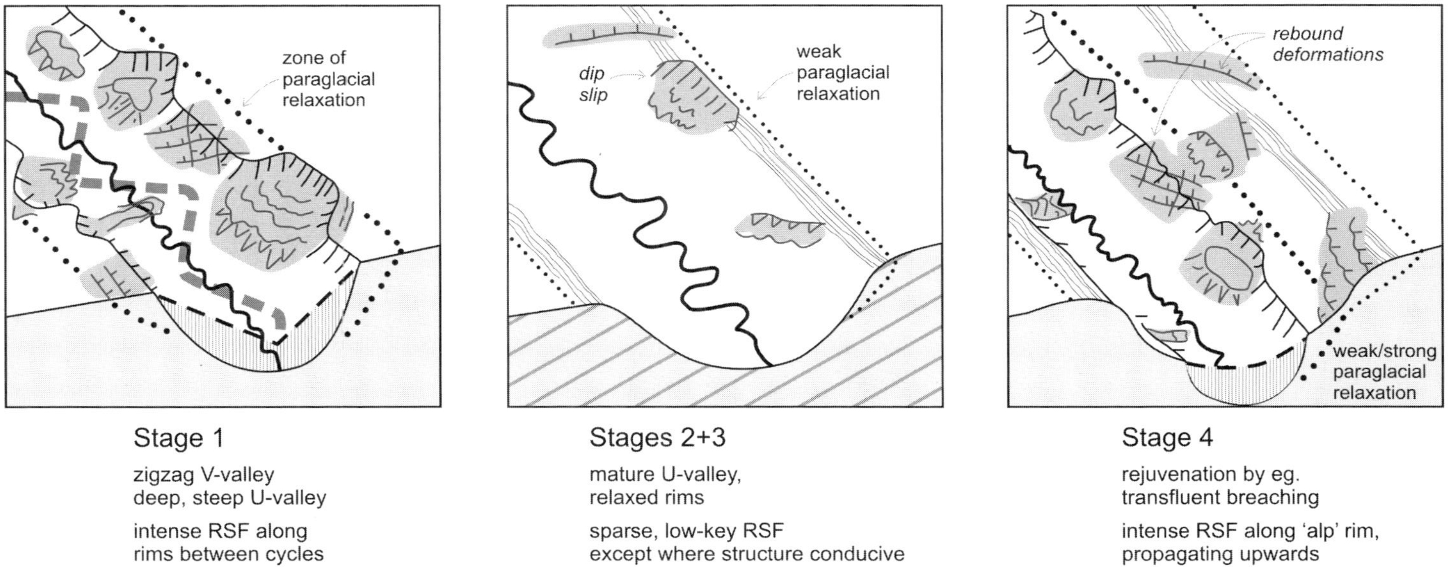

Fig. 18. Model of evolving paraglacial trough-widening effects over time. RSF occurs in 'zones of paraglacial relaxation' of varying scale. Each stage may well persist through a number of glacial–paraglacial cycles. The two processes alternately undo each other's efforts in terms of preferred slope angle. *Stage 1: fluvial V to glacial U profile conversion.* During early glacial stages, as original fluvial valleys adapt to ice discharge, RSF is likely to be intense, especially where glacial erosion is concentrated. In fallible lithologies erosion will be endemic, with all valley sides affected. In massive lithologies such as granite, where the unit of rock-mass failure seldom achieves the RSF threshold size of 0.01 km^2/100 000 m^3, hyperabundant minor failure is likely. *Stage 2: preferential trough enlargement.* As trough shape and size approach peak efficiency for catchment ice discharge, concentrated glacial erosion will diminish, as will the scope for paraglacial RSF. In structurally controlled terrain, RSF will occur preferentially on failure planes dipping valleyward, promoting trough asymmetry and lateral displacement of divides. *Stage 3: maturity.* Where valleys have adjusted to ice discharge and 'stress-hardened' slopes have regained quasi-stability, RSF will become sparse. This fits Scottish evidence for mature glaciated valleys with relaxed profiles such as Dochart–Tay (Jarman 2003*a*). In valleys long adapted to ice discharge, bulk erosion over a glacial cycle has become insufficient to daylight new failure planes, or to provoke stresses during isostatic rebound in excess of shear strengths. Some main valleys are essentially pre-glacial forms with rather limited modification by ice, notably where they follow ancient lineaments rather than dendritic or zig-zag courses. *Stage 4: rejuvenation.* Late Quaternary RSF activity may be a response to: (a) *widespread glacial reincision* where wholesale glacial erosion and ensuing glacio-isostatic rebound maintain available relief sufficiently that glaciers are constantly reincising their floors, perhaps creating 'alps' with a fresh zone of paraglacial relaxation (e.g. the Randa RSF locus, Eberhardt *et al.* 2004); and (b) *locally concentrated erosion*, where a glacier is exploiting a linear weakness, such as fault crush, or where shifting icesheds/dispersal routes and ice streaming promote glacial transfluence and breaching. Augmentation of local catchment ice by several-fold is possible: this will render the existing trough underfit and promote its rapid enlargement.

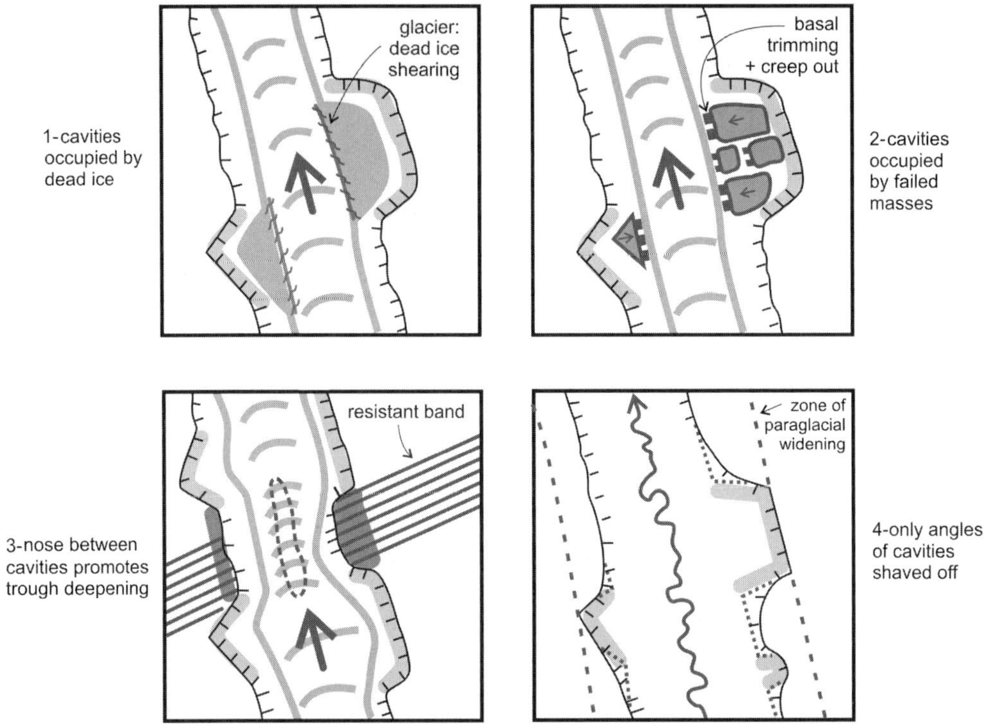

Fig. 19. Four possible reasons why the 'zone of paraglacial relaxation' may not be fully exploited for trough widening during the next glaciation. In (1) cavities are occupied by dead ice, whether endo- or exogenic (the classic glacier image is of uniform width). Likewise in (2) where failed masses remain largely in their cavities, they might simply be bypassed by the next glaciers; trimming of the failed toe may only promote limited further creep in the next interstadial, thus requiring many cycles for complete evacuation. It is not clear why slope segments between failure bays are eroded more rapidly than ordinary open slopes: channelling of ice through a constriction (3) might promote acceleration or overdeepening rather than enhanced widening. And where noses between RSF cavities are broad (4), or in resistant rock, the 'channel-smoothing' contribution to valley-widening must be restricted to rubbing off their angular mouths.

significantly exposed to future glacial attack, especially in resistant bedrock. Wholesale trough widening by this means requires that, over multiple glacial cycles, cumulative RSF incidence must affect almost the entire valley side. If structural strength/fallibility is consistent along a valley side, then this may be possible: geotechnical studies of RSFs generally find that slopes were close to critical thresholds, and only required small triggering forces to fail (e.g. Bjerrum & Jørstad 1968; Holmes 1984). This would imply that adjacent sectors of valley wall are ripe for failure in the next cycle. However, it is equally likely that variations in rock strength and discontinuity configuration will promote survival of inter-cavity bluffs. The evidence of incipient failures (Table 3) suggests that RSF propagates from itself rather than in fresh locations. Indeed, RSF may promote initiation of corries and side bays as much as widening of troughs (e.g. Fig. 10), if not to the extent suggested by Turnbull & Davies (2006) that most corries originate from (seismically-triggered) RSF cavities.

Any measures derived from extant RSF bites are thus no more than a broad indication of their actual contribution to valley widening during recent glacial cycles, and a tentative guide to their overall contribution during the Quaternary. The combined D_{BAY} and incipient D_{RIM} measures (typically 200–500 m) suggest the depth of a 'zone of paraglacial relaxation' along trough walls (Fig. 18) within which RSF may occur over a long period, and which will migrate upslope at a widely varying pace.

Overwidening, slope reduction, scarp retreat and paleic relief elimination

It is conventionally assumed that glacial trough size correlates with ice discharge, and that troughs cease to widen once equilibrium is attained, as attested by declining gross glacial erosion over the Quaternary (Sugden & John 1976). However, some glacial troughs display unusually high width : depth ratios, and Bentley & Dugmore (1998) proposed RSF cyclicity as a possible factor. This would be a classic positive feedback loop, with alternating glacial and paraglacial erosional processes widening valleys beyond the standard parabola. A further positive feedback arises if the overwidened valleys promote ice streaming, which draws in transfluent ice via glacial breaches (ice piracy).

In reality, troughs that appear overwidened in relation to ice catchment can be found in weak and massive lithologies alike, and with RSF both sparse (e.g. Rondane, Norway) and prolific (e.g. northern Iceland). Conversely, narrow profiles can occur in Scottish metasediments whether or not RSF is abundant. The three valleys in north Norway (Fig. 12) are variously parabolic with a few RSFs, shovel-shaped with abundant RSF, and maturely overwide with negligible RSF. With complex preglacial landscape and glaciological interactions, the role of RSF is hard to detect and quantify. What this feedback loop does highlight is that RSF is in itself a slope reduction process (e.g. Fig. 17). Only in tandem with glacial erosion does it become a scarp-retreat process, which is what 'overwidening' amounts to.

The corollary of trough widening is encroachment into the preglacial upland surface or 'paleic relief' (Gjessing 1967). Many RSFs bite significantly, whether by actual cavity or incipient failure, into paleic surface rims. Paleic relief is extensive in the selectively eroded eastern Highlands, and although many extant RSFs encroach into it (e.g. Figs 8 and 15) their percentage impact is small. Conversely, in Linton Zones III–IV (Fig. 6), where paleic remnants become vestigial (Godard 1965), RSF is a major contributor to their elimination (e.g. Figs 10 and 16). In north Norway several paleic surface blocks are undergoing significant attrition by RSF (Fig. 12). Identifying evolving RSF impacts in these migrating zones might assist in reconstructing the pre-Quaternary extent and character of upland relief.

Conclusions

- Some quantification of RSF contribution to trough widening is possible, but deriving a process rate is presently unrealistic. It is easy and instructive to measure RSF breadth and the proportion of valley sides affected. However, measuring 'depth of bite' is compromised by problems of planes of reference, irregular configurations and non-evacuated cavities. Map-based results obtained here show order-of-magnitude RSF impacts. Geotechnical evidence of failure cavity shape and depth would identify failure surface/zone behaviour, and help calibrate a slope-stress model for glacial troughs. Analysing glacier–failed mass interactions might clarify the probable 'yield' (or conversion rate into future widening) for RSFs of different types over glacial–paraglacial cycles.
- RSF is a high-magnitude–low-frequency contributor to trough widening of locally considerable impact. In the Scottish Highlands, the average RSF makes a significantly deeper 'bite' into trough walls than any other glacial or para/periglacial process. Average RSF breadths across the five sample areas range 270–600 m. In two of the main clusters RSF affects 9% and 14% of total valley wall, rising to 26–28% in three valleys in Knoydart and the Luss Hills. Average maximum contour recess (D_{MAX}) ranges from 40 to 45 m across the four largest areas. This suggests the scale on which metasedimentary structures in older ranges respond to deglaciation stresses. The D_{MAX} value must not be misrepresented as a real measure of RSF depth or as a widening rate per glacial cycle. It is merely a spot measure of the scale of visible RSF cavities. It may exaggerate the 'degree of purchase' offered to the next valley glacier in exploiting failed trough walls, if the average value across the whole failure is low. It may understate the impact where bedrock cavities are concealed by residual failed masses and subsequent infill.
- The 'zone of paraglacial relaxation' is best approximated by the D_{BAY} measure that ranges from 170 to 400 m, in conjunction with cases of rim bite and incipient failure, with D_{RIM} values of over 100 m (maximally 550 m).
- A model of RSF is developed for glacial–paraglacial cycles over the Quaternary (Fig. 18). The absence of such a model in the literature reflects: a historic focus on glacial trough deepening rather than widening; a lack of awareness of paraglacial RSF as a process; and the lack of evidence from ranges at earlier stages of glaciation. The model shows that RSF is a key process within a glacial–paraglacial cycle with powerful feedbacks. Glacial troughs may be widened by RSF to more than the normal parabolic profile, enabling ice piracy with consequences for ice-sheet profiles and elevations, ice dispersal patterns, and landscape evolution. RSF intensity would have been greatest as fluvial valleys underwent adaptation to ice discharge, and diminished as trough walls became stress-hardened, except where undergoing rejuvenated incision, for example in response to glacial breaching.

- The zone of paraglacial relaxation will vary over space and time, and will not be fully exploited in any one cycle. Unknown factors include: the ability of glaciers to evacuate failed debris masses or exploit valley-wall cavities; the effects of long glacials as against short stadials; and how mountain slopes respond to glacio-isostatic compression and rebound. RSF in the Scottish Highlands may be primarily a paraglacial response to the last main (late Devensian) deglaciation, with the Loch Lomond Stadial as a possibly anomalous event complicating interpretation of extant RSF incidence.
- Within the zone of paraglacial relaxation, RSF incidence is localized and unlikely to affect entire valley sides. The failure of some valley walls and not others may be due to intact slopes persisting as strong points. The extent to which the rate of trough widening by glaciers can be accelerated by prior RSF pock-marking the walls is thus debatable. Present RSF sparsity may indicate maturity where it was previously prevalent; or that structure or lithology are unconducive to RSF. Alternatively, troughs may have widened too slowly for RSF ever to have been significant, notably in areas with capacious preglacial valleys. Wholesale trough widening is therefore a complex and little-understood process to which RSF makes a significant contribution. The model suggests why this contribution has been greater in the past, and plan-depth data from extant Scottish RSFs can offer a pointer to its efficacy.

I would like to thank G. Holmes for the RSF database in his unpublished thesis; J. Gordon for access to air photographs at Scottish Natural Heritage; the British Geological Survey (Edinburgh) for access to unpublished field mapping; J.L. Sollid for providing the geomorphological map of Kåfjord; and the editors and referees for their advice and encouragement.

References

ARSENAULT, A. M. & MEIGS, A. J. 2005. Contribution of deep-seated bedrock landslides to erosion of a glaciated basin in southern Alaska. *Earth Surface Processes and Landforms*, **30**, 1111–1125.

AUGUSTINUS, P. C. 1995. Glacial valley cross-profile development: the influence of *in situ* rock stress and rock mass strength, with examples from the Southern Alps, New Zealand. *Geomorphology*, **14**, 87–97.

BALLANTYNE, C. K. 1997. Report on field meeting. The Quaternary of Islay and Jura. *Quaternary Newsletter*, **83**, 31–34.

BALLANTYNE, C. K. 2002. Paraglacial geomorphology. *Quaternary Science Reviews*, **21**, 1935–2017.

BALLANTYNE, C. K. 2007*a*. Beinn Alligin. *In*: COOPER, R. (ed.) *Mass Movement in Great Britain*. Geological Conservation Review, **33**. Joint Nature Conservation Committee, Peterborough, 111–116.

BALLANTYNE, C. K. 2007*b*. Coire Gabhail. *In*: COOPER, R. (ed.) *Mass Movement in Great Britain*. Geological Conservation Review, **33**. Joint Nature Conservation Committee, Peterborough, 132–136.

BALLANTYNE, C. K. & HARRIS, C. 1994. *The Periglaciation of Great Britain*. Cambridge University Press, Cambridge.

BENN, D. I. & EVANS, D. J. A. 1998. *Glaciers and Glaciation*. Wiley, Chichester.

BENTLEY, M. & DUGMORE, A. J. 1998. Landslides and the rate of glacial trough formation in Iceland. *Quaternary Proceedings*, **6**, 11–15.

BJERRUM, L. & JØRSTAD, F. A. 1968. *Stability of Rock Slopes in Norway*. Norwegian Technical Institute, Publication, **69**.

CLAYTON, K. M. 1974. Zones of glacial erosion. *Institute of British Geographers Special Publication*, **7**, 163–176 (includes unpublished material by D. L. Linton).

CRUDEN, D. M. & HU, X. Q. 1993. Exhaustion and steady-state models for predicting landslide hazards in the Canadian Rocky Mountains. *Geomorphology*, **8**, 279–285.

DAWSON, A. G. 1977. A fossil lobate rock glacier in Jura. *Scottish Journal of Geology*, **13**, 37–42.

DEHLS, J. F., OLESEN, O., OLSEN, L. & BLIKRA, L. H. 2000. Neotectonic faulting in northern Norway; the Stuoragurra and Nordmannvikdalen postglacial faults. *Quaternary Science Reviews*, **19**, 1447–1460.

EBERHARDT, E., STEAD, D. & COGGAN, J. S. 2004. Numerical analysis of initiation and progressive failure in natural rock slopes – the Randa rockslide. *International Journal of Rock Mechanics and Mining Sciences*, **41**, 69–87.

EVANS, I. S. 1997. Process and form in the erosion of glaciated mountains. *In*: STODDART, D. R. (ed.) *Process and Form in Geomorphology*. Routledge, London, 145–174.

EVANS, S. G., MUGNOZZA, G. S., STROM, A. & HERMANNS, R. L. (eds) 2006. *Landslides From Massive Rock Slope Failure*. Springer, Dordrecht.

FENTON, C. H. 1991. *Neotectonics and Paleoseismicity in NW Scotland*. PhD thesis, University of Glasgow.

GJESSING, J. 1967. Norway's paleic surface. *Norsk Geografisk Tidsskrift*, **21**, 69–132.

GLASSER, N. F. & HALL, A. M. 1997. Calculating glacial erosion rates in NE Scotland. *Geomorphology*, **20**, 29–48.

GODARD, A. 1965. *Recherches de Géomorphologie en Écosse du Nord-Ouest*. Université de Strasbourg, Publications de la Faculté des Lettres, Fondation Baulig.

GORDON, J. E. 1977. Morphometry of cirques in the Kintail–Affric–Cannich area of northwest Scotland. *Geografiska Annaler*, **59A**, 177–194.

GORDON, J. E., BIRNIE, R. V. & TIMMIS, R. 1978. A major rockfall and debris slide on the Lyell Glacier, South Georgia. *Arctic and Alpine Research*, **10**, 49–60.

HALL, A. M. & JARMAN, D. 2004. Quaternary landscape evolution – plateau dissection by transfluent glacial breaching. *In*: LUKAS, S., MERRITT, J. & MITCHELL, W. (eds) *Central Grampian Highlands – Field Guide*. Quaternary Research Association, London, 26–40.

HARBOR, J. M. 1992. Numerical modelling of the development of U-shaped valleys by glacial erosion.

Bulletin of the Geological Society of America, **104**, 1364–1375.

HEBDON, N. J., ATKINSON, T. C., LAWSON, T. J. & YOUNG, I. R. 1997. Rate of glacial valley deepening during the late Quaternary in Assynt, Scotland. *Earth Surface Processes and Landforms*, **22**, 307–315.

HEWITT, K. 1988. Catastrophic landslide deposits in the Karakoram Himalaya. *Science*, **242**, 64–67.

HOLMES, G. 1984. *Rock-slope Failure in Parts of the Scottish Highlands*. PhD thesis, University of Edinburgh.

HOVIUS, N. & STARK, C. P. 2006. Landslide-driven erosion and topographic evolution of active mountain belts. *In*: EVANS, S. G., MUGNOZZA, G. S., STROM, A. & HERMANNS, R. L. (eds) *Landslides From Massive Rock Slope Failure*. Springer, Dordrecht, 573–590.

HUTCHINSON, J. N. 1988. General report: morphological and geotechnical parameters of landslides in relation to geology and hydrogeology. *In*: *Proceedings of the fifth Internationl Symposium on Landslides, Lausanne, Volume* 1. Balkema, Rotterdam, 3–35.

JARMAN, D. 2002. Rock slope failure and landscape evolution in the Caledonian Mountains, as exemplified in the Abisko area, northern Sweden. *Geografiska Annaler*, **84A**, 213–224.

JARMAN, D. 2003*a*. Paraglacial landscape evolution – the significance of rock slope failure. *In*: EVANS, D. J. A. (ed.) *Western Highland Boundary – Field Guide*. Quaternary Research Association, London, 50–68.

JARMAN, D. 2003*b*. Tullich Hill rock slope failures, Glen Douglas. *In*: EVANS, D. J. A. (ed.) *Western Highland Boundary – Field Guide*. Quaternary Research Association, London, 200–208.

JARMAN, D. 2003*c*. The Glen Shiel rock slope failure cluster. *In*: TIPPING, R. (ed.) *Glen Affric and Glen Shiel: Field Guide*. Quaternary Research Association, London, 165–183.

JARMAN, D. 2004*a*. Rock slope failures of the Gaick Pass. *In*: LUKAS, S., MERRITT, J. & MITCHELL, W. (eds) *Central Grampian Highlands – Field Guide*. Quaternary Research Association, London, 103–111.

JARMAN, D. 2004*b*. The Cobbler – a mountain shaped by rock slope failure. *Scottish Geographical Journal*, **120**, 227–240.

JARMAN, D. 2006. Large rock slope failures in the Scottish Highlands: characterisation, causes and spatial distribution. *Engineering Geology*, **83**, 161–182.

JARMAN, D. 2007*a*. Introduction to the mass movements in the older mountain areas of Britain. *In*: COOPER, R. (ed.) *Mass Movements in Great Britain*. Geological Conservation Review, **33**, Joint Nature Conservation Committee, Peterborough, 33–55.

JARMAN, D. 2007*b*. Sgurr na Ciste Duibhe. *In*: COOPER, R. (ed.) *Mass Movements in Great Britain*. Geological Conservation Review, **33**. Joint Nature Conservation Committee, Peterborough, 62–69.

JARMAN, D. 2007*c*. Druim Shionnach. *In*: COOPER, R. (ed.) *Mass Movements in Great Britain*. Geological Conservation Review, **33**. Joint Nature Conservation Committee, Peterborough, 70–74.

JARMAN, D. 2007*d*. Glen Ample. *In*: COOPER, R. (ed.) *Mass Movements in Great Britain*. Geological Conservation Review, **33**. Joint Nature Conservation Committee, Peterborough, 82–91.

JARMAN, D. 2008. The Roy–Lochy rock slope failure cluster: implications for glacial breaching, ice movements, and Parallel Road dislocations. *In*: PALMER, A. P., LOWE, J. J. & ROSE, J. (eds) *The Quaternary of Glen Roy and Vicinity – Field Guide*. Quaternary Research Association, London, 98–104.

JARMAN, D. & BALLANTYNE, C. K. 2002. Beinn Fhada, Kintail: a classic example of paraglacial rock slope deformation. *Scottish Geographical Journal*, **118**, 59–68.

JARMAN, D. & LUKAS, S. 2007. Ben Hee. *In*: COOPER, R. (ed.) *Mass Movements in Great Britain*. Geological Conservation Review, **33**. Joint Nature Conservation Committee, Peterborough, 99–107.

JARVIS, J. J. 1985. *Large-scale Toppling Failure in Metamorphic Rock Slopes*. PhD thesis, University of London.

KORUP, O., CLAGUE, J. L., HERMANNS, R. L., HEWITT, K., STROM, A. L. & WEIDINGER, J. T. 2007. Giant landslides, topography, and erosion. *Earth and Planetary Science Letters*, **261**, 578–589.

KVERNDAL, A. I. & SOLLID, J. L. 1993. Late Weichselian glaciation and deglaciation in NE Troms, north Norway. *Norsk Geografisk Tidsskrift*, **47**, 163–177.

LI, Y., HARBOR, J. *ET AL*. 2005. Ice sheet erosion patterns in valley systems in northern Sweden investigated using cosmogenic nuclides. *Earth Surface Processes and Landforms*, **30**, 1039–1049.

LINTON, D. L. & MOISLEY, H. A. 1960. The origin of Loch Lomond. *Scottish Geographical Magazine*, **76**, 26–37.

SIGURÐSSON, O. & WILLIAMS, R. S., JR. 1991. Rockslides on the terminus of Jökulsárgilsjökull, southern Iceland. *Geografiska Annaler*, **73A**, 129–140.

SISSONS, J. B. 1976. A remarkable protalus rampart in Wester Ross. *Scottish Geographical Magazine*, **92**, 182–190.

SUGDEN, D. E. & JOHN, B. S. 1976. *Glaciers and Landscape*. Arnold, London.

TOLGENSBAKK, J. & SOLLID, J. L. 1988. *Kåfjord*, Kvartaergeologi og geomorfologi, 1:50 000, 1634 II. Geografisk Institutt, Universitet i Oslo.

TURNBULL, J. M. & DAVIES, T. R. H. 2006. A mass movement origin for cirques. *Earth Surface Processes and Landforms*, **31**, 1129–1148.

WATTERS, R. J. 1972. *Slope Stability in the Metamorphic Rock of the Scottish Highlands*. PhD thesis, University of London.

Rockfall talus slopes and associated talus-foot features in the glaciated uplands of Great Britain and Ireland: periglacial, paraglacial or composite landforms?

PETER WILSON

Environmental Sciences Research Institute, School of Environmental Sciences, University of Ulster at Coleraine, Cromore Road, Coleraine, Co. Londonderry BT52 1SA, Northern Ireland, UK (e-mail: P.Wilson@ulster.ac.uk)

Abstract: The traditional interpretation of talus slopes and talus-foot landforms in the glaciated uplands of Great Britain and Ireland has been that they are periglacial landforms associated with freeze–thaw activity and permafrost. Since about 1990 some reassessment of this widely held view has occurred, and paraglacial rockfall and rock-slope failure are now considered to have played a significant role in the development of some talus landforms; in certain cases a wholly paraglacial origin is advocated. In order to determine formative processes, critical site-specific evidence (morphological and sedimentological) needs to be obtained. This will enable models of the deglacial–post-glacial evolution of these landscapes to be proposed and allow the palaeoenvironmental significance of the landforms to be established. Distinguishing between a periglacial and paraglacial origin might be assisted by application of cosmogenic isotope surface-exposure dating, which may demonstrate a Holocene age for a particular landform and thus rule out a permafrost-related origin. However, there will be instances where the application of dating will not differentiate, as in the case of Late Glacial landforms that could be either periglacial or paraglacial. It is likely that equifinality applies with respect to these landforms and a composite origin for them should also be considered. This latter issue is one that has not previously been given much consideration, probably because of the inherent difficulties in recognizing the products of different processes in landforms for which exposures of their constituent materials are rare.

Rockfall talus is common on many steep hillslopes in the glaciated uplands of Great Britain and Ireland (Lewis 1985; Ballantyne & Harris 1994). Talus slopes usually have an upper rectilinear segment in the range 33°–40° and a marked basal concavity of approximately 5–30°. The size and shape of constituent clasts normally exhibits downslope sorting, and a preferred downslope orientation of clasts has also been recorded (Andrews 1961; Statham 1973; Ballantyne 1984). In terms of planform morphology, both talus cones (fans) and talus sheets have been recognized (Andrews 1961; Wilson 1990*a*), and talus development has taken place on both massive and fissile rock types (Ball 1966; Statham 1976). Many talus slopes are now essentially relict features, because rockfall activity is currently non-existent or intermittent, and they support either a partial or complete vegetation cover. It is generally believed that talus accumulation occurred in the interval *c.* 18–10 ka cal BP, between deglaciation following the Last Glacial Maximum (LGM) and the early part of the Holocene (Andrews 1961; Ball 1966; Ballantyne & Eckford 1984; Wilson 1990*a*; Salt & Ballantyne 1997). On the basis of palaeoclimatic evaluations, talus development has been associated with freeze–thaw cycles acting on rockwalls during this period (Andrews 1961; Ball 1966; Tufnell 1969; Clayton 1981; Ballantyne & Kirkbride 1987). Therefore, the traditional view of talus slopes is that they developed under periglacial conditions and they are regarded as a type of periglacial debris accumulation (Lewis 1985; Ballantyne & Harris 1994). Presently, talus modification by gullying, debris flows and slides, and snow avalanching is widespread (Luckman 1992; Hinchliffe & Ballantyne 1999; Curry & Morris 2004; Anderson & Harrison 2006).

At the foot of many talus slopes there are linear, arcuate or lobate extensions of coarse rock debris. These accumulations have been categorized as either protalus rock glaciers or protalus ramparts based on their morphological and sedimentological characteristics and their distance from the talus foot (e.g. Dawson 1977; Colhoun 1981; Gray 1982; Ballantyne & Kirkbride 1986; Wilson 1990*b*, *c*, 1993; Anderson *et al.* 2001; Harrison & Anderson 2001; Ballantyne 2002*a*). In most cases development of these landforms is ascribed to the Loch Lomond Stade (LLS; 12.9–11.5 ka cal BP) rather than the period of deglaciation that followed the LGM because: (1) they only occur beyond the mapped limits of LLS glaciers, the inference being that they are contemporaneous with those glaciers; (2) protalus rock glaciers are regarded as landforms that require aggrading as opposed to degrading

From: KNIGHT, J. & HARRISON, S. (eds) *Periglacial and Paraglacial Processes and Environments.*
The Geological Society, London, Special Publications, **320**, 133–144.
DOI: 10.1144/SP320.9 0305-8719/09/$15.00 © The Geological Society Publishing House 2009.

permafrost in order to move downslope from the talus foot; and (3) protalus ramparts require the growth of steeply inclined firn fields. The latter two points strongly support the assertion that a period of renewed cooling was required following deglaciation in order to facilitate development of those landforms (Ballantyne & Kirkbride 1986). The only period for which there is geochronological and litho- and biostratigraphical evidence of pronounced renewed cooling in upland areas of Great Britain and Ireland following LGM deglaciation is the LLS (e.g. Gray & Coxon 1991; Pennington 1996). The severe climate of the LLS (Atkinson *et al.* 1987) is thought to have been conducive to high rates of debris production as a result of freeze–thaw activity and permafrost aggradation (Ballantyne & Harris 1994; Isarin 1997). Thus, as with talus, it has become conventional to regard many talus-foot features as essentially periglacial landforms.

A reassessment

Within the last two decades the results of numerous studies, both in the British Isles and elsewhere, have led to a reassessment of the climatic (periglacial) hypothesis of talus development. Based on estimates of talus accumulation rates and/or rockwall retreat rates, it has been suggested that present-day rates of rockfall are too low to account for the volumes of talus that have accumulated since local deglaciation (André 1997; Luckman & Fiske 1997; Hinchliffe & Ballantyne 1999; Curry & Morris 2004; Anderson & Harrison 2006). It is considered that higher rates of talus accumulation were operative throughout the period of ice wastage and immediately afterwards, and were followed by a decline in rates as climate ameliorated (Ballantyne & Harris 1994). This is a reasonable proposition: rockfall activity from newly exposed cliffs is likely to have been enhanced because of permafrost degradation, freeze–thaw cycles and an abundance of meltwater. High rates of rockfall and large-scale rock-slope failure due to stress release have also been recorded on recently deglaciated terrain (e.g. Augustinus 1995; Holm *et al.* 2004; Korup *et al.* 2004; Matthews & Shakesby 2004; Arsenault & Meigs 2005), and talus accumulation has been a consequence of these processes. Such observations have resulted in some reassessment of talus origin. Talus slopes are now no longer viewed as being of wholly periglacial origin, but as either paraglacial landforms or composite landforms of periglacial–paraglacial derivation (André 1997; Luckman & Fiske 1997; Ballantyne 2002*b*; Wilson 2005; Sass 2006).

There has also been reassessment of the periglacial origin for some of the protalus rock glaciers and protalus ramparts found in Great Britain and Ireland. Based on detailed morphological and sedimentological analyses, several of these landforms are now considered to be products of large-scale rock-slope failures that produced protalus rampart and rock glacier 'mimics' (cf. Whalley & Martin 1992). Consequently, a paraglacial origin has been advocated (Ballantyne 1986, 1999; Sandeman & Ballantyne 1996; Curry *et al.* 2001; Wilson 2004). It seems, therefore, that equifinality may apply with respect to these features.

The paraglacial concept

When first introduced in the early 1970s, the term 'paraglacial' was used in the context of fluvial-system response to the large quantities of glacial sediment available for reworking during and after deglaciation, and the rapid system adjustments from glacial to non-glacial conditions (Ryder 1971; Church & Ryder 1972). Over the past 35 years usage of the term has been extended, and it now encompasses themes as diverse as adjustments of mountain rockwalls and coastal environments (e.g. Wyrwoll 1977; Forbes & Syvitski 1994). This has led to redefinition of the term as 'non-glacial Earth surface processes, sediment accumulations, landforms, landsystems and landscapes that are directly conditioned by glaciation and deglaciation' (Ballantyne 2002*b*, p. 1938). This broadening of the definition recognizes that glaciation and deglaciation impact on a range of geomorphological processes and landscape components, rather than on the fluvial system alone. There is now an acceptance that paraglacial adjustments affect all landscapes that have experienced glaciation. With respect to the glaciated landscapes of Great Britain and Ireland, the challenge for geomorphologists is to determine to which landforms the paraglacial tag can be justifiably attached.

Indicators for periglacial–paraglacial talus and talus-foot landforms

Determining the conditions under which talus slopes and talus-foot landforms developed in the glaciated uplands of Great Britain and Ireland is not an easy task, but is one that must be attempted in order that meaningful models for the deglacial–post-glacial evolution of these landscapes can be constructed. This is not the first time that upland landforms have been considered in the context of a periglacial–paraglacial debate (cf. Harrison 1996), and it is unlikely to be the last. The areas of discussion outlined later direct attention to some of the problems of and the possible solutions to recognition of talus landforms and their interpretation. It is not intended to polarize the periglacial–paraglacial debate; a composite origin for some (perhaps many) talus landforms may apply especially as both effects are likely to have operated throughout the Late Glacial period.

Age determinations

Establishing the age of talus landforms may provide an opportunity of clarifying the processes responsible for their formation. However, for talus landforms that developed between approximately 18 and 11.5 ka cal BP it may not be possible to distinguish a periglacial from a paraglacial origin because we do not actually know the amplitude of the fluctuations in intensity of upland periglacial processes across this interval and therefore we cannot easily discount this option. Even during the Windermere Interstade (*c.* 16.5–12.9 ka cal BP: Coope & Pennington 1977; Pennington 1978) there is the possibility that upland areas experienced some degree of periglacial activity and permafrost. Nevertheless, the application of cosmogenic isotope surface-exposure dating to the upper faces of large boulders from landform surfaces offers some potential (Cockburn & Summerfield 2004), but the uncertainties associated with cosmogenic dating (often ± 1–2 ka over the Late Glacial period) mean that we might not be able to separate talus landforms of Windermere Interstade age from those that accumulated immediately prior to and immediately after this interval. In some circumstances cosmogenic dates falling entirely within the Holocene stage (<11.5 ka cal BP) may enable a greater degree of confidence to be assigned to non-periglacial processes as the formative mechanisms for talus landforms because freeze–thaw processes have been much less severe and climatic conditions were not conducive for permafrost.

At many sites in Great Britain and Ireland, large-scale rock-slope failures have created major undercliff landforms composed of disrupted and displaced bedrock masses, and/or coarse rock debris. Although only two failures – at The Storr, Isle of Skye, Scotland (Ballantyne *et al.* 1998) and Beinn Alligin, Wester Ross, Scotland (Ballantyne & Stone 2004) – have been dated directly, a paraglacial origin has been proposed for several of them (e.g. Shakesby & Matthews 1996; Sellier & Lawson 1998; Hutchinson & Millar 2001; Jarman 2006; Wilson & Smith 2006).

Cosmogenic ^{10}Be ages have demonstrated that the Beinn Alligin 'ice-cored rock glacier' did not develop when glacier ice occupied the site (Sissons 1975, 1976*a*; Whalley 1976) but was formed about 4 ka BP. The feature, composed of Precambrian Torridon sandstone, is interpreted by Ballantyne & Stone (2004) as an excess-runout rock-avalanche deposit that probably resulted from a combination of paraglacial stress release and seismic activity. Cosmogenic dating has enabled the rock glacier hypothesis for the origin of this landform to be falsified. A steep talus cone composed of rockslide debris occurs at the foot of the rock-avalanche scar and must therefore be contemporary with or younger than the rock-avalanche deposit.

At The Storr cosmogenic ^{36}Cl exposure dates of *c.* 6.5 ka BP constrain a large rock-slope failure in Tertiary basalts to the mid-Holocene (Ballantyne *et al.* 1998). Since then, a talus slope has developed beneath the headscarp. The talus could have formed rapidly as the oversteepened failure scarp adjusted to altered stress levels. A periglacial origin for some of the talus cannot be rejected completely, even though mid- to late-Holocene freeze–thaw activity is likely to have been of reduced severity relative to that of the Late Glacial period, and granular disintegration of the rockwall has probably also made a contribution (Hinchliffe & Ballantyne 1999). In this example, cosmogenic dating has shown that the talus is of undoubted Holocene age, but the processes responsible for its accumulation remain to be established.

Where suitable materials are available, other techniques (e.g. optically stimulated luminescence (OSL) and ^{14}C dating) also offer potential for age determination of talus landforms. Some use has been made of ^{14}C dating in order to determine the timing of episodes of talus reworking (e.g. Hinchliffe 1999; Curry & Black 2002), but as yet no absolute ages have been obtained that relate to sub-talus landsurfaces.

Talus slopes

It is unlikely that talus-slope gradients and clast shape, fabric and sorting will enable differentiation between rockfalls resulting from freeze–thaw or stress-release processes: irrespective of how rockfalls are generated, talus characteristics will be indistinguishable. In contrast, abundant boulders of sizes considerably greater than those that are normally associated with local talus slopes may be suggestive of paraglacial rockfall inputs (Sandeman & Ballantyne 1996; Sass 2006). It is difficult to provide a definition for what is meant by 'abundant' and 'large', individual site assessment is needed in relation to neighbouring sites. Particle size on talus slopes is a function of joint spacing on the free-face (Wilson 1990*a*) and rock hardness: size is likely to be greater where joints are more widely spaced and where the rock type is relatively resistant to fragmentation following rockfall impacts. These properties need taking into account and it should be remembered that most boulders at the talus foot were probably slightly to significantly larger when they were released from the face, and underwent some size reduction as a result of impact fragmentation.

Critical evidence for talus formation by paraglacial processes has rarely if ever been sought, although such processes have frequently been inferred intuitively and/or on estimated rates of talus accumulation (Ballantyne & Eckford 1984; Ballantyne & Kirkbride 1987; Ballantyne & Harris 1994; Boardman 1996; Hinchliffe & Ballantyne 1999; Curry & Morris 2004; Anderson & Harrison 2006). The free-face or ground directly above it may yield pertinent morphological information, but these areas are seldom examined in detail, usually because they are difficult to access.

At Wasdale in the Lake District (England), the Wast Water screes extend for 2.5 km along the south side of the lake as talus sheets and cones that have developed from Ordovician volcanic lavas and tuffs (Fig. 1). Andrews (1961) and Huddart (2002) favoured the periglacial hypothesis for talus formation at this site, while Boardman (1996) considered that rockfall associated with unloading of the rockwalls was likely to have been important. Apparently, no examination of the top of the free-face for signs of former rockwall instability had taken place, in spite of its accessibility. Recently, Wilson (2005) reported that three areas of disrupted bedrock occur along the scarp edge. At these sites well-vegetated, slightly sinuous and low-amplitude (<2 m) ridges and depressions, spaced 2–20 m apart, and flights of low-gradient (<10°) benches up to 50 m wide are aligned (sub)-parallel to the scarp edge for distances of 150–300 m and extend up to 130 m back from the edge (Fig. 2). Wilson (2005) interpreted the ridges as large rock slabs created by tensional spreading of the scarp, and the benches as slight downslope displacements of large bedrock masses. At the talus foot below two of these sites there are numerous boulders exceeding 2 m in length. It was proposed that these concentrations of boulders resulted from large-scale rock-slope failures, rather than intermittent rockfall activity (cf. Sandeman & Ballantyne 1996), and that the scarp-edge ridges and depressions represented residual components of the failures that had generated the boulders. It was not possible to determine how much of the scarp had failed or the style(s) of failure. No partially intact rock masses occur on the talus, this may be because: (1) all the slope failures disintegrated to boulder-sized debris; (2) any rock masses that did not disintegrate have since been buried by further talus accumulation; and/or (3) large rock masses descended to below lake level. Rock slabs and tension cracks have been reported from behind the headscarps of rock-slope failures elsewhere in Great Britain (Johnson & Vaughan 1989; Sandeman & Ballantyne 1996; Shakesby & Matthews 1996; Jarman 2003; Wilson 2007). The field evidence strongly suggests that paraglacial rock-slope failures have contributed material to the Wast Water screes, but exactly how much remains to be established.

Fig. 1. Part of the Wast Water screes, Lake District (England). A concentration of large (>2 m) boulders occurs just above the water line, suggesting a rock-slope failure contribution to talus accumulation.

Fig. 2. Vegetated ridges and depressions attributed to tensional spreading on the cliff top above the talus shown in Figure 1.

There are other Lake District sites where paraglacial rock-slope failures have probably made significant inputs to talus slopes. Talus below Bowder Crag in Borrowdale includes numerous large (2–8 m-long) boulders (Fig. 3), and at the talus foot stands the 18.6 m-long 1274 t Bowder Stone (Smith 2002) (Fig. 4). The crag has widely spaced joints and several wedge-shaped failure scars, and the ground above the crag consists of broad steps and broken outcrops suggestive of arrested large-scale rock-slope failures. As with the Wasdale site, Bowder Crag is also in Ordovician volcanic lavas and tuffs.

Recent work by Sass (2006) and Sass & Krautblatter (2007) has demonstrated the value of geophysical techniques for assessment of the structure

Fig. 3. Boulders in the talus below Bowder Crag, Lake District. Each of the four large boulders in the upper middle part of the picture exceeds 2 m in length; many of the foreground boulders exceed a length of 1 m.

Fig. 4. The Bowder Stone, Lake District. The *a*-axis of the boulder is aligned horizontally away from the end that is in view and is 18.6 m in length. The vertical height of the boulder as seen in the photograph is 8.18 m

and composition of talus. In particular, ground-penetrating radar images of alpine talus slopes have revealed the presence of large boulders below the surface of the talus and these have been interpreted as paraglacial in origin. The technique appears to offer great promise with respect to understanding talus-slope development, but it has yet to be applied to sites in Great Britain and Ireland.

Protalus rock glaciers

The idea that rock glaciers could result from large-scale slope failures was proposed in the early twentieth century (Howe 1909). This issue was taken up by Johnson (1983, 1984*a*, *b*), who discussed examples from the Yukon in which various forms of high-magnitude, low-frequency events, involving the failure of unconsolidated talus and glacial sediments on steep valley sides, were shown to have created talus-foot landforms that had close morphological and topographical similarity with protalus rock glaciers. Support for this discovery has since been provided by Shakesby *et al.* (1987), Vick (1987), Luckman & Fiske (1997) and Ballantyne (2002*b*), who have shown that debris input from slope-failure mechanisms can produce landforms that are morphologically similar to protalus rock glaciers formed by creep of a talus–interstitial ice (permafrost) mixture.

Several of the protalus rock glaciers identified in Great Britain and Ireland have been reassessed and are now considered to represent debris landforms created by paraglacial rock-slope failures (Sandeman & Ballantyne 1996; Ballantyne 1999; Wilson 2004) (Fig. 5). Several lines of evidence support these claims: (1) hillside scars are present above some of the debris features and are interpreted as rock-slope failure source areas; (2) large areas of disrupted and displaced bedrock attributed to slope failure occur on the adjacent hillsides; and (3) the constituent material of the debris landforms often has a considerable component that is an order of magnitude coarser than that on the backing talus, suggesting it was produced by rock-slope failures rather than discrete and intermittent rockfalls. A further line of evidence in support of a rock-slope failure origin for talus and talus-foot debris accumulations is the presence of tension cracks along the scarp edge, as described at Wasdale, but their absence cannot be taken to indicate that failure has not occurred. For example, coarse debris of Precambrian quartzite forms a massive ridge below talus slopes on the flanks of Muckish Mountain, Co. Donegal (Ireland) (Fig. 6). Wilson (1990*c*) considered the ridge to be a talus-foot rock glacier but, for reasons given above, has reinterpreted it as a product of large-scale rock-slope failure (Wilson 2004). However, ridges and tension furrows are absent from the scarp edge above the talus (Fig. 7), suggesting that if failure did occur it did not leave residual evidence in the form of rock slabs partly attached to the scarp. This illustrates that not all situations are identical; local geological factors probably account for some of the differences in the characteristics of scarps from which rock-slope failure debris derives.

Fig. 5. This debris landform in Strath Nethy, Cairngorms (Scotland) was categorized by Sisson (1979, p. 79) as a 'boulder deposit associated with former snow beds'. Sandeman & Ballantyne (1996, p. 140) termed it a 'talus rock glacier', but recognized that sliding failures from the granite cliffs upslope provided much of the debris. The Cairngorm rock glaciers are currently the subject of a cosmogenic isotope surface-exposure dating programme (C.K. Ballantyne pers. comm. 2007).

Being able to distinguish between true (permafrost-related) protalus rock glaciers and similar landforms resulting from large-scale slope failure is important because the former features have been used to estimate rates of rockwall retreat and/or to draw palaeoclimate inferences for the LLS, in which they are thought to have formed (Dawson 1977; Sissons 1980; Chattopadhyay 1984; Wilson 1990*b*). These estimates and inferences do not apply if the features are slope-failure deposits because: (1) they probably represent short-lived high-magnitude events; (2) they did not

Fig. 6. Quarried debris ridge at Muckish, Co. Donegal (Ireland). The vertical height of the quarried face at its left-hand side is approximately 45 m. Wilson (1990*c*) regarded this feature to be a relict talus-foot rock glacier, but has since considered it to result from paraglacial rock-slope failure (Wilson 2004).

Fig. 7. View along the crest of the rockwall that overlooks the debris accumulation shown in Figure 6. Note the absence of evidence for tensional spreading (cf. Fig. 2).

necessarily develop in the LLS; and (3) they may have no palaeoclimatic significance. An enduring problem is determining whether paraglacial slope-failure debris was modified by permafrost creep in a contemporary or subsequent periglacial climate. Holocene-age dates (cf. Ballantyne & Stone 2004; Ballantyne *et al.* 1998) will enable this possibility to be excluded, Late Glacial age dates will not.

Protalus ramparts

A number of relict protalus ramparts have been identified in Great Britain and Ireland based on their morphological similarities with actively forming features in arctic and alpine environments (Colhoun 1981; Ballantyne & Kirkbride 1986; Anderson *et al.* 2001; Ballantyne 2002*a*). Some features earlier regarded as ramparts have been reinterpreted as either moraines or rock-slope failure deposits (Ballantyne 1986; Shakesby & Matthews 1993; Curry *et al.* 2001; Wilson 2004) because of their excessive size and because they do not 'fit' with the morphometrical criteria established by Ballantyne & Benn (1994) for differentiating ramparts from other talus-foot landforms.

It does not seem to have been previously considered in any detail that relict talus-foot ramps and ridges that conform to the constraints outlined by Ballantyne & Benn (1994) are anything other than protalus ramparts. Possible alternative origins were recognized by Ballantyne & Benn (1994, p. 152) when they stated that 'Ramps or ridges of debris that fall *within* the critical distance (30–70 m) for protalus rampart development may have some other origin', and Shakesby (1997, p. 414) stated that 'all possible origins ... should be considered before accepting a rampart origin... '. However, alternative modes of formation are not always critically evaluated.

Whilst it is not doubted that relict protalus ramparts could have formed in the generally accepted manner, whereby debris falls from cliffs and then slides, bounces and/or rolls across a perennial firn field to accumulate at its downslope margin (Rapp 1960; Ballantyne 1987; Perez 1988; Hall & Meiklejohn 1997), not all talus-foot debris ramps and ridges need to have formed in this way. If large-scale slope failures are capable of producing talus-foot debris accumulations that mimic the morphology of protalus rock glaciers, then there is no reason why smaller-scale slope failures should not produce features that resemble protalus ramparts. The development of such ramparts by paraglacial slope failure need not involve a firn field. For example, the Baosbheinn (Scotland) protalus rampart (Sissons 1976*b*) has been reinterpreted by Ballantyne (1986) as the product of rock-slope failure(s) that descended across the surface of a former firn field. Whilst rock-slope failure has undoubtedly occurred at this site, a firn field does not necessarily have to have been present in order to account for the planform morphology assumed by the debris accumulation (cf. Curry *et al.* 2001).

Almost all the relict protalus ramparts described in the literature are of classical form, being linear or arcuate ridges of predominantly coarse rock debris separated from the talus by a shallow depression. Morphologically less distinct features occur at the foot of many talus slopes, but these have not received the same amount of attention as the former features. Thus, there is a bias in the literature towards ramparts that display classical morphology, probably because they have proved easiest to recognize. It may be the case that these features *are* of periglacial origin, but inputs of paraglacial rockfall and rockslide debris cannot be excluded entirely, and the less distinct features may be of wholly paraglacial origin. Rampart clast sizes and geomorphological evidence for rock-slope failure on and above the rockwall source areas may provide clues.

If ramparts are essentially paraglacial rather than periglacial features then, as with protalus rock glaciers, they need not have developed during the LLS, as is usually inferred, and no palaeoclimatic inferences can be drawn because they are not climate-dependent landforms. Again, absolute age dating may be able to assist; Late Glacial ages will not distinguish periglacial from paraglacial debris, Holocene dates may be taken to indicate a non-periglacial origin.

Summary

Talus slopes and talus-foot landforms in Great Britain and Ireland have stimulated a considerable amount of research. Prior to about 1990, most of these features were considered to be of purely periglacial origin but since then a reassessment of some of them has occurred and paraglacial rockfall and rock-slope failure are recognized as having played a significant part in talus-landform development. In some cases, paraglacial processes are thought to have been of greater importance than processes associated with permafrost and freeze–thaw activity (Wilson 2004). However, there is still much uncertainty as to the respective roles of periglacial and paraglacial processes, and difficulties in identifying specific processes remain.

At present the shift towards invoking a paraglacial origin for some talus-related landforms continues. It needs to be remembered that talus can and does accumulate under periglacial conditions (McCarroll *et al.* 1998, 2001). Absolute age dating using cosmogenic isotopes may help to resolve the origin of some landforms, but is unlikely to do so for all. Equifinality seems likely to apply in respect of these features and a composite origin for some cannot be excluded and seems highly probable. Geomorphologists have tended to overlook the issue of composite landforms, probably because of the inherent difficulties in recognizing the products of different processes in landforms for which exposures of their constituent materials are rare and also because of the apparent absence of modern analogues. Relict talus-related landforms continue to present a major challenge to geomorphologists because their formative processes impact directly on their palaeoenvironmental significance.

References

Anderson, E. & Harrison, S. 2006. Late Quaternary paraglacial sedimentation in the Macgillycuddy's Reeks, southwest Ireland. *Irish Geography*, **39**, 69–77.

Anderson, E., Harrison, S. & Passmore, D. 2001. A Late-Glacial protalus rampart in Macgillycuddy's Reeks, south-west Ireland. *Irish Journal of Earth Sciences*, **19**, 43–50.

André, M.-F. 1997. Holocene rockwall retreat in Svalbard: a triple-rate evolution. *Earth Surface Processes and Landforms*, **22**, 423–440.

Andrews, J. T. 1961. The development of scree slopes in the English Lake District and central Quebec-Labrador. *Cahiers de Géographie de Quebec*, **10**, 219–230.

Arsenault, A. M. & Meigs, A. J. 2005. Contribution of deep-seated bedrock landslides to erosion of a glaciated basin in southern Alaska. *Earth Surface Processes and Landforms*, **30**, 1111–1125.

Atkinson, T. C., Briffa, K. R. & Coope, G. R. 1987. Seasonal temperatures in Britain during the past 22,000 years, reconstructed using beetle remains. *Nature*, **325**, 587–592.

Augustinus, P. 1995. Rock mass strength and the stability of some glacial valley slopes. *Zeitschrift für Geomorphologie*, **39**, 55–68.

Ball, D. F. 1966. Late-Glacial scree in Wales. *Biuletyn Peryglacjalny*, **15**, 151–163.

Ballantyne, C. K. 1984. The Late Devensian periglaciation of upland Scotland. *Quaternary Science Reviews*, **3**, 311–343.

Ballantyne, C. K. 1986. Protalus rampart development and the limits of former glaciers in the vicinity of Baosbheinn, Wester Ross. *Scottish Journal of Geology*, **22**, 13–25.

Ballantyne, C. K. 1987. Some observations on the morphology and sedimentology of two active protalus ramparts, Lyngen, northern Norway. *Arctic and Alpine Research*, **19**, 167–174.

Ballantyne, C. K. 1999. Maximum altitude of Late Devensian glaciation on the Isle of Mull and Isle of Jura. *Scottish Journal of Geology*, **35**, 97–106.

Ballantyne, C. K. 2002*a*. The Conachair protalus rampart, St. Kilda. *Scottish Geographical Journal*, **118**, 343–350.

Ballantyne, C. K. 2002*b*. Paraglacial geomorphology. *Quaternary Science Reviews*, **21**, 1935–2017.

Ballantyne, C. K. & Benn, D. I. 1994. Glaciological constraints on protalus rampart development. *Permafrost and Periglacial Processes*, **5**, 145–153.

BALLANTYNE, C. K. & ECKFORD, J. D. 1984. Characteristics and evolution of two relict talus slopes in Scotland. *Scottish Geographical Magazine*, **100**, 20–33.

BALLANTYNE, C. K. & HARRIS, C. 1994. *The Periglaciation of Great Britain*. Cambridge University Press, Cambridge.

BALLANTYNE, C. K. & KIRKBRIDE, M. P. 1986. The characteristics and significance of some Lateglacial protalus ramparts in upland Britain. *Earth Surface Processes and Landforms*, **11**, 659–671.

BALLANTYNE, C. K. & KIRKBRIDE, M. P. 1987. Rockfall activity in upland Britain during the Loch Lomond Stadial. *Geographical Journal*, **153**, 86–92.

BALLANTYNE, C. K. & STONE, J. O. 2004. The Beinn Alligin rock avalanche, NW Scotland: cosmogenic ^{10}Be dating, interpretation and significance. *The Holocene*, **14**, 448–453.

BALLANTYNE, C. K., STONE, J. O. & FIFIELD, L. K. 1998. Cosmogenic Cl-36 dating of postglacial landsliding at The Storr, Isle of Skye, Scotland. *The Holocene*, **8**, 347–351.

BOARDMAN, J. 1996. *Classic Landforms of the Lake District*. The Geographical Association, Sheffield.

CHATTOPADHYAY, G. P. 1984. A fossil valley-wall rock glacier in the Cairngorm mountains. *Scottish Journal of Geology*, **20**, 121–125.

CHURCH, M. & RYDER, J. M. 1972. Paraglacial sedimentation: a consideration of fluvial processes conditioned by glaciation. *Geological Society of America Bulletin*, **83**, 3059–3071.

CLAYTON, K. 1981. Explanatory description of the landforms of the Malham area. *Field Studies*, **5**, 389–423.

COCKBURN, H. A. P. & SUMMERFIELD, M. A. 2004. Geomorphological applications of cosmogenic isotope analysis. *Progress in Physical Geography*, **28**, 1–42.

COLHOUN, E. A. 1981. A protalus rampart from the western Mourne Mountains, Northern Ireland. *Irish Geography*, **14**, 85–90.

COOPE, G. R. & PENNINGTON, W. 1977. The Windermere Interstadial of the Late Devensian. *Philosophical Transactions of the Royal Society*, **280**, 337–340.

CURRY, A. M. & BLACK, R. 2002. Structure, sedimentology and evolution of rockfall talus, Mynydd Du, south Wales. *Proceedings of the Geologists' Association*, **114**, 49–64.

CURRY, A. M. & MORRIS, C. J. 2004. Lateglacial and Holocene talus slope development and rockwall retreat on Mynydd Du, UK. *Geomorphology*, **58**, 85–106.

CURRY, A. M., WALDEN, J. & CHESHIRE, D. A. 2001. The Nant Ffrancon 'protalus rampart': evidence for Late Pleistocene paraglacial landsliding in Snowdonia, Wales. *Proceedings of the Geologists' Association*, **112**, 317–330.

DAWSON, A. G. 1977. A fossil lobate rock glacier in Jura. *Scottish Journal of Geology*, **13**, 37–42.

FORBES, D. L. & SYVITSKI, J. P. M. 1994. Paraglacial coasts. *In*: CARTER, R. W. G. & WOODROFFE, C. D. (eds) *Coastal evolution: Late Quaternary shoreline morphodynamics*. Cambridge University Press, Cambridge, 373–424.

GRAY, J. M. 1982. The last glaciers (Loch Lomond Advance) in Snowdonia, N. Wales. *Geological Journal*, **17**, 111–133.

GRAY, J. M. & COXON, P. 1991. The Loch Lomond Stadial glaciation in Britain and Ireland. *In*: EHLERS, J., GIBBARD, P. L. & ROSE, J. (eds) *Glacial Deposits in Great Britain and Ireland*. Balkema, Rotterdam, 89–105.

HALL, K. & MEIKLEJOHN, I. 1997. Some observations regarding protalus ramparts. *Permafrost and Periglacial Processes*, **8**, 245–249.

HARRISON, S. 1996. Paraglacial or periglacial? The sedimentology of slope deposits in upland Northumberland. *In*: ANDERSON, M. G. & BROOKS, S. M. (eds) *Advances in Hillslope Processes*. Wiley, Chichester, 1197–1218.

HARRISON, S. & ANDERSON, E. 2001. A late Devensian rock glacier in the Nantlle Valley, North Wales. *Glacial Geology and Geomorphology*. Available from http://boris.qub.ac.uk/ggg/papers/full/2001/rp012001/rp01.html.

HINCHLIFFE, S. 1999. Timing and significance of talus slope reworking, Trotternish, Skye, northwest Scotland. *The Holocene*, **9**, 483–494.

HINCHLIFFE, S. & BALLANTYNE, C. K. 1999. Talus accumulation and rockwall retreat, Trotternish, Isle of Skye, Scotland. *Scottish Geographical Journal*, **115**, 53–70.

HOLM, K., BOVIS, M. & JAKOB, M. 2004. The landslide response of alpine basins to post-Little Ice Age glacial thinning and retreat in southwestern British Columbia. *Geomorphology*, **57**, 201–216.

HOWE, E. 1909. Landslides in the San Juan Mountains, Colorado. *United States Geological Survey Professional Paper*, **67**, 31–40.

HUDDART, D. 2002. Wasdale screes. *In*: HUDDART, D. & GLASSER, N. F. (eds) *Quaternary of Northern England*. Geological Conservation Review Series, **25**. Joint Nature Conservation Committee, Peterborough, 343–348.

HUTCHINSON, J. N. & MILLAR, D. L. 2001. The Craig Goch landslide dam, Meirionnydd. mid Wales. *In*: WALKER, M. J. C. & MCCARROLL, D. (eds) *The Quaternary of West Wales: Field Guide*. Quaternary Research Association, London, 113–125.

ISARIN, R. F. B. 1997. Permafrost distribution and temperatures in Europe during the Younger Dryas. *Permafrost and Periglacial Processes*, **8**, 313–333.

JARMAN, D. 2003. Paraglacial landscape evolution – the significance of rock slope failure. *In*: EVANS, D. J. A. (ed.) *The Quaternary of the Western Highland Boundary: Field Guide*. Quaternary Research Association, London, 50–68.

JARMAN, D. 2006. Large rock slope failures in the Highlands of Scotland: characterisation, causes and spatial distribution. *Engineering Geology*, **83**, 161–182.

JOHNSON, P. G. 1983. Rock glaciers. A case for a change in nomenclature. *Geografiska Annaler*, **65A**, 27–34.

JOHNSON, P. G. 1984*a*. Paraglacial conditions of instability and mass movement: a discussion. *Zeitschrift für Geomorphologie*, **28**, 235–250.

JOHNSON, P. G. 1984*b*. Rock glacier formation by high-magnitude low-frequency slope processes in the southwest Yukon. *Annals of the Association of American Geographers*, **74**, 408–419.

JOHNSON, R. H. & VAUGHAN, R. D. 1989. The Cowms Rocks landslide. *Geological Journal*, **24**, 359–370.

KORUP, O., MCSAVENEY, M. J. & DAVIES, T. R. H. 2004. Sediment generation and delivery from large historic landslides in the Southern Alps, New Zealand. *Geomorphology*, **61**, 189–207.

LEWIS, C. A. 1985. Periglacial features. *In*: EDWARDS, K. J. & WARREN, W. P. (eds) *The Quaternary history of Ireland*. Academic Press, London, 95–113.

LUCKMAN, B. H. 1992. Debris flows and snow avalanche landforms in the Lairig Ghru, Cairngorm Mountains, Scotland. *Geografiska Annaler*, **74A**, 109–121.

LUCKMAN, B. H. & FISKE, C. J. 1997. Holocene development of coarse debris landforms in the Canadian Rockies. *Paläoklimaforschung*, **19**, 283–297.

MATTHEWS, J. A. & SHAKESBY, R. A. 2004. A twentieth-century neoparaglacial rock topple on a glacier foreland, Ötztal Alps, Austria. *The Holocene*, **14**, 454–458.

MCCARROLL, D., SHAKESBY, R. A. & MATTHEWS, J. A. 1998. Spatial and temporal patterns of late Holocene rockfall activity on a Norwegian talus slope: a lichenometric and simulation-modeling approach. *Arctic and Alpine Research*, **30**, 51–60.

MCCARROLL, D., SHAKESBY, R. A. & MATTHEWS, J. A. 2001. Enhanced rockfall activity during the Little Ice Age: further lichenometric evidence from a Norwegian talus. *Permafrost and Periglacial Processes*, **12**, 157–164.

PENNINGTON, W. 1978. Quaternary geology. *In*: MOSELEY, F. (ed.) *Geology of the Lake District*. Yorkshire Geological Society, Occasional Publication, **3**, 207–225.

PENNINGTON, W. 1996. Limnic sediments and the taphonomy of Lateglacial pollen assemblages. *Quaternary Science Reviews*, **15**, 501–520.

PEREZ, F. L. 1988. Debris transport over a snow surface: a field experiment. *Revue de Geomorphologie Dynamique*, **37**, 81–101.

RAPP, A. 1960. Recent development of mountain slopes in Kärkevagge and surroundings, northern Scandinavia. *Geografiska Annaler*, **42**, 65–200.

RYDER, J. M. 1971. The stratigraphy and morphology of paraglacial alluvial fans in south-central British Columbia. *Canadian Journal of Earth Science*, **8**, 279–298.

SALT, K. E. & BALLANTYNE, C. K. 1997. The structure and sedimentology of relict talus, Knockan, Assynt, N. W. Scotland. *Scottish Geographical Magazine*, **113**, 82–89.

SASS, O. 2006. Determination of the internal structure of alpine talus deposits using different geophysical methods (Lechtaler Alps, Austria). *Geomorphology*, **80**, 45–58.

SASS, O. & KRAUTBLATTER, M. 2007. Debris flow-dominated and rockfall-dominated talus slopes: genetic models derived from GPR measurements. *Geomorphology*, **86**, 176–192.

SANDEMAN, A. F. & BALLANTYNE, C. K. 1996. Talus rock glaciers in Scotland: characteristics and controls on formation. *Scottish Geographical Magazine*, **112**, 138–146.

SELLIER, D. & LAWSON, T. J. 1998. A complex slope failure on Beinn nan Cnaimhseag, Assynt, Sutherland. *Scottish Geographical Magazine*, **114**, 85–93.

SHAKESBY, R. A. 1997. Pronival (protalus) ramparts: a review of forms, processes, diagnostic criteria and palaeoenvironmental implications. *Progress in Physical Geography*, **21**, 394–418.

SHAKESBY, R. A. & MATTHEWS, J. A. 1993. Loch Lomond Stadial glacier at Fan Hir, Mynydd Du (Brecon Beacons), South Wales: critical evidence and palaeoclimatic implications. *Geological Journal*, **28**, 69–79.

SHAKESBY, R. A. & MATTHEWS, J. A. 1996. Glacial activity and paraglacial landsliding in the Devensian Lateglacial: evidence from Craig Cerrig-gleisiad and Fan Dringarth, Forest Fawr (Brecon Beacons), South Wales. *Geological Journal*, **31**, 143–157.

SHAKESBY, R. A., DAWSON, A. G. & MATTHEWS, J. A. 1987. Rock glaciers, protalus ramparts and related phenomena, Rondane, Norway: a continuum of large-scale talus-derived landforms. *Boreas*, **16**, 305–317.

SISSONS, J. B. 1975. A fossil rock glacier in Wester Ross. *Scottish Journal of Geology*, **11**, 83–86.

SISSONS, J. B. 1976*a*. A fossil rock glacier in Wester Ross. *Scottish Journal of Geology*, **12**, 178–179.

SISSONS, J. B. 1976*b*. A remarkable protalus rampart complex in Wester Ross. *Scottish Geographical Magazine*, **92**, 182–190.

SISSONS, J. B. 1979. The Loch Lomond Advance in the Cairngorm Mountains. *Scottish Geographical Journal*, **95**, 66–82.

SISSONS, J. B. 1980. The Loch Lomond Advance in the Lake District, northern England. *Transactions of the Royal Society of Edinburgh: Earth Sciences*, **71**, 13–27.

SMITH, R. A. 2002. The Bowder Stone, Grange-in-Borrowdale, Cumbria. *Proceedings of the Cumberland Geological Society*, **6**, 525–539.

STATHAM, I. 1973. Scree slope development under conditions of surface particle movement. *Transactions of the Institute of British Geographers*, **59**, 41–53.

STATHAM, I. 1976. A scree slope rockfall model. *Earth Surface Processes*, **1**, 43–62.

TUFNELL, L. 1969. The range of periglacial phenomena in northern England. *Biuletyn Peryglacjalny*, **19**, 291–323.

VICK, S. G. 1987. Significance of landsliding in rock glacier formation and movement. *In*: GIARDINO, J. R., SHRODER, J. F. & VITEK, J. D. (eds) *Rock Glaciers*. Allen & Unwin, Boston, 239–263.

WHALLEY, W. B. 1976. A fossil rock glacier in Wester Ross. *Scottish Journal of Geology*, **12**, 175–178.

WHALLEY, W. B. & MARTIN, H. E. 1992. Rock glaciers: II models and mechanisms. *Progress in Physical Geography*, **16**, 127–186.

WILSON, P. 1990*a*. Clast size variations on talus: some observations from northwest Ireland. *Earth Surface Processes and Landforms*, **15**, 183–188.

WILSON, P. 1990*b*. Characteristics and significance of protalus ramparts and fossil rock glaciers on Errigal Mountain, Co. Donegal. *Proceedings of the Royal Irish Academy*, **90B**, 1–21.

WILSON, P. 1990*c*. Morphology, sedimentological characteristics and origin of a fossil rock glacier on Muckish Mountain, north-west Ireland. *Geografiska Annaler*, **72A**, 237–247.

WILSON, P. 1993. Description and origin of some talus-foot debris accumulations, Aghla Mountains, Co. Donegal, Ireland. *Permafrost and Periglacial Processes*, **4**, 231–244.

WILSON, P. 2004. Relict rock glaciers, slope failure deposits, or polygenetic features? A re-assessment of some Donegal debris landforms. *Irish Geography*, **37**, 77–87.

WILSON, P. 2005. Paraglacial rock-slope failures in Wasdale, western Lake District, England: morphology, styles and significance. *Proceedings of the Geologists' Association*, **116**, 349–361.

WILSON, P. 2007. The Kirkby Fell rock-slope failure, Malham, Yorkshire Dales. *North West Geography*, **7**, 1–9.

WILSON, P. & SMITH, A. 2006. Geomorphological characteristics and significance of late Quaternary paraglacial rock-slope failures on Skiddaw group terrain, Lake District, northwest England. *Geografiska Annaler*, **88A**, 237–252.

WYRWOLL, K.-H. 1977. Causes of rock-slope failure in a cold area: Labrador–Ungava. *Geological Society of America Reviews in Engineering Geology*, **3**, 59–67.

Paraglacial adjustment of the fluvial system to Late Pleistocene deglaciation: the Milfield Basin, northern England

DAVID G. PASSMORE[1]* & CLIVE WADDINGTON[2]

[1]*School of Geography, Politics and Sociology, Newcastle University, Newcastle upon Tyne NE1 7RU, UK*

[2]*Archaeological Research Services Ltd, Angel House, Portland Square, Bakewell, Derbyshire DE45 1HB, UK*

**Corresponding author (e-mail: d.g.passmore@newcastle.ac.uk)*

Abstract: Landform–sediment assemblages in the middle reaches of the River Till in the Milfield Basin, northern England, provide a comparatively rare example of a fluvial system emerging from drainage of a Late Pleistocene ice-dammed lake. This paper reviews the chronology and sequencing of Late Pleistocene lake drainage and early Holocene valley-floor development using new geomorphological, palaeoenvironmental and radiocarbon data, and considers the results in the context of paraglacial models of landscape response. The balance of currently available evidence suggests drainage of the proglacial lake occurred some time between the end of the Dimlington Stadial and the relatively mild climate of the Windermere Interstadial. Fluvial downcutting through glaciodeltaic and glaciolacustrine sediments was associated with recoupling of the fluvial sediment system to lower reaches of the Till and paraglacial development of inset fluvial terraces and valley widening prior to establishment of early Holocene channel systems at least 13–15 m below the equivalent glaciodeltaic surface. This short-lived phase of high paraglacial sediment yield was followed by relatively abrupt relaxation of coarse-sediment reworking as Holocene channels became largely decoupled from Late Pleistocene sand and gravel terraces in the basin. The combination of a bedrock barrier at the basin outlet and relatively gentle valley gradients in the basin has promoted a tendency towards Holocene floodplain alluviation (in central parts of the basin) with little net change in channel elevation until recent historic times. Paraglacial landscape modifications will have continued to exert an influence on the fluvial system during the Holocene, especially with regard to fine sediment yields from localized erosion of glaciolacustrine deposits, but this has most probably diminished considerably with time as sediment supplies to the basin increasingly reflect the impact of anthropogenic catchment disturbance and reworking of Holocene valley-floor deposits.

Fluvial systems in formerly glaciated catchments of NW Europe are recognized as having responded rapidly to climate changes spanning the Late Glacial (Late Weichselian–Devensian) period (Vandenburghe 2003), and these are often discussed in the context of a paraglacial (cf. Church & Ryder 1972; Ballantyne 2002) model of landscape development. In these scenarios high rates of sediment supply during and immediately following a glaciation typically promote aggradation by braided, meltwater-fed channels, while transitions to interstadial conditions are often marked by fluvial downcutting as glacially derived sediment sources are stabilized by soil and vegetation development (e.g. Ballantyne 2002; Antoine *et al.* 2003). Subsequent patterns of Holocene valley-floor development may, however, continue to be conditioned by sediment yields derived from reworked glacial and paraglacial deposits (see review by Ballantyne 2002). Some recent studies have demonstrated that episodes of fluvial downcutting at the onset of the Late Glacial period and at the Late Glacial–Holocene transition may be accomplished over centennial timescales (e.g. Antoine *et al.* 2003; Pastre *et al.* 2003), but it is frequently the case that fluvial archives lack adequate dating controls with which to constrain phases of valley incision and early Holocene alluviation. This is especially true in upland contexts where post-glacial erosion and reworking of older valley fills is liable to have removed the sedimentary record of the earliest Holocene (Lewin *et al.* 2005).

In their review of the UK Holocene fluvial archive, Lewin *et al.* (2005) identify flood-basin settings as offering the highest potential for preserving dateable Late Glacial and early Holocene deposits, and this has proved to be the case in the wide, low-relief valley floor of the River Till in the Milfield Basin, northern England (Fig. 1). This setting has preserved dateable palaeochannel (Passmore *et al.*

From: KNIGHT, J. & HARRISON, S. (eds) *Periglacial and Paraglacial Processes and Environments*. The Geological Society, London, Special Publications, **320**, 145–164.
DOI: 10.1144/SP320.10 0305-8719/09/$15.00

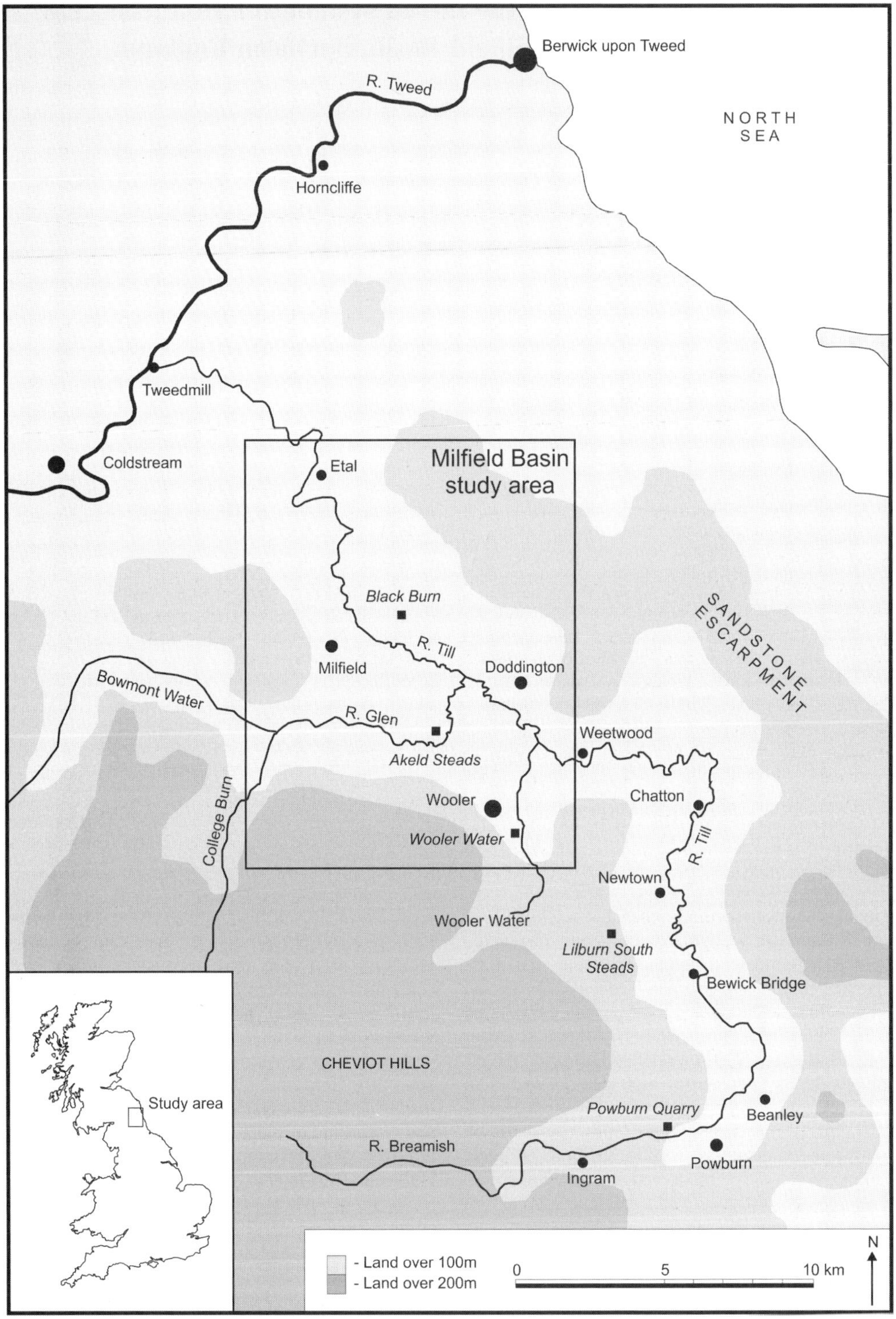

Fig. 1. Map of the River Breamish–Till and Lower Tweed valleys showing relief, the location of the Milfield Basin study area and the sites mentioned in text.

2002) and flood-basin (Tipping 1998) sequences that permit evaluation of the character and elevation of early Holocene channel and floodplain environments that are inset below Late Glacial glaciofluvial, glaciolacustrine and glaciodeltaic terraces. However, dating controls on the Late Glacial sequences have hitherto been limited, and this has led to conflicting interpretations of the chronology and pattern of deglaciation and fluvial-system development (Clapperton 1971*a*; Payton 1980, 1988, 1992; Tipping 1998). Debate has focused in particular on evidence for the development and subsequent drainage of an extensive proglacial lake in the basin that, as has been demonstrated elsewhere (e.g. Campy *et al.* 1988; Ballantyne 2002), is liable to have exerted a major control on river base levels and paraglacial sediment supply.

A reconsideration of the deglaciation history of the Milfield Basin has recently been facilitated by the multidisciplinary Till–Tweed Geoarchaeology Project (Passmore & Waddington 2009). In assessing the landform, sediment and archaeological associations in the valleys of the River Till and Lower Tweed, this project has undertaken systematic high-resolution geomorphological mapping and sediment coring of the Milfield Basin valley floor (see also Passmore *et al.* 2002, 2006) that has yielded new insights into the Late Glacial and early Holocene history of valley-floor development. This paper presents some new data from the Till–Tweed project with the aims of: (i) evaluating competing models of Late Glacial basin development, particularly with regard to the chronology and character of proglacial lake drainage; and (ii) refining the chronology of Late Glacial and early Holocene incision in the basin and its tributary valleys, and the development of accommodation space for Holocene fluvial activity. These aims are reviewed in the context of models of paraglacial landsystem development (cf. Ballantyne 2002).

Background to the study area

The Milfield Basin forms a physically distinct area of low-lying [below 70 m OD (ordnance datum)] Late Devensian glacial drift and Holocene alluvial deposits extending over 15 km^2 in the middle reaches of the River Till valley, north Northumberland (Figs 1–3). The basin is underlain by Carboniferous Cementstones and flanked to the west by the rounded domes of the Cheviot Hills, a complex of Devonian volcanic rocks predominantly comprising ashes, pyroclasts and andesitic lavas, and a later Devonian granite intrusion. To the east of the basin the landscape is dominated by gently curving escarpments with west-facing craggy ridges formed by Carboniferous Fell Sandstones. Soil cover in the region includes well-drained brown earths in many of the lowland settings, with gleys on intermediate slopes, and podzols and peats developed in upland locations (Payton 1980). Draining the basin is the gently meandering River Till, the second largest tributary of the Tweed, with a catchment area of 650 km^2. In central parts of the basin the Till traverses a broad expanse of low-relief Holocene alluvium between 34 and 36 m OD, has a gradient of 0.001 m m^{-1}, and conveys a fine gravel and sand bedload (Figs 1, 3 and 4). It is joined by the confluent rivers Glen and Wooler Water, and also the artificially straightened Humbleton Burn, a minor tributary that joins the Till from the southern flanks of the Cheviots (Figs 1 and 3). Prominent flood embankments built during the early nineteenth century (Archer 1992) confine the Till and its tributary channels. Below the confluence with the Glen, and downvalley to Etal, the Till has a gradient of 0.0005 m m^{-1} and becomes progressively confined to a valley floor up to 1 km wide that lies between upstanding Late Devensian terraces (Figs 3 and 4). Lower reaches of the Till, between Etal and the confluence with the Lower Tweed at Tweedmill, occupy a narrow and deeply entrenched bedrock gorge cut through Fell Sandstone and Cementstone with a relatively steep overall gradient of 0.002 m m^{-1} (Figs 1 and 4).

Quaternary history

Northumberland has a record of glacial research that extends back to William Buckland's identification of large 'moraines' on the eastern flanks of the Cheviots in the mid-nineteenth century (Boylan 1981). The broad pattern of ice-sheet flows over Northumbria was established by Raistrick (1931) and it is generally understood that during the Last Glacial Maximum, most of the Cheviot–lower Tweed area was overrun by ice flows originating from the Southern Uplands and Solway Firth to the west (Douglas 1991). On the basis of the distribution of Cheviot and non-Cheviot erratics, Clapperton (1970) has also argued that easterly flowing ice was deflected around the northern and southern margins of the Cheviot Hills by the presence of a local ice cap, although the nature and extent of the latter is uncertain (Lunn 2004). Particular attention has focused on two extensive areas of glaciofluvial sand and gravel in the Cheviot–lower Tweed region; a northerly deposit that stretches between Cornhill in the Tweed valley around the northern flank of the Cheviots (termed the Cornhill 'kettle moraine' by Sissons 1967), and to the south between Wooler and the Breamish valley at Powburn (Fig. 2). These characteristic hummocky landscapes comprise ice-contact and proglacial features including eskers, kames, kettle holes and

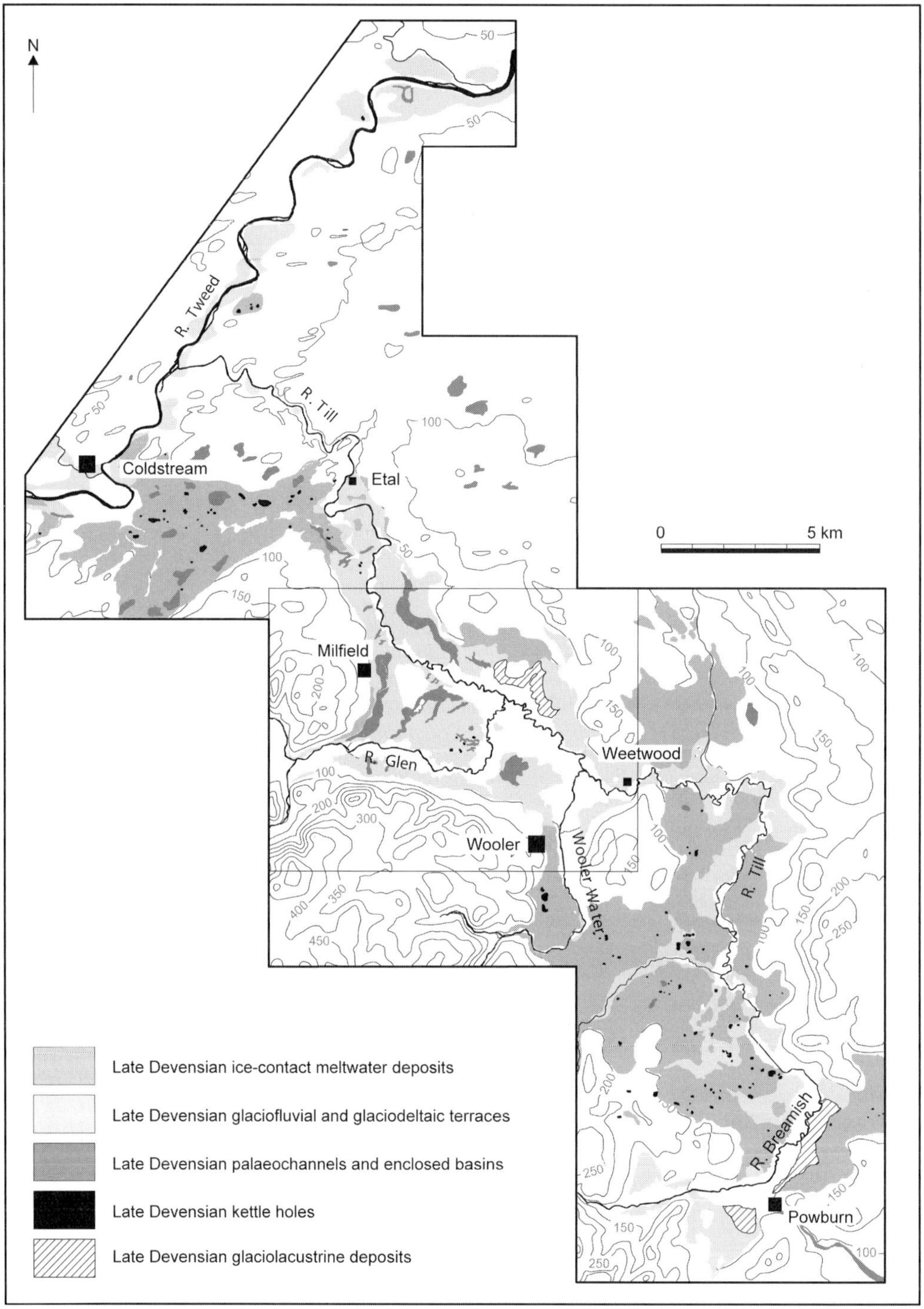

Fig. 2. Map of the River Breamish–Till and Lower Tweed valley showing major depositional landform assemblages associated with Late Devensian deglaciation. The box shows the location of Figure 3.

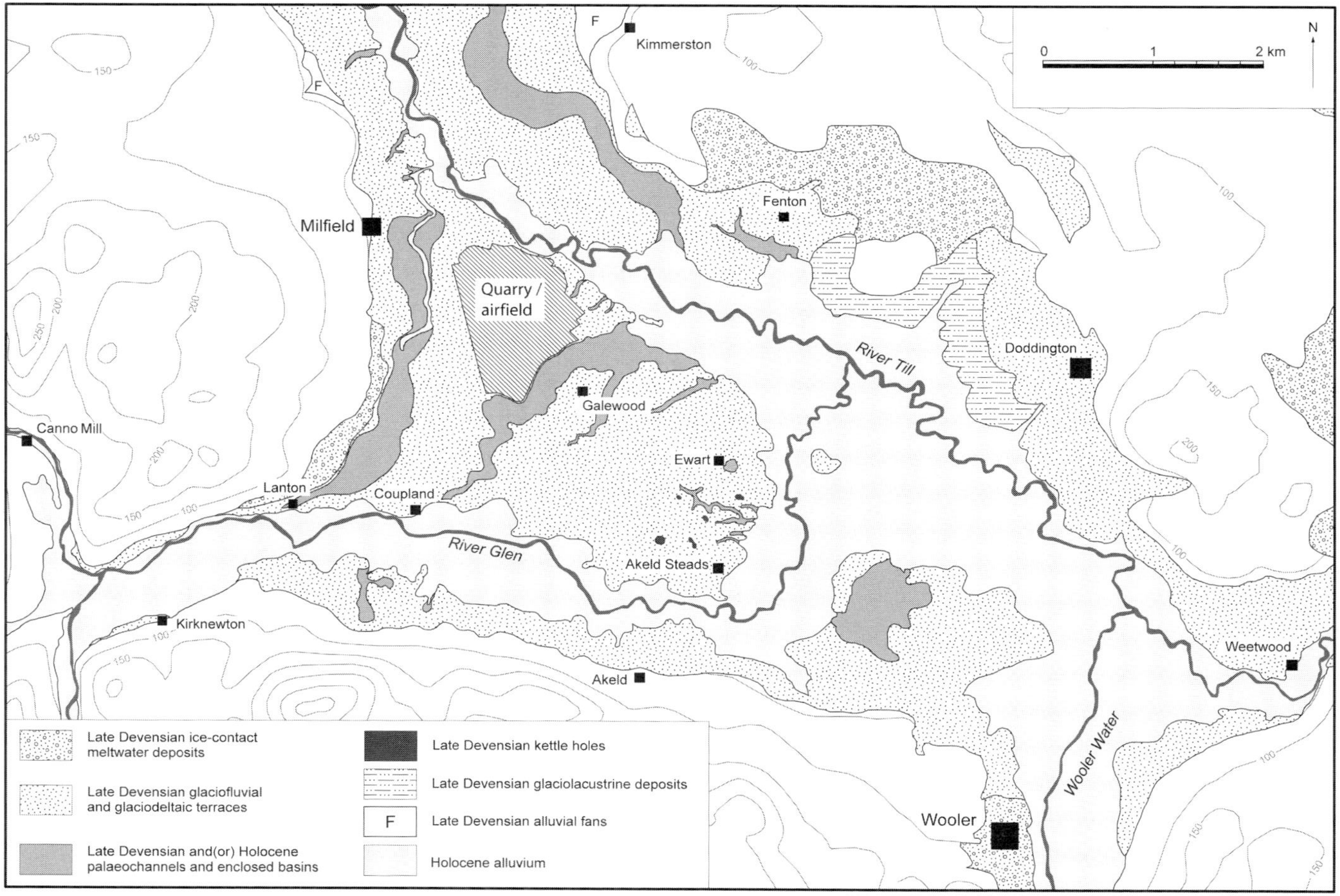

Fig. 3. Map of the Milfield Basin showing Late Devensian depositional landform assemblages associated with deglaciation, Holocene alluvium, and the site of the Milfield airfield and aggregate quarry.

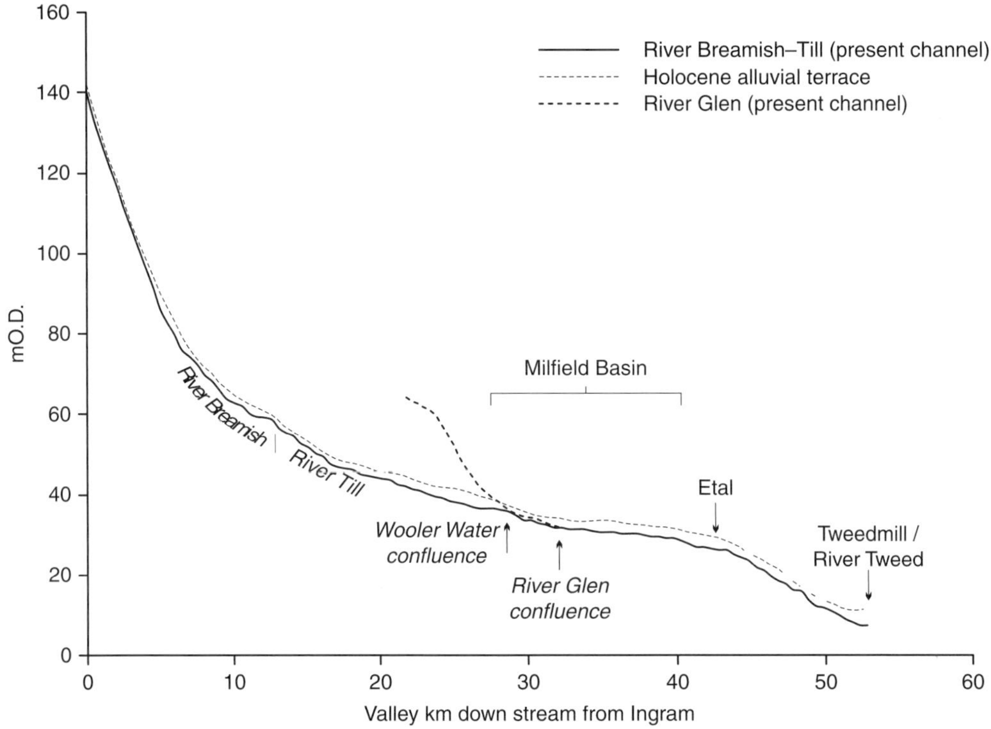

Fig. 4. Long-profile of the present River Breamish–Till channel and typical elevation of the flanking Holocene alluvial surface. Also shown are the confluences with the tributary rivers Glen and Wooler Water and the River Tweed at Tweedmill.

terraced outwash associated with downwasting and retreating ice at the end of the Dimlington Stadial (Douglas 1991; Lunn 2004). These deposits, together with extensive and complex systems of meltwater channels on the northern and eastern flanks of the Cheviot Hills, have been documented by Kendall & Muff (1901, 1903), Carruthers *et al.* (1932), Common (1954), Derbyshire (1961) and, especially, Clapperton (1970, 1971*a*, *b*).

More recent investigations of the glacial and deglacial history of the region have focused on Late Devensian slope deposits in the Cheviot Hills (Douglas & Harrison 1985, 1987; Harrison 1996, 2002), periglacial modification of soils and sediments in the region of the Milfield Basin (Payton 1980, 1988, 1992), and deposits infilling a kettle hole in ice-contact sand and gravel deposits near Lilburn South Steads, approximately 5 km SW of Chatton (Jones *et al.* 2000) (Fig. 1). However, secure chronological controls for regional Late Quaternary landform and sedimentary sequences are still generally lacking. Previously published radiocarbon dates on Late Devensian deposits are confined to a glaciolacustrine sequence in the eastern part of the Milfield Basin (a single date reported by Payton 1988 and 1992 – discussed later), while radiocarbon dating of the Lilburn Steads kettle-fill sequence presented problems of interpretation due to the assays returning ages between 1000 and 2500 years older than biostratigraphical evidence (including pollen, ostracod and molluscan data) from the site (Jones *et al.* 2000). Such discrepancies have been noted in similar geological and geomorphic contexts elsewhere in the UK (e.g. Walker *et al.* 1993) and may be attributable to contamination of samples with older carbon from detrital inwash and hard-water errors. Accordingly, Jones *et al.* (2000) resorted to correlation of biostratigraphic associations to constrain the lower part of the Lilburn Steads kettle-fill sequence to the period spanning the Windermere Interstadial, Loch Lomond (Younger Dryas) Stadial and the Early Holocene to approximately 8000 cal BP.

Methods

The Till–Tweed Project has adopted a range of geoarchaeological methods (Passmore *et al.* 2006; Passmore & Waddington 2009) that have centred on a programme of geomorphological mapping and survey of Late Devensian and Holocene landform assemblages in the Milfield Basin. This aspect of the project has been facilitated by LiDAR (Light Detection and Ranging) data commissioned from the Environment Agency, supported by analysis of historic and modern Ordnance Survey maps, regional geological maps, published articles on the soils (Payton 1980, 1992) and geomorphology (including Clapperton 1971*a*, *b*; Payton 1980; Tipping 1994*a*, *b*, 1998) of the study area and a programme of field visits (Passmore & Waddington 2009). Detailed sedimentological, palaeoecological and geochronological investigations in the study area have focused on sediment cores extracted from two coring transects and discrete palaeochannel and flood-basin features across the Milfield Basin valley floor between Weetwood and Milfield (Fig. 1). All cores/sections were logged for colour, texture, bedding structures and inclusions, while organic-rich sedimentary sequences were sampled as continuous bodies (either in cores or monolith tins) and removed intact to the laboratory for storage and subsampling for radiocarbon and palaeoecological analyses. All dates quoted below (unless stated otherwise) are calibrated date ranges (95% confidence intervals) calculated by the maximum intercept method (Stuiver & Reimer 1986), using the program OxCal v.3.5 (Bronk Ramsey 1995, 1998, 2001) and the INTCAL98 dataset (Stuiver *et al.* 1998). Palaeoecological methods are described fully in Passmore *et al.* (2006) and Passmore & Waddington (2009) and are not repeated here.

Late Devensian glaciolacustrine, glaciofluvial and glaciodeltaic landform–sediment associations

Late Devensian glaciolacustrine deposits

Thick deposits of laminated silts, clays and occasionally fine sands identified and mapped in the Milfield Basin (Gunn 1895; Clapperton 1971*a*; Payton 1980) and the Breamish–Till valley between Beanley and Chatton (Clapperton 1971*a*) have been interpreted as glaciolacustrine sediments associated with the development of localized proglacial lakes during the later stages of Late Devensian deglaciation (Fig. 2). Clapperton's (1971*a*) account of the Milfield Basin proglacial lake deposits builds on earlier work by Gunn (1895), who described these sediments as reaching a thickness of at least 22 m in the southern part of the basin. Lake development is believed to have arisen through damming of the basin's northern outlet by a combination of stagnant Tweed valley ice and a bedrock barrier at Etal (Clapperton 1971*a*; Payton 1980) (Fig. 2). Payton's (1980) investigation of the Milfield Basin soils, while acknowledging the need for more detailed geomorphological analysis of the former lake sediments, described glaciolacustrine deposits as reaching surface elevations of 42 m OD on the margins of the basin, where they are locally overlain by alluvial sediments and soliflucted slope deposits (Fig. 3). Published dating control on the latter stages of glaciolacustrine sedimentation is derived from radiocarbon dating of a buried humic gley soil developed in laminated silts and clays at Black Burn, located in a small tributary valley on the eastern flanks of the basin (Payton 1988, 1992) (Fig. 1). The buried topsoil (bApg) horizon of this soil at approximately 37 m OD yielded a ^{14}C date of *c.* 13544–13129 cal BP (HAR-4308; Table 1) (Payton 1992). Overlying the buried soil are about 2 m of laminated silts and clays that are interpreted by Payton (1988) as lacustrine sediment characterized by biogenic rather than density-graded laminations.

Late Devensian glaciofluvial and glaciodeltaic terraces

Glaciofluvial and glaciodeltaic outwash deposits are extensively developed around the margins of the Milfield Basin (Fig. 3). A broad fan-shaped expanse of sand and gravel, up to 10 m thick, spreads north and east into the basin from an apex at the mouth of the River Glen valley at Lanton (Fig. 3), and has been described by Clapperton (1971*a*) as an outwash delta built out into a large proglacial lake that filled the basin during deglaciation. The surface of this feature has a maximum elevation of 69 m OD at its apex, and slopes north and east to margins at 40–42 m OD; to the south and east the surface terminates in a locally well-defined terrace margin that rises 5–10 m above the Holocene alluvial valley floor and the present rivers Till and Glen (Fig. 5). Underlying sediments of this terrace complex have been exposed by aggregate quarrying SE of Milfield (Fig. 3), and have been described by Payton (1980) as comprising cross-bedded sands and cross-laminated sands and silts representing glaciodeltaic foreset beds, unconformably overlain by up to 2 m of plane-bedded sandy gravels deposited by subaerial braided river channels. The erosional contact recorded between these foreset and topset beds lies at about 45 m OD, and is interpreted by Payton (1980) as offering a minimum estimate of the former proglacial lake level in the basin. Inset some 2–3 m into the glaciodeltaic terrace surface are channelized depressions with broad, low-relief floors that have been described as former courses of the proto-river Glen (Payton 1980) (Figs 3

Table 1. *^{14}C dates and calibration details for selected samples (see text for context and calibration details)*

Laboratory code	Sample reference	Core	Material	^{14}C Age (BP)	Calibrated date range (95% confidence)
SUERC-9080	GW90 (90 cm)	Galewood 1	Wood fragment	11 490 ± 35	14 381–13 998 cal BP
SUERC-9081	GW115 (115 cm)	Galewood 1	Wood fragment	12 280 ± 40	13 415–13 255 cal BP
BETA-125959	MSH1-14 (291–299 cm)	MSH1-14 (Humbleton Burn)	Silty peat	11 740 ± 70	13 775–13 420 cal BP
HAR-4308*	Black Burn M13 (322 cm)	Black Burn	Peat	11 460 ± 100	13 544–13 129 cal BP
SUERC-522		MSH1-21	Wood fragment	4700 ± 55	5582–5317 cal BP

*After Payton (1988, 1992).

and 5). One such channel to the north of Galewood (here termed the Galewood depression; Fig. 3) traverses the central part of the terrace and has been found to be locally infilled with up to 130 cm of fine sediment, including approximately 25–30 cm of humified peaty silt at the base of the sequence where it overlies coarse sands and gravels (Passmore & Waddington 2009).

No direct dating controls have been previously reported for the glaciodeltaic terrace, and the argument for its deposition having occurred prior to the Late Glacial (Windermere) Interstadial rests on morphostratigraphic relationships with the glaciolacustrine sequence in the basin and geomorphological evidence of periglacial processes including cryoturbation structures, fine-sediment capping on larger clasts and abundant cropmark evidence of polygonally patterned ground on surfaces between 40 and 50 m OD (Payton 1992). Recent geoarchaeological investigations associated with the nearby Cheviot Quarry (Passmore & Waddington 2009; Johnson & Waddington 2008) have subsequently dated the upper and lower levels of the basal peaty sediment infilling the Galewood Depression to the period *c.* 14 381–13 998 cal BP (SUERC-9080; Table 1) and *c.* 13 415–13 255 cal BP (SUERC-9081; Table 1), respectively. Dating sediment accumulation in the abandoned channel at Galewood to the Late Glacial (Windermere) Interstadial provides further support for the assignment of the glaciodeltaic terrace to the later stages of the Dimlington Stadial.

A suite of sand and gravel terraces lie below the margins of the glaciodeltaic terrace in the Milfield Basin and have been interpreted by Clapperton (1971*a*) as glaciofluvial terraces that represent fluvial reworking of glaciodeltaic sediments. They are most extensively developed on both sides of the Till valley floor in the northern part of the basin between Milfield and Etal (Figs 3 and 5), and have surfaces some 3–4 m below the glaciodeltaic terrace. Mapping and sediment coring undertaken for the Milfield Basin project has identified a further group of low-lying sand and gravel terraces in the southern and western parts of the basin in areas that have been previously mapped as Holocene alluvium (Payton 1980, 1992; Tipping 1998) (Figs 3 and 6). The largest terrace in this group lies inset below glaciofluvial and glaciodeltaic deposits between Akeld Steads and Turvelaws, and has a gently undulating surface 1–2 m above the main Holocene alluvial surface. This terrace is traversed by the artificially straightened course of the Humbleton Burn, a small tributary of the present River Glen that rises on the northern flanks of the Cheviot Hills (Fig. 6), and has been shown by sediment cores and archaeological test pits (Transect MSH2; Fig. 7; Passmore & Waddington 2009) to comprise between 2 and 3 m of well- to poorly-bedded

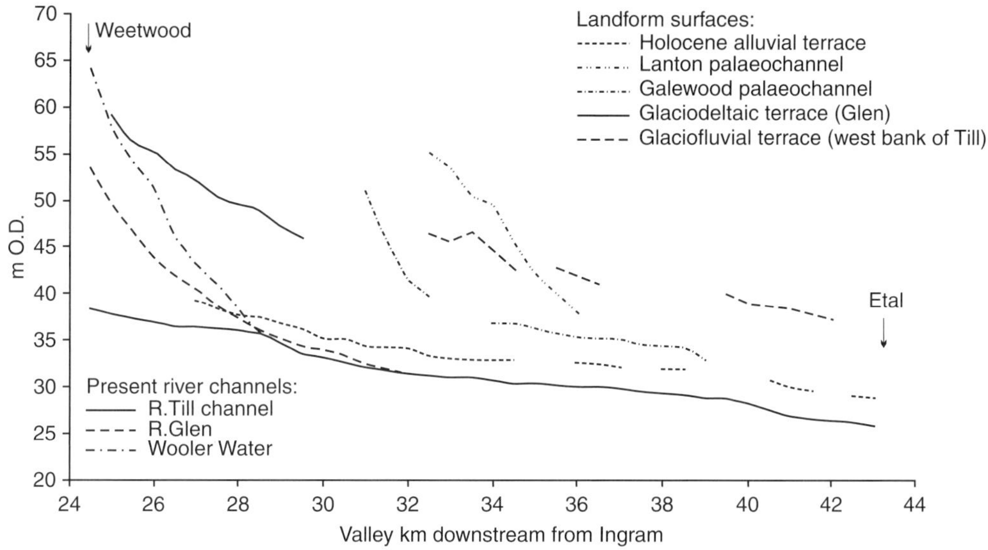

Fig. 5. Long-profiles of the present rivers Till, Wooler Water and Glen in the Milfield Basin between Weetwood and Etal. Also shown are surface profiles of Holocene alluvium, glaciodeltaic terraces (flanking the River Glen) and inset palaeochannels, and glaciofluvial terrace deposits downvalley of the Till–Glen confluence.

inorganic gravelly sands and silts. These fluvial sediments typically overlie inorganic and finely laminated light blue and grey silts and clays (Fig. 7) that are interpreted as truncated glaciolacustrine deposits. A smaller terrace remnant of this assemblage is evident to the east of Bridge End, where it forms a localized low-relief sand and gravel surface that is surrounded by Holocene alluvium (Fig. 6).

Both terrace units have gently dipping margins that are onlapped by Holocene silts and clays, but to the east of Akeld Steads their higher elevations have been shown by aerial photographs to exhibit a well-developed pattern of polygonal ice-wedge casts (T. Gates pers. comm.) (Fig. 8). Accordingly, these terraces are interpreted as reworked glaciodeltaic and glaciofluvial sediments that were deposited as a low-angle fan during incision of the River Glen through the main delta surface and underlying glaciolacustrine sediments following drainage of the proglacial lake (see the next section); abandonment of the terrace surfaces must have occurred prior to periglacial modification during the Dimlington and/or Loch Lomond Stadial.

Late Devensian and Holocene fluvial sequences in the Milfield Basin

Central parts of the Milfield Basin

Sedimentary sequences underlying the central part of the Milfield Basin have been investigated along sediment-coring transect MSH1, extending for some 2.7 km between Bridge End and a small crossing over the river Till to the SW of Doddington (Fig. 6). Selected sediment core logs and the surface profile of MSH1 are illustrated in Figure 9; this shows the alluvial surface to rise gently from a low of 34.5 m OD at Humbleton Burn, in the central part of the basin, to 35–36 m OD in the vicinity of the rivers Glen and Till, respectively, on the west and east side of the valley floor. Transects revealed alluvial sedimentary sequences to achieve depths between 200 and 500 cm (Fig. 9) and to overlie inorganic blue/pink, finely laminated fine sands, silts and clays that are consistent with accounts of glaciolacustrine deposits described by Clapperton (1971*a*) and Payton (1980, 1988). A full description of the Holocene alluvial sequence can be found in Passmore & Waddington (2009); here we focus on sediment cores taken in the vicinity of Humbleton Burn in the central part of the basin (cores MSH1-1, MSH1-14 and MSH1-21, Fig. 9), and which are notable for featuring beds of peat and/or organic fine sands, silts and clays at depths between 250 and 350 cm below the surface (*c*. 31–32 m OD). These organic-rich sediments are interpreted as buried shallow channel fills or flood-basin depressions that have no modern surface expression. A radiocarbon assay of approximately. 13 775–13 421 cal BP (Beta-125959; Table 1) from peat obtained from core MSH1-14 at 291–299 cm suggests this organic deposit dates to the Late Glacial (Windermere) Interstadial (Fig. 9). Pollen counts from organic-rich fine sediments

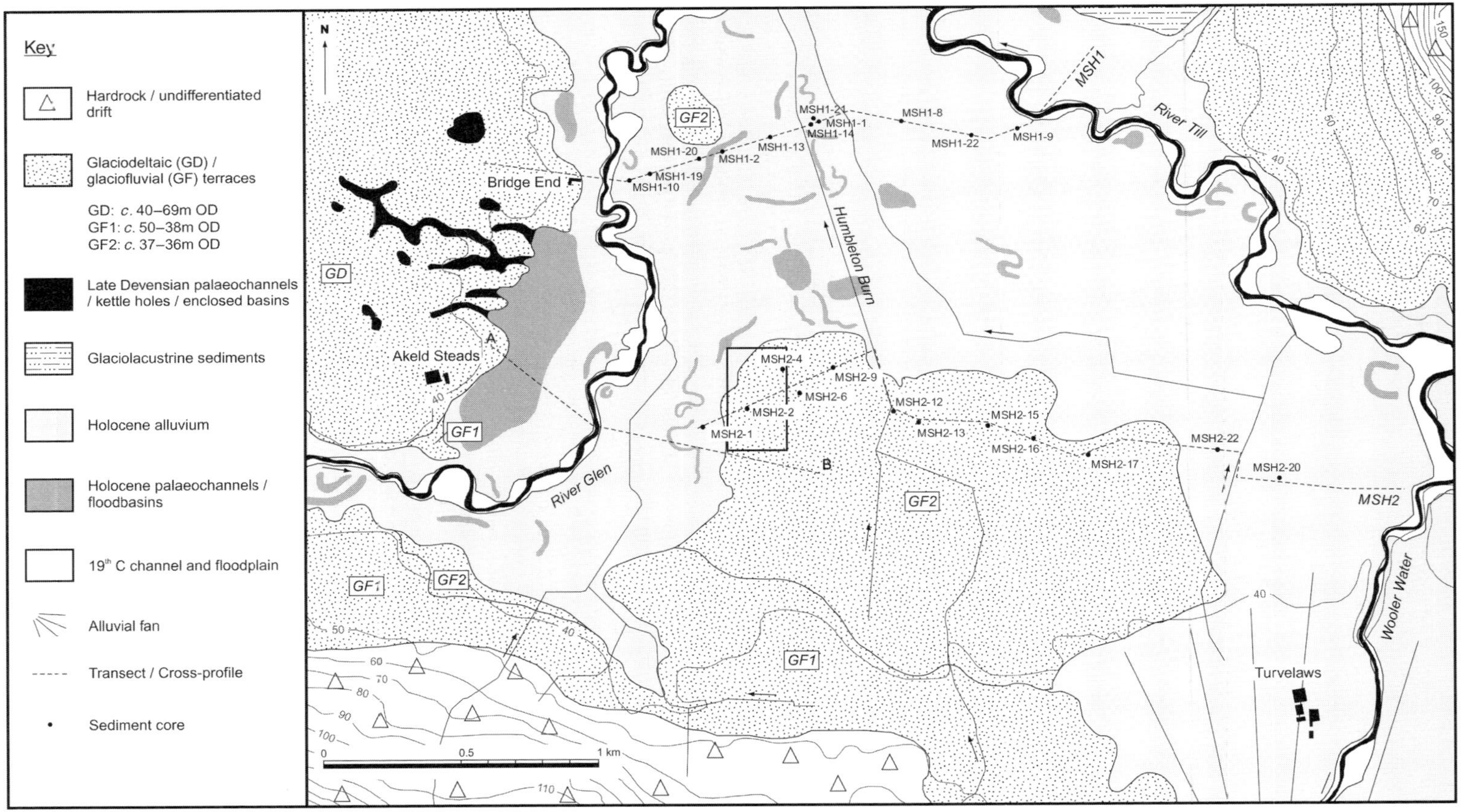

Fig. 6. Simplified geomorphological map of the valley floor between Akeld Steads and Turvelaws (Milfield Basin) showing transects MSH1 and MSH2, cross-profile A–B, major palaeochannels and selected sediment-core locations; the box shows the location of the aerial photograph in Figure 8.

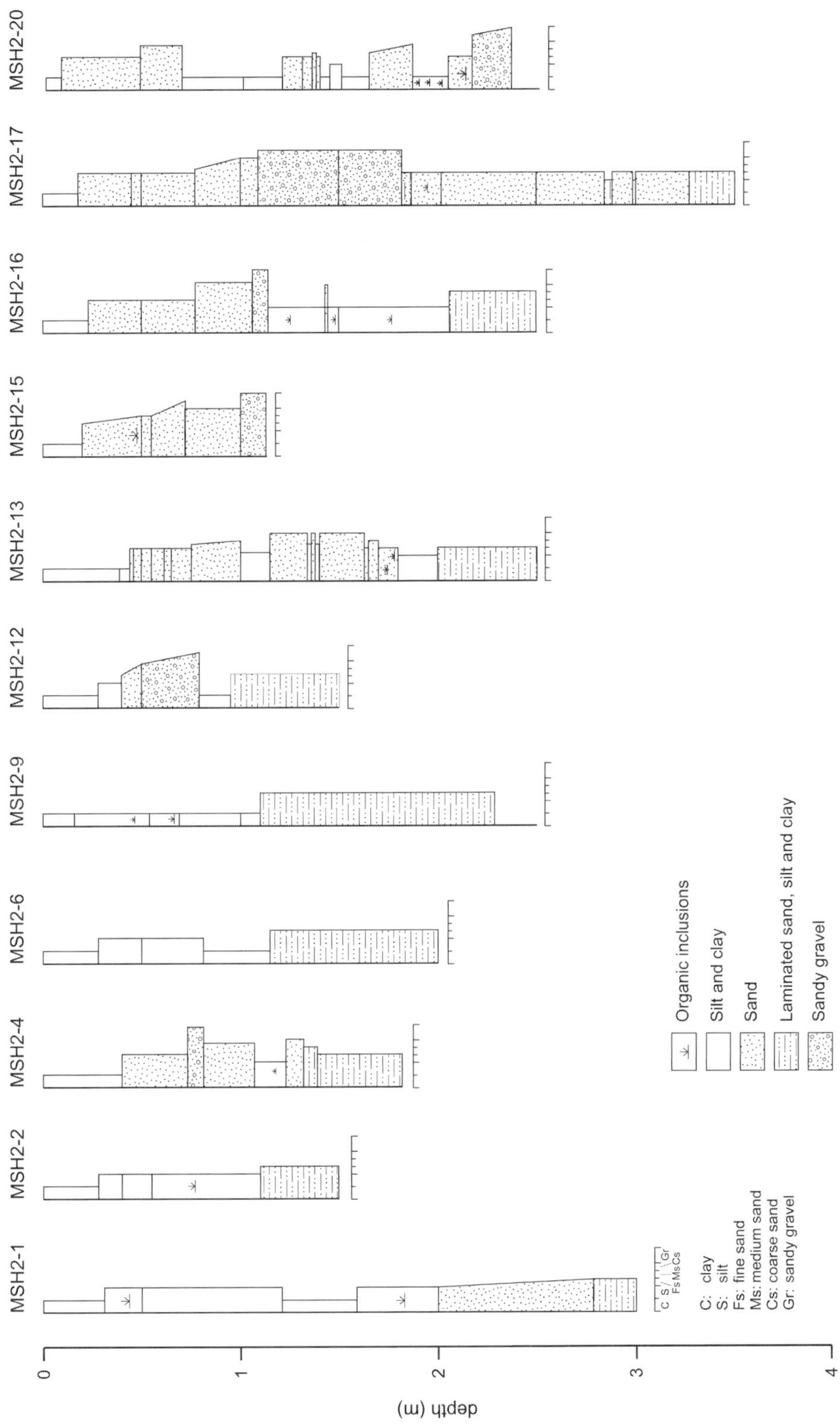

Fig. 7. Selected sediment-core logs from Transect MSH2 (Milfield Basin) (see Fig. 6 for core locations).

Fig. 8. Aerial photograph of a low-relief terrace surface east of Akeld Steads showing cropmark evidence of polygonal ice-wedge casts; see Figure 6 for the location of the photograph (© T. Gates).

between 310 and 330 cm in the nearby core MSH1-21, characterized by a hazel–birch–juniper scrub (including the arctic–alpine dwarf-shrub *Betula nana* L.), grasses (Poaceae) and *Filipendula*, bear comparison with very early Holocene vegetation assemblages recorded in the kettle hole fill at Lilburn South Steads, 9.7 km to the SE (Jones *et al.* 2000). Confirmation of this age estimate is currently problematic, however, since MSH1-21 has been radiocarbon dated to *c.* 5582–5317 cal BP at 322 cm (SUERC-522; Table 1; see Fig. 9). This date is believed to be in error as a result of sample contamination during the coring exercise, and hence the interpretation of a Late Glacial context for this deposit remains provisional, pending further palaeoecological and radiocarbon analysis.

Lower reaches of the River Glen

Previous work by Borek (1975) and Tipping (1994*a*, 1998) has documented a 3.5 m sequence of flood-basin peat and interbedded fine-grained alluvium at Akeld Steads, located in the lower reaches of the River Glen and lying immediately adjacent to the upstanding (*c.* 45 m OD) glaciodeltaic terrace margin (Figs 6 and 10). To the SE of the River Glen the modern floodplain surface lies inset 1 m below the lowest-elevation Late Devensian fan terrace described earlier. The organic-rich sedimentary sequence at Akeld Steads underlies the modern alluvial surface at 36 m OD, and spans the period between *c.* 11 600 (31.5 m OD) and *c.* 2800 cal BP (35 m OD; Tipping 1998). Geomorphological mapping and sediment coring undertaken for this project has demonstrated that this flood-basin fill extends for some 800 m along the glaciodeltaic terrace margin, and has a laterally persistent peaty infill sequence between 345 and 120 cm thick; in the downstream limits of the flood-basin these peaty sediments directly overlie blue/grey finely laminated sands, silts and clays to a recorded depth of 200 cm, and which are

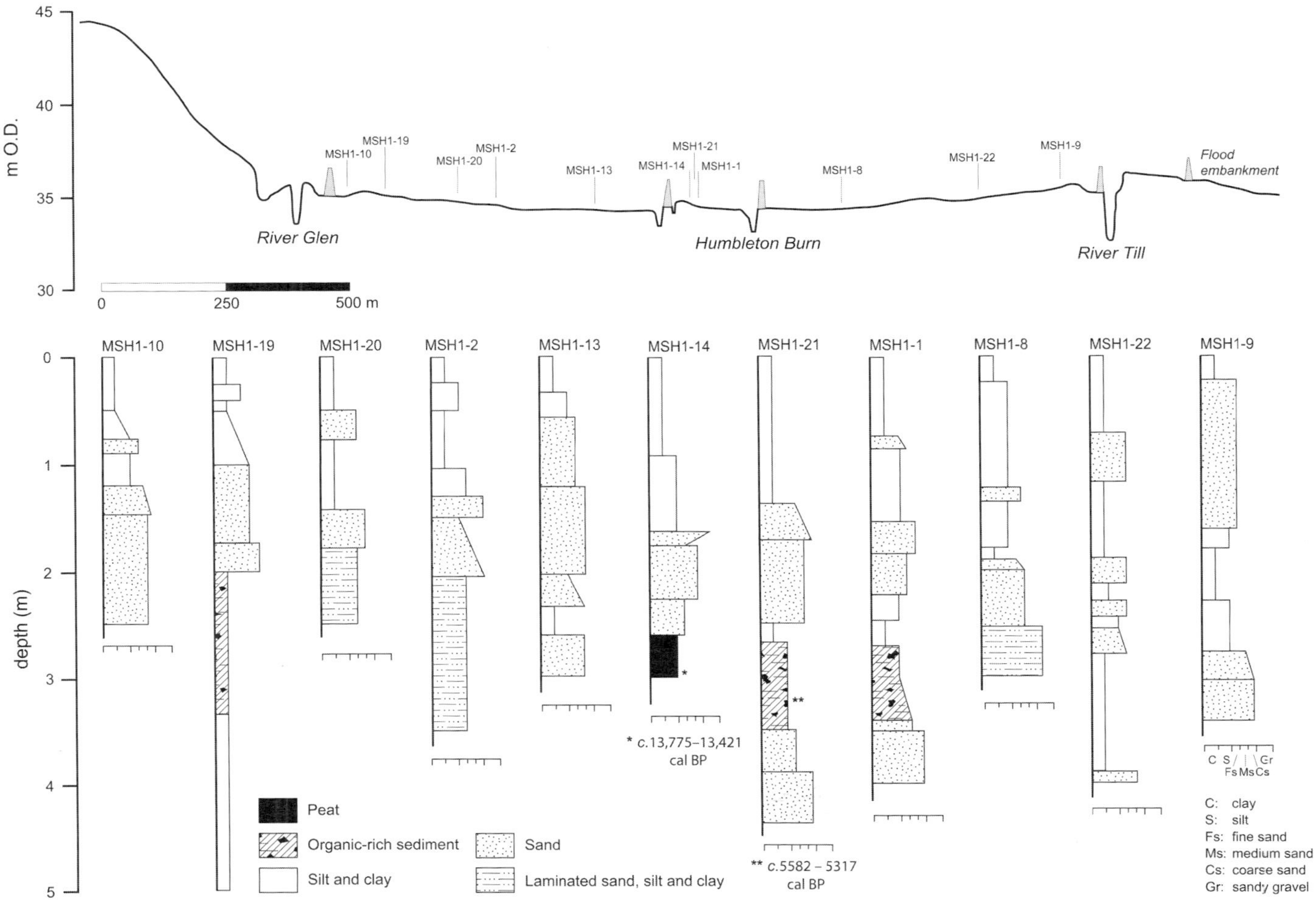

Fig. 9. Cross-profile of the valley floor of the Milfield Basin along Transect MSH1 (derived from LiDAR data) showing locations and logs of selected sediment cores.

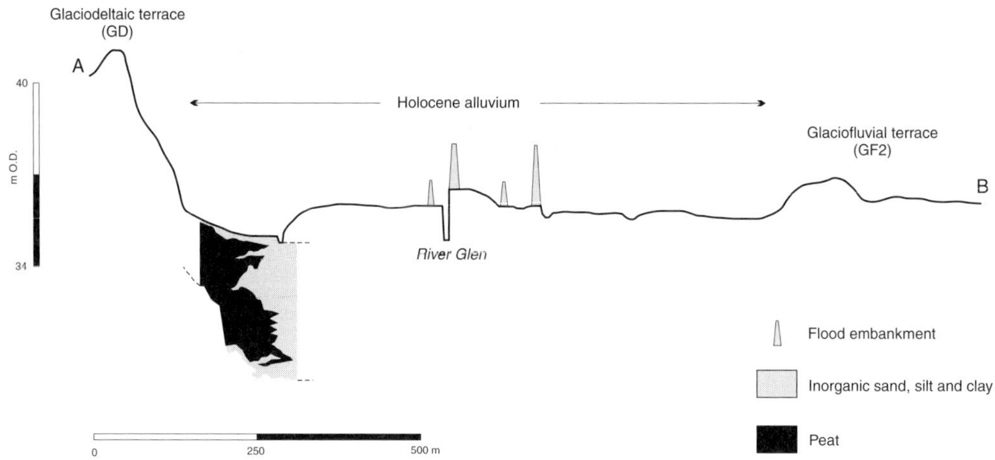

Fig. 10. Cross-profile A–B of the valley floor of the Milfield Basin extending SE from Akeld Steads showing a summary of the Holocene sedimentary sequence described by Tipping (1994*a*, 1998).

interpreted as glaciolacustrine deposits (Passmore & Waddington 2009).

Wooler Water

Late Devensian and Holocene sedimentary sequences in the valley floor of the Wooler Water upstream of Wooler have been the subject of investigations by Clapperton *et al.* (1971) and Tipping (1992, 1994*b*), while local channel and floodplain adjustments to recent historic aggregate extraction have also been reported by Sear & Archer (1998). On the western side of the valley an extensive spread of kame, esker and kettle-hole deposits lie up to 175 m OD, and are truncated to the east by a broad gravel terrace some 10 m thick and 500 m wide that has been termed the Haugh Head Terrace (Tipping 1994*b*). This surface can be traced downstream through Wooler and out into the Milfield Basin as a broad, low-relief fan that merges with the Holocene alluvial surface and, to the NW, the low-level glaciofluvial terrace surface developed along the southern margins of the basin (Passmore & Waddington 2009) (Fig. 6). However, and in contrast to the latter glaciofluvial deposits, aerial photographs of the Wooler Water fan exhibit no evidence of periglacial modification (T. Gates pers. comm.). Downstream of Wooler, the present channel of the Wooler Water lies inset approximately 1 m below the fan surface, but upstream in the vicinity of Earle Mill (1 km south of Wooler; Fig. 1) some 8 m of channel incision has occurred since the 1960s in response to aggregate extraction of channel and floodplain sand and gravel (Sear & Archer 1998); here, channel-bed surveys by local water authorities indicate the 1966 river bed to lie within 1 m of the adjacent Haugh Head terrace surface.

Dating controls for the valley-fill sequence in the Wooler Water are derived from a 2 m-thick peat bed buried beneath some 3.5 m of fluvial gravels (termed the Earle Mill Terrace by Tipping 1994*b*). Upper levels of this peat have been dated to approximately 4000 cal BP (SRR-3658; Tipping 1992, 1994*b*), but pollen evidence from the lower levels of the sequence suggests the inception of peat development is likely to have begun during the Early Holocene (Clapperton *et al.* 1971). The chronology and character of Late Devensian and Holocene fluvial valley-floor development has been subject to differing interpretations. Clapperton *et al.* (1971) acknowledged that deposition of the Haugh Head Terrace may have commenced during regional deglaciation, but they interpret the Earle Mill peat deposit as a kettle-hole fill that was buried by late Holocene (post-*c.* 5700 cal BP) gravel aggradation associated with the development of the Haugh Head Terrace. Tipping's (1994*b*, 1998) re-evaluation of the site argued that the peat bed had developed on a poorly-drained alluvial-valley floor cut into the Haugh Head Terrace, and that the overlying gravels were associated with a discrete episode of post-4000 cal BP fluvial deposition (termed the Earle Mill Terrace) that aggraded to within 1–2 m of the Haugh Head surface. In this model the Haugh Head aggradation is viewed primarily as a response to increased discharge and sediment loads during the Loch Lomond Stadial; the terrace surface was subsequently abandoned by valley-floor entrenchment before or during the earliest Holocene. Assessment of these competing interpretations is complicated by the recent history of gravel extraction, and Sear & Archer (1998) regard the Earle Mill terrace and two lower fluvial units as artifacts of post-1960 flooding and gravel mining.

Further assessment of the fluvial history will be difficult in view of extensive post-extraction landscaping (see Passmore & Waddington 2009) of the site, but here it is noted that both Clapperton *et al.* (1971) and Tipping (1994*b*, 1998) envisaged net valley-floor entrenchment some time between regional deglaciation and the earliest Holocene. At Earle Mill this incision attained at least 7 m, but relatively flat terrace remnants on the adjacent truncated kamiform complex at 73 m OD suggest net fluvial incision since about 15 000 cal BP may have been as much as 15 m.

Discussion

Following Church & Ryder (1972), thick deposits of glaciolacustrine, glaciodeltaic and glaciofluvial sediment infilling the valley floor of the Milfield Basin may be interpreted as a paraglacial sequence comprising reworked glacigenic material from upstream tributary and trunk stream valleys. It is noted, however, that some more recent studies have preferred to exclude the deposits of glacial meltwater streams from their classifications of paraglacial landform assemblages (see the discussion in Ballantyne 2002). Irrespective of their classification, it is argued later in this section that in the context of paraglacial landsystem responses the timing and extent of subsequent incision and lateral reworking of these meltwater-derived deposits has implications for coupling of the fluvial sediment system, downstream sediment yields, and also the establishment of accommodation space and boundary conditions for post-glacial fluvial-system development.

Lake drainage and establishment of secondary paraglacial systems in the Milfield Basin and lower reaches of the Till

Proglacial lake impoundment and delta formation in the Milfield Basin constitutes the final phase of meltwater drainage described in Clapperton's (1971*a*) geomorphological synthesis of regional deglaciation, and would appear to be consistent with the typical deglaciation transition described by Ballantyne (2002) whereby proximal, ice-contact lakes evolve to distal glacier lakes in which sediment influx is dominated by meltwater rivers draining retreating glacier margins. In the Milfield Basin, however, distal margins of the lake may have remained in contact with an ice margin for much of the lake history, since Clapperton (1971*a*) envisaged a gradual reduction in lake levels controlled by outflow beneath stagnant ice to the north of Etal and incision of the rock-cut meanders of the present River Till between Etal and the confluence with the Tweed (Figs 1 and 4). This process would have been complete by the time of final ice melt, by which time erosion of the Etal gorge had been accomplished to a level approximating the modern channel elevation.

Subsequent work by Payton (1980, 1988) has argued, however, for at least two episodes of lake formation and drainage that collectively spanned the Windermere Interstadial and Loch Lomond Stadial; this model rests on the Windermere Interstadial ^{14}C age date obtained on the buried soil developed on glaciolacustrine deposits at Black Burn and its subsequent burial by a further 2 m of lacustrine sediments. This interpretation could not be reconciled with base-level changes controlled by erosion of the bedrock barrier at Etal and hence was attributed to changes in discharge and(or) climate within a closed basin (Payton 1988; see also the discussion in Tipping 1998). In this scenario lake-level rises post-13 300 cal BP are argued to reflect climatic deterioration immediately prior to or during the Loch Lomond Stadial. This assumption, in combination with a re-evaluation of the geomorphological context of the Earle Mill peat bed near Wooler, prompted Tipping (1998) to assign the major phase of channel entrenchment and valley-floor widening at both Akeld Steads and in the Wooler Water to the late Loch Lomond Stadial and(or) earliest Holocene.

Figure 11 plots the age and elevation (m OD) for Late Devensian and Holocene active channel-bed and flood-basin peat surfaces recorded in the valleys of the Wooler Water (at Earle Mill), Glen (at Akeld Steads) and Till (at Thirlings and in central parts of the Milfield Basin at Humbleton Burn), together with the minimum elevation of proglacial lake levels and the Black Burn palaeosol dated by Payton (1980, 1988). The combination of post-war gravel extraction and landscaping renders it difficult to establish the number and scale of valley-floor incision phases in the Wooler Water between about 15 000 and 10 000 cal BP. However, new radiocarbon dates and geomorphological evidence yielded by the Till–Tweed Project would appear to be inconsistent with the presence of a post-13 300 cal BP proglacial lake in the Milfield Basin on the following grounds. First, the Windermere Interstadial date reported here for organic-rich wetland deposits at approximately 39 m OD in the Galewood palaeochannel is incompatible with a further phase of glaciolacustrine sedimentation to elevations of *c.* 37–39 m OD; the Galewood deposits lie between 90 and 115 cm below the modern surface, are buried by non-lacustrine

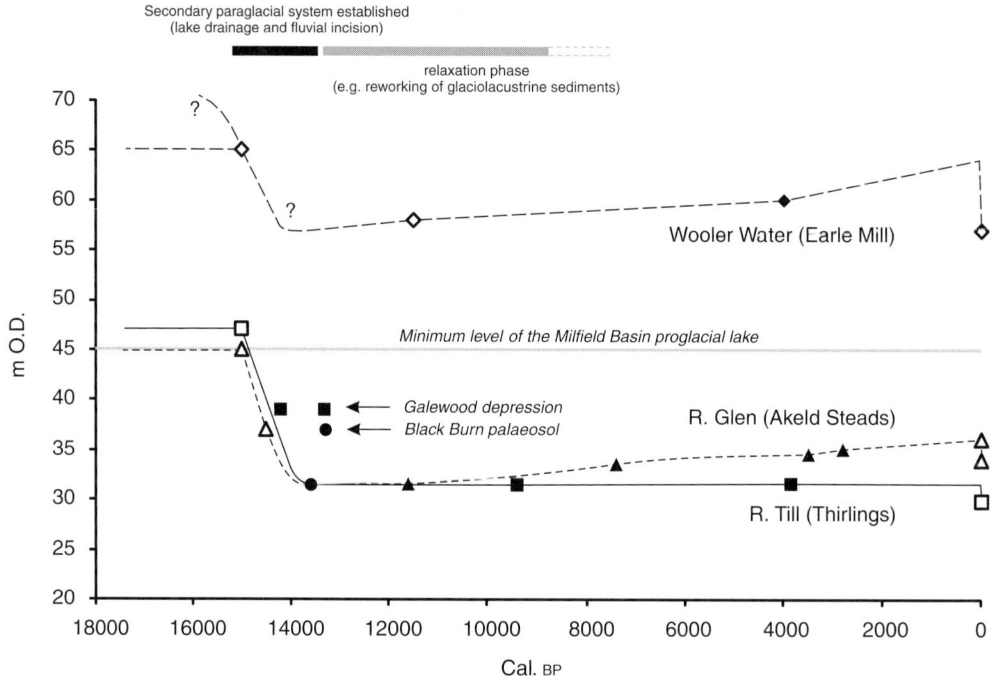

Fig. 11. Age–elevation plots for active channel and floodplain environments in the rivers Wooler Water (at Earle Mill), Glen (at Akeld Steads) and Till (at Thirlings) (filled symbols indicate radiocarbon dated levels, while open symbols are dates obtained by geomorphological inference or historical data; see the text for details). Also shown are elevations of the proglacial lake (estimate of minimum lake surface – see the text for details) and the Black Burn palaeosol (dated by Payton 1980, 1988), and an index of paraglacial activity (see text for details).

sediment and show no evidence of subsequent erosion. Secondly, peaty deposits preserved at 31.5 m OD in the central Milfield Basin have also been provisionally dated to the Windermere Interstadial and these are overlain by 2.5 m of fine-grained alluvium (Fig. 9). An interstadial landsurface at this elevation would imply that a contemporary drainage outlet for the basin below 31.5 m OD had been established well before the Loch Lomond Stadial. Thirdly, evidence for polygonal ice-wedge formation on the surface of low-lying (37 m OD) glaciofluvial fan deposits to the east of Akeld Steads (Fig. 8) is also incompatible with a high (+40 m OD) lake-level stand during the Loch Lomond Stadial. Accordingly, the balance of evidence would suggest a near-complete or total drainage of the lake and fluvial incision to at least 31.5 m OD (in central parts of the basin) to have been accomplished some time within the approximately 2–3 ka between the disappearance of Late Devensian ice and the later part of the Windermere Interstadial. This revised model of lake drainage requires a re-evaluation of the sediments burying the Black Burn palaeosol described by Payton (1988); here it is suggested that these overlying deposits most probably accumulated in a small, localized valley-floor depression during the Loch Lomond Stadial.

Abandonment of the main glaciodeltaic terrace surface in the basin is associated with the development of downvalley sand and gravel terraces, incised palaeochannel belts, and a low-elevation sand and gravel fan at the former mouth of the River Glen, all of which point to an episodic lowering of lake levels and channel-bed elevations rather than a single event. This landform assemblage probably reflects adjustment of basin drainage to erosion of the rock barrier at Etal, but possibly also the reorganization of major drainage routeways through the basin as river channels adjusted to lowering of base levels. The landform assemblage is described here as being of glaciofluvial origin on the grounds that meltwater is likely, at least in part, to have been feeding the drainage system. However, the sequence of events is consistent with the establishment of a secondary paraglacial system (*sensu* Ballantyne 2002) whereby *in situ* glacigenic sediments and paraglacial sediment stores from within and upstream of the basin are remobilized by fluvial processes.

Proglacial lake drainage of the Milfield Basin will have acted to recouple the valley sediment system to lower reaches of the Till downstream of the ice–rock barrier at Etal (Figs 2 and 4) and induce a surge in paraglacial sediment yields. Sediment delivery during this period will have included coarse sediments reworked from glaciodeltaic and glaciofluvial terraces, and, especially during the latter stages of fluvial downcutting, a significant proportion of fine-grained sediment derived from extensive and readily erodible glaciolacustrine deposits. Relatively rapid and large-scale Late Glacial reworking of glaciolacustrine deposits has been evidenced in several recent studies (e.g. Campy *et al.* 1998; Ballantyne 2002), and the low relief of the central part of the Milfield Basin most probably attests to extensive lateral migration of trunk channels and tributary streams in the deglaciation phase prior to the Windermere Interstadial, and possibly through the Loch Lomond Stadial and early Holocene periods.

Holocene valley-floor development and relaxation of the paraglacial system

At the onset of the Holocene period channel-bed elevations of the Glen and Till were incised some 13–15 m below the adjacent glaciodeltaic terrace surface (Figs 8, 9 and 11) and some localities in the basin, most notably at Akeld Steads (Tipping 1998), were established as flood basins. The development of early Holocene flood basins has been recorded in similarly low-relief valley floors elsewhere in the UK (e.g. Parker & Robinson 2003) and may be linked to the development of levées comprising coarse, reworked Lateglacial sediments; this has been interpreted by Lewin *et al.* (2005) as a potential legacy of paraglacial activity. The pattern and chronology of Holocene fluvial valley-floor development will be fully described elsewhere (Passmore & Waddington 2009), but here we note that, until recent historic times, Holocene fluvial activity has exhibited a tendency to valley-floor aggradation in lower reaches of the Glen (*c.* 4.5 m), Wooler Water (*c.* 4–5 m) and Humbleton Burn (2.5 m, Fig. 11). Aggradation is liable to have been promoted by the reduction in valley gradients encountered by streams entering the basin, whereas at Thirlings, in the central part of the basin below the confluence with the Glen, the Till appears to have experienced little net change in channel-bed elevation (Fig. 11). The Holocene fluvial record of the Milfield Basin therefore bears comparison with similarly low-gradient valley settings in the lower reaches of northern British rivers (e.g. Passmore *et al.* 1992) rather than the typically incised fluvial terrace sequence of steeper upland valley floors (e.g. Macklin & Lewin 1989; Tipping 1995; Passmore & Macklin 2000), and this most probably reflects the long-term base-level control exerted by the Etal rock barrier (Fig. 4).

Up to 5 m of Holocene fluvial sediment infills the central part of the present Milfield Basin (Figs 7 and 9) and the unusually long record of Holocene alluviation extends back to the earliest Holocene (Tipping 1998; Passmore & Waddington 2009). Holocene alluvial surfaces typically lie inset up to 10 m below the upstanding Late Glacial sand and gravel terraces, and in some localities palaeochannels appear to be cut into, or lie immediately adjacent to, glaciodeltaic and glaciofluvial terrace bluffs (Fig. 3). In general, however, it is likely that the lateral margins of the Holocene valley floor were largely established during the Late Glacial period of downcutting and lateral reworking, and that these have persisted until present times with only minor trimming and modification. Indeed, the tendency towards early to mid Holocene erosion and elimination of older fluvial units and paraglacial sediment stores observed in Lewin *et al.*'s (2005) overview of UK fluvial histories does not appear to be characteristic of the valley-fill sequence in the Milfield Basin; in particular, extensive deposits of coarse-grained glacigenic and paraglacial sediment have entered a phase of longer-term storage that are substantially decoupled from channel and floodplain processes under climatic regimes that have prevailed during the Holocene.

Fine-grained sediment delivery to Holocene floodplains within and downstream of the basin, by contrast, is likely to have included some proportion of material derived from deposits emplaced during the preceding glacial and deglaciation period, and this is consistent with the concept of an extended paraglacial cycle operating on a subcatchment scale (e.g. Ballantyne 2002). Glaciolacustrine sediments, for example, may be locally observed in eroding banks of the present Glen and Till. However, from mid-Holocene times the impact of human activity on catchment soil and vegetation cover (Tipping 1992, 1998; Passmore & Waddington 2009) will have been exercising an influence on water and sediment yields to the basin while channel and floodplain development will have increasingly been associated with reworking of previously emplaced Holocene alluvium. Accordingly, and especially in terms of coarse-sediment yields, the Milfield Basin is liable to have experienced a relatively short phase of high paraglacial sediment yields that had substantially relaxed by early Holocene times, and possibly as early as the Windermere Interstadial.

For much of the Holocene period, therefore, fluvial systems in the Milfield Basin have developed in the context of a comparatively small and fine-sediment-dominated proportion of paraglacial sediment and, following Lewin *et al.*'s (2005) arguments, may be more appropriately considered as 'post-paraglacial'.

Conclusions

Fluvial systems emerging from the rapid drainage of temporary ice and moraine-dammed proglacial lakes have been widely documented in recent historic times, and may be considered in the context of a paraglacial landscape response to glacier retreat. These scenarios are less well recorded in the landform and sedimentary record of Quaternary deglaciations, but have been shown to provide localized opportunities for the deposition and subsequent incision of thick sedimentary sequences over relatively short timescales. Landform–sediment assemblages in the Milfield Basin provide a comparatively rare example of such an event in northern Britain, and our ongoing geomorphological and palaeoenvironmental investigations in the area are beginning to refine our understanding of the chronology and character of deglaciation, and the paraglacial response of the River Till and its principal tributaries. While further work will be required to augment the chronology explored here, the balance of currently available evidence would support a sequence of paraglacial fluvial response that has been associated with the drainage of a large proglacial lake some time between deglaciation and the relatively mild climate of the Windermere Interstadial. Fluvial downcutting through glaciodeltaic and glaciolacustrine sediments was associated with paraglacial development of inset fluvial terraces and valley widening prior to the establishment of early Holocene channel systems at least 13–15 m below the equivalent glaciodeltaic surface. Lake drainage and channel incision will also have recoupled the fluvial-sediment system to lower reaches of the Till and generated a short-lived phase of relatively high paraglacial sediment yield.

An important control on both lake drainage and subsequent Holocene channel and floodplain development has been exercised by a bedrock barrier at the basin outlet, and the relatively gentle valley gradients in the basin have promoted a tendency towards Holocene floodplain alluviation (in central parts of the basin) with little net change in channel elevation until recent historic times. Extended paraglacial activity will have continued to exert an influence on the fluvial system during the Holocene, but this has most probably diminished considerably with time as sediment supplies to the basin increasingly reflect the impact of anthropogenic catchment disturbance and reworking of Holocene valley-floor deposits.

The authors would like to thank English Heritage for supporting work in the Milfield Basin, and the many landowners and farmers who granted access to land. The review comments of R. Tipping and J. Kemp greatly helped us improve the manuscript.

References

ANTOINE, P., MUNAUT, A.-V., LIMONDIN-LOZOUET, N., PONEL, P., DUPLERON, J. & DUPLERON, M. 2003. Response of the Selle River to climatic modifications during the Lateglacial and Early Holocene (Somme Basin–Northern France). *Quaternary Science Reviews*, **22**, 2061–2076.

ARCHER, D. 1992. *Land of Singing Waters: Rivers and Great Floods of Northumbria*. Spredden Press, Stocksfield.

BALLANTYNE, C. K. 2002. Paraglacial geomorphology. *Quaternary Science Reviews*, **21**, 1935–2017.

BOREK, M. J. E. 1975. *Pollen Analysis and Vegetational History of the Akeld Basin*. MSc thesis, University of Durham.

BOYLAN, P. J. 1981. The role of William Buckland (1784–1856) in the recognition of glaciation in Great Britain. *In*: NEALE, J. & FLENLEY, J. (eds) *The Quaternary in Britain*. Pergamon, Oxford, 1–8.

BRONK RAMSEY, C. 1995. Radiocarbon calibration and analysis of stratigraphy: the OxCal program. *Radiocarbon*, **37**, 425–430.

BRONK RAMSEY, C. 1998. Probability and dating. *Radiocarbon*, **40**, 461–474.

BRONK RAMSEY, C. 2001. Development of the radiocarbon calibration program OxCal. *Radiocarbon*, **43**, 355–363.

CAMPY, M., BUONCRISTIANI, J. F. & BICHET, V. 1998. Sediment yield from glacio-lacustrine calcareous deposits during the postglacial period in the Combe d'Ain (Jura, France). *Earth Surface Processes and Landforms*, **23**, 429–444.

CARRUTHERS, R. G., BURNETT, G. A. & ANDERSON, W. 1932. *The Geology of the Cheviot Hills*. HMSO, London.

CHURCH, M. & RYDER, J. M. 1972. Paraglacial sedimentation: a consideration of fluvial processes conditioned by glaciation. *Geological Society of America Bulletin*, **83**, 3059–3071.

CLAPPERTON, C. M. 1970. The evidence for a Cheviot ice cap. *Transactions of the Institute of British Geographers*, **50**, 115–127.

CLAPPERTON, C. M. 1971*a*. The pattern of deglaciation in north Northumberland. *Transactions of the Institute of British Geographers*, **53**, 67–78.

CLAPPERTON, C. M. 1971*b*. The location and origin of glacial meltwater phenomena in the eastern Cheviot Hills. *Proceedings of the Yorkshire Geological Society*, **38**, 361–380.

CLAPPERTON, C. M., DURNO, S. E. & SQUIRES, R. H. 1971. Evidence for the Flandrian history of the Wooler Water, Northumberland, provided by pollen analysis. *Scottish Geographical Journal*, **87**, 14–20.

COMMON, R. 1954. The geomorphology of the east Cheviot area. *Scottish Geographical Journal*, **70**, 124–138.

DERBYSHIRE, E. 1961. Subglacial col gullies and the deglaciation of the North-East Cheviots. *Transactions and Papers (Institute of British Geographers)*, **29**, 31–46.

DOUGLAS, T. 1991. Glacial deposits of Northumbria. *In*: EHLERS, J., GIBBARD, P. L. & ROSE, J. (eds) *Glacial Deposits in Great Britain and Ireland*. Balkema, Rotterdam, 169–174.

DOUGLAS, T. D. & HARRISON, S. 1985. Periglacial landforms and sediments in the Cheviots. *In*: BOARDMAN, J. (ed.) *Field Guide to the Periglacial Landforms of Northern England*. Quaternary Research Association, Cambridge, 68–76.

DOUGLAS, T. D. & HARRISON, S. 1987. Late Devensian slope deposits in the Cheviot Hills. *In*: BOARDMAN, J. (ed.) *Periglacial Processes and Landforms in Britain and Ireland*. Cambridge University Press, Cambridge, 237–244.

GUNN, W. 1895. *The Geology of Part of Northumberland, Including the Country Between Wooler and Coldstream* (explanation of quarter-sheet 110S.W., new series, sheet 3). HMSO, London.

HARRISON, S. 1996. Paraglacial or periglacial? The sedimentology of slope deposits in upland Northumberland. *In*: ANDERSON, M. G. & BROOKS, S. M. (eds) *Advances in Hillslope Processes*. Wiley, Chichester, 1197–1218.

HARRISON, S. 2002. Lithological variability of Quaternary slope deposits in the Cheviot Hills, UK. *Proceedings of the Geologists' Association*, **113**, 121–138.

JOHNSON, B. & WADDINGTON, C. 2008. Prehistoric and Dark Age settlement remains from Cheviot Quarry, Milfield Basin, Northumberland. *Archaeological Journal*, **165**, 101–258.

JONES, R. L., KEEN, D. H. & ROBINSON, J. E. 2000. Devensian Lateglacial and early Holocene floral and faunal records from NE Northumberland. *Proceedings of the Yorkshire Geological Society*, **53**, 97–110.

KENDALL, P. F. & MUFF, H. B. 1901. Evidence of ancient glacier-dammed lakes in the Cheviots. *Geological Magazine N.S.*, **8**, 513–515.

KENDALL, P. F. & MUFF, H. B. 1903. The evidence for glacier-dammed lakes in the Cheviot Hills. *Transactions of the Edinburgh Geological Society*, **8**, 226–230.

LEWIN, J., MACKLIN, M. G. & JOHNSTONE, E. 2005. Interpreting alluvial archives: sedimentological factors in the British Holocene fluvial record. *Quaternary Science Reviews*, **24**, 1873–1889.

LUNN, A. G. 2004. *Northumberland*. Harper Collins, London.

MACKLIN, M. G. & LEWIN, J. 1989. Sediment transfer and transformation of an alluvial valley floor: the River South Tyne, Northumbria, UK. *Earth Surface Processes and Landforms*, **14**, 233–246.

PARKER, A. G. & ROBINSON, M. A. 2003. Palaeoenvironmental investigations on the middle Thames at Dorney, UK. *In*: HOWARD, A. J., MACKLIN, M. G. & PASSMORE, D. G. (eds) *Alluvial Archaeology in Europe*. Swets & Zeitlinger, Lisse, 43–60.

PASSMORE, D. G. & MACKLIN, M. G. 2000. Late Holocene channel and floodplain development in a wandering gravel-bed river: The River South Tyne at Lambley, northern England. *Earth Surface Processes and Landforms*, **25**, 1237–1256.

PASSMORE, D. G. & WADDINGTON, C. 2009. *Managing an Archaeological Landscape. Till–Tweed Studies, North-East England*, Volume 1. Oxbow, Oxford.

PASSMORE, D. G., MACKLIN, M. G., STEVENSON, A. C., O'BRIEN, C. F. & DAVIES, B. A. S. 1992. A Holocene alluvial sequence in the lower Tyne valley, northern Britain: a record of river response to environmental change. *The Holocene*, **2**, 138–147.

PASSMORE, D. G., VAN DER SCHRIEK, T. & WADDINGTON, C. 2006. Enhancing the evaluation and management of river valley archaeology; geoarchaeology in the Till–Tweed catchment, northern England. *Archaeological Prospection*, **13**, 269–281.

PASSMORE, D. G., WADDINGTON, C. & HOUGHTON, S. J. 2002. Geoarchaeology of the Milfield Basin, northern England; towards an integrated archaeological prospection, research and management framework. *Archaeological Prospection*, **9**, 71–91.

PASTRE, J. -F., LIMONDIN-LOZOUET, N., LEROYER, C., PONEL, P. & FONTUGNE, M. 2003. River system evolution and environmental changes during the Lateglacial in the Paris Basin (France). *Quaternary Science Reviews*, **22**, 2177–2188.

PAYTON, R. 1980. The soils of the Milfield Plain, Northumberland. *North of England Soils Discussion Group Proceedings*, **16**, 1–52.

PAYTON, R. 1988. *The Characteristics and Genesis of Fragipans in British Soils*. PhD thesis, University of Newcastle upon Tyne.

PAYTON, R. 1992. Fragipan formation in argillic brown earths (fragiadalfs) of the Milfield Plain, North-East England. *Journal of Soil Science*, **43**, 621–644.

RAISTRICK, A. 1931. The glaciation of Northumberland and Durham. *Proceedings of the Geologists' Association*, **42**, 281–291.

SEAR, D. A. & ARCHER, D. 1998. Effects of gravel extraction on stability of gravel-bed rivers: the Wooler Water, Northumberland, UK. *In*: KLINGEMAN, P. C., BESCHTA, R. L., KOMAR, P. D. & BRADLEY, J. B. (eds) *Gravel-bed Rivers in the Environment*. Water Resources Publications, LLC, Highlands Ranch, CO, 415–432.

SISSONS, J. B. 1967. *The Evolution of Scotland's Scenery*. Oliver & Boyd, Edinburgh.

STUIVER, M. & REIMER, P. J. 1986. A computer program for radiocarbon age calculation. *Radiocarbon*, **28**, 1022–1030.

STUIVER, M., REIMER, P. J. *ET AL*. 1998. INTCAL98 Radiocarbon age calibration, 24,000–0 cal BP. *Radiocarbon*, **40**, 1041–1083.

TIPPING, R. M. 1992. The determination of cause in the creation of major prehistoric valley fills in the Cheviot Hills, Anglo-Scottish Border. *In*: NEEDHAM, S. & MACKLIN, M. C. (eds) *Alluvial Archaeology in Britain*. Oxbow, Oxford, 111–121.

TIPPING, R. M. 1994*a*. Akeld Steads and the evolution of the Milfield plain. Geomorphology and Late Quaternary Evolution of the Cheviot Hills. *In*: HARRISON, S. & TIPPING, R. M. (eds) *The Geomorphology and Late Quaternary Evolution of the Cheviot Hills*.

British Geomorphological Research Group, London, 80–87.

TIPPING, R. M. 1994*b*. The Harthope valley upstream of Skirl Naked (the Wooler Water at Wooler). *In*: HARRISON, S. & TIPPING, R. M. (eds) *The Geomorphology and Late Quaternary Evolution of the Cheviot Hills*. British Geomorphological Research Group, London, 77–80.

TIPPING, R. M. 1995. Holocene evolution of a lowland Scottish landscape: Kirkpatrick Fleming III. Fluvial history. *The Holocene*, **5**, 184–195.

TIPPING, R. M. 1998. The chronology of late Quaternary fluvial activity in part of the Milfield Basin, northeast England. *Earth Surface Processes and Landforms*, **23**, 845–856.

VANDENBERGHE, J. 2003. Climate forcing of fluvial system development: an evolution of ideas. *Quaternary Science Reviews*, **22**, 2053–2060.

WALKER, M. J. C., COOPE, G. R. & LOWE, J. J. 1993. The Devensian (Weichselian) Lateglacial palaeoenvironmental record from Gransmoor, East Yorkshire, England. *Quaternary Science Reviews*, **12**, 659–680.

The limitations of Quaternary lithostratigraphy: an example from southern Ireland

JASPER KNIGHT

Department of Geography, University of Exeter, Cornwall Campus, Penryn, Cornwall TR10 9EZ, UK (e-mail: j.knight@exeter.ac.uk)

Abstract: The lithostratigraphy of Quaternary extra-glacial (cold-climate) sediments is described from five sites along the south coast of Ireland. These sediments are compared to a regional-scale lithostratigraphic framework. This comparison enables consideration of the applicability of standard lithostratigraphic principles and practices to the Quaternary record. It is concluded that a strict lithostratigraphic approach is not favourable towards interpreting the age or depositional process/environment of Quaternary sediments, either in isolation or as part of a sedimentary succession. The concepts of time transgression and lateral equivalence are useful interpretive tools that can support lithostratigraphic procedures as applied to the Quaternary record.

Lithostratigraphy, as one of the fundamental tools of the discipline of Geology, is central to understanding the relative ordering and age of events, and, by interpretation of sedimentological characteristics, reconstruction of past climates (Salvador 1994; Bowen 1999). Both these components of lithostratigraphy (relative age and climatic interpretation) are particularly important in the understanding of Quaternary sediment successions. This is because well-preserved Quaternary sediments, in basinal and ice-marginal settings (e.g. Eyles & McCabe 1989), have potential to inform on the dynamic behaviour and timing of ice-sheet oscillations, and changes in vegetation, temperature, precipitation and sea-level, and at high resolution. These principles of lithostratigraphy have been used in the British Isles to subdivide and correlate Quaternary sediment successions, and to fit them within the global marine isotope stage (MIS) time frame (Mitchell *et al.* 1973; Bowen 1999). In the western and southern British Isles, located outside ice-sheet limits, the Quaternary record is dominated by extra-glacial (cold-climate) landforms and sediments (Stephens 1970; Mitchell 1977; Ballantyne & Harris 1994; Bates *et al.* 2003; Murton *et al.* 2003). These are significant because cold-climate processes have, during the Quaternary, operated on a larger spatial scale and with a longer temporal persistence than glacial processes themselves (cf. Karte & Liedtke 1981; Ballantyne & Harris 1994). The extra-glacial Quaternary record is, therefore, potentially longer and has fewer time gaps than the record inside ice-sheet margins (e.g. Stephens 1970), which has significance for reconstructing longer-term and regional-scale Quaternary climate (Kirkby 1995; Renssen *et al.* 2000).

Using standard lithostratigraphic procedures in order to identify the relative ordering and age of events in unglaciated areas of the British Isles is not straightforward, for two reasons. First, in these areas loose sediments have been formed, moved downslope under gravity, deposited and remobilized many times during the Quaternary. The extant sedimentary record is, therefore, a palimpsest of cycles of slope activity (DeWolf 1988; Knight 2005*a*). In addition, because sediment supply is strongly controlled by local factors (including rock type, slope, aspect, microclimate, etc.), the stratigraphic elements identified at any one site may be difficult to place in a regional lithostratigraphic framework. Second, there is a general absence of dated or dateable materials in extra-glacial areas with which to support a regional lithostratigraphic framework, or to link it to a MIS time frame.

Other, alternative approaches have also been used in the British Isles in order to evaluate the relative age and stratigraphic position of Quaternary landforms and sediments. The degree of weathering and landscape modification by periglacial processes has been used as a proxy indicator of the relative age and/or severity of periglacial conditions (e.g. Finch & Synge 1966; Williams 1968; Stephens 1970; Hoare & McCabe 1981; Kirkby 1995). This assumption has been used to suggest that southernmost Britain was not glaciated during cold Quaternary MIS but was affected by repeated periglaciation (Mitchell 1977; Warren 1987; Ballantyne & Harris 1994; Bates *et al.* 2003). In southern Ireland it was

From: KNIGHT, J. & HARRISON, S. (eds) *Periglacial and Paraglacial Processes and Environments*.
The Geological Society, London, Special Publications, **320**, 165–180.
DOI: 10.1144/SP320.11 0305-8719/09/$15.00

argued that the South of Ireland End Moraine (SIEM) marks the boundary between young and fresh glacial landscapes to the north, and old and degraded non-glacial landscapes to the south (Charlesworth 1928; Farrington 1947; Synge 1970, 1979). Recent work, however, has shown that: (1) the SIEM is a recessional rather than maximal ice limit (Ballantyne *et al.* 2006; McCabe 2008); (2) cold-climate landforms and structures are present both within and outside of the SIEM (Lewis 1978); and (3) fresh-looking glacial landforms and sediments are common outside of the SIEM limit (Lewis 1978; McCabe 1998; Bowen *et al.* 2002; Ó Cofaigh & Evans 2007).

The foregoing discussion highlights some potential issues in the use of extra-glacial lithostratigraphic patterns in reconstructing the relative ordering of Quaternary events and past climates. A strict lithostratigraphic approach can result in some stratigraphic elements being assigned a certain age based on the age of adjacent elements, rather than on objective field evidence. In addition, a lithostratigraphic framework often has the undesired outcome of establishing an interpretive paradigm for the entire sediment sequence (not just certain stratigraphic elements) that is subsequently difficult to overcome (Warren 1985).

A more useful way forward, and one discussed in this paper, is to: (1) use the lithostratigraphic approach as a working framework against which to compare sediment stratigraphies from different locations in order to understand the regional continuity of, and variability within, individual sediment units; and (2) consider the detailed sedimentology of these units, including their primary sedimentary structures, in order to make inferences of process, environment and climate independent of the interpretation of surrounding units. A key advantage of this approach is that it does not require *a priori* radiometric dating, but uses lithostratigraphy as a relative-age tool to help understand temporal variations in sediment deposition processes and climatic setting.

Aims and structure

This paper uses the two-stage procedure, identified earlier, in order to aid interpretation of the Quaternary sediment record of five sites in Counties Kerry and Cork, southernmost Ireland (Fig. 1). In detail, this paper: (1) describes the regional framework of Quaternary sediments in southern Ireland; (2) compares evidence from five field sites to this regional framework; and (3) considers the extent to which the regional framework can account for local sediment stratigraphies and the detailed sedimentology seen at the sites.

Lithostratigraphic framework of Quaternary sediments in southernmost Ireland

The regional-scale Quaternary lithostratigraphy of southernmost Ireland was constructed by Wright & Muff (1904), who identified seven main stratigraphic elements (Fig. 2). The sediment units that make up six of these elements are generally tabular, flat-lying, have clear and planar bounding contacts, are sedimentologically distinct and are of regional extent. This has favoured their use in a regional stratigraphic context (Farrington 1965), and the Wright & Muff framework has persisted almost unchanged to present (Stephens 1970; Mitchell *et al.* 1973; Warren 1985; McCabe 1999). Subsequently, these stratigraphic elements have been attributed to certain MIS (e.g. Gallagher & Thorp 1997; McCabe 1999). Briefly, these seven stratigraphic elements comprise the following.

(1) A rock platform (marked element 1 on Fig. 2) is present intermittently along the southern and eastern Ireland coast, and lies at 2–5 m above Irish ordnance datum (OD) (Stephens 1957; Farrington 1966; Synge 1981; Hoare 1991; Gallagher & Thorp 1997). Farrington (1966) argued that there are in fact two platforms, with an elevation difference of 1–5 ft (0.3–1.5 m), formed during different sea-level stages. Morphologically, the platform is seaward-dipping, cuts across bedrock bedding planes, and is polished and striated when preserved beneath a glacial sediment cover. Although the elevation of the platform is above present mean sea level, it is lower than many other interglacial beaches (Charlesworth 1963; Devoy 1983), and cannot really be called 'raised' as it falls within the range of present-day storm surges (Bartholdy & Aagaard 2001).

(2) The platform is overlain by 'stratified pebbly raised-beach sand, lying among sub-angular blocks of rock' (Wright & Muff 1904) (element 2 on Fig. 2). The rock blocks have been interpreted as eroded sea stacks, cliff rockfall or periglacial shattered boulders (Farrington 1966). The 'beach' deposit itself, typically 0.5–1.0 m thick, comprises sorted, laterally variable sand and gravel beds, and contain reworked erratics (Hoare 1991). A beach interpretation of these sediments hinges on their relationship to the underlying rock platform, and implies that they are conformable and contemporaneous, but it is also possible that they are of different ages (Gallagher & Thorp 1997).

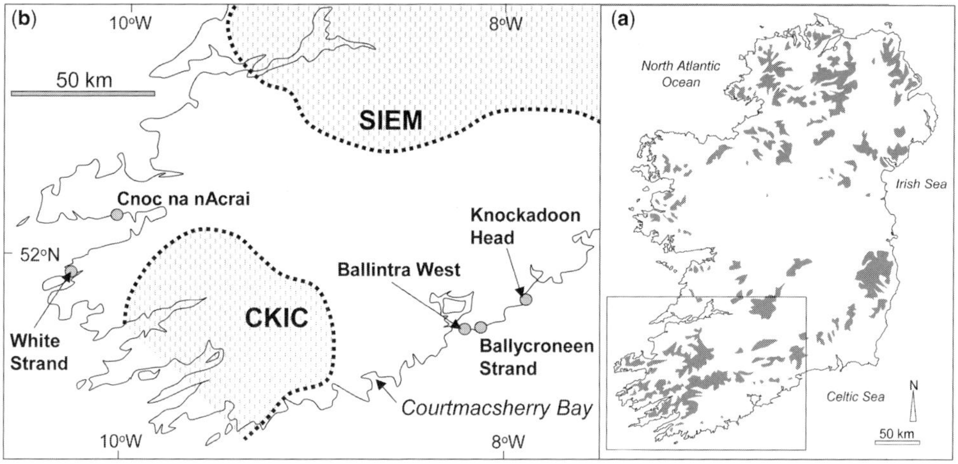

Fig. 1. (**a**) Location of the study area in southernmost Ireland (boxed), with areas over 200 m elevation (shaded). (**b**) Location of the five study sites and general (minimal) Devensian ice limits in southern Ireland, terminating at the South of Ireland End Moraine (SIEM), and the extent of the separate Cork–Kerry Ice Cap (CKIC) (after Stephens *et al.* 1975).

(3) Aeolian sand discontinuously caps the 'beach' deposit (element 3 on Fig. 2). Although common regionally, the sand is rarely more than 10–20 cm thick, and often forms intraformational lenses and stringers within element 4, below. At Fethard (County Wexford), the sand has been dated using the infrared stimulated luminescence method to 128 610 ± 16 795 years BP (Gallagher & Thorp 1997), corresponding to MIS 5–6, but the age error intercepts a date on underlying gravels (element 2) at this site. At other locations, cool–temperate organic deposits occupy this stratigraphic position (Devoy 1983; Heijnis *et al.* 1993). The aeolian sand unit probably formed under a periglacial,

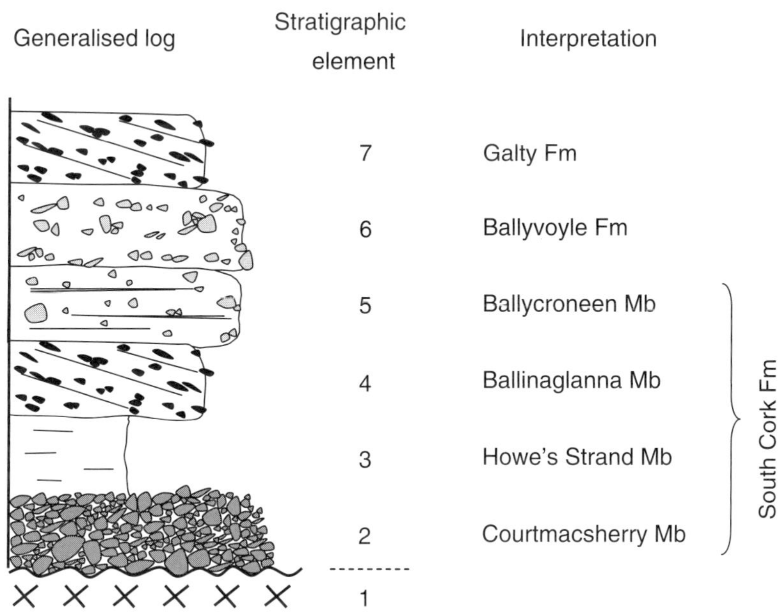

Fig. 2. Quaternary stratigraphic elements in southernmost Ireland, identified after Wright & Muff (1904) with formations and members named after McCabe (1999).

Fig. 4. (**a**) At Ballycroneen Strand the bedrock surface (not seen) is overlain by rounded beach gravels (element 2), sharply overlain in turn (at the position of the trowel) by matrix-supported sandy diamicton (element 4). (**b**) Diamictic breccia at Ballycroneen Strand (element 4), comprising angular local mudstone clasts set in a silt-rich matrix. Note the deformation around these larger clasts of smaller components within the matrix.

thick) comprising angular and flat-lying local clasts (element 4). The diamicton is variably clast- to matrix-supported and contains laterally continuous and undulating sand stringers, present particularly in the lower part of the section, that have a vertical spacing of 20–50 cm. Laterally, decimetre (dm)-scale stratification is present within the diamicton, with massive, matrix-dominated and granular gravels interbedded with clast-supported to openwork angular breccia containing local clasts up to 70 cm in diameter. These clasts are flat-lying, sometimes stacked and imbricated, and are in places clearly separated by a pebble-dominated matrix in which pebble long-axes follow clast margins (Fig. 4b).

The diamicton is sharply overlain by a laterally continuous and flat-lying stratified coarse sand unit (1 m thick; element 4 or 7) containing isolated, angular clasts. The overlying unit (element 5 or 6 or 7) contains massive and clast-supported pebbles that are flat-lying and edge-rounded, and has a deformed upper contact. Overlying this is a massive diamicton (1 m thick; element 6) that contains angular clasts, and massive sand unit (element 7).

Ballintra West

This site (50°47.5′N, 08°11′W) is located 10 km SE of Cobh, County Cork, where Quaternary sediments are exposed in a cliff section. The bedrock upper surface varies in elevation across the site (element 1). Where this bedrock surface is located at height in the cliff face (*c.* 4–6 m above beach level), it is onlapped by a massive gravel unit (element 2). This unit (<1.5 m thick) has a sharp upper boundary, and comprises well-rounded pebbles and boulders of varying lithology that are chaotically organized and occasionally interlocking. This unit is also discontinuous across the exposure and only preserved where bedrock hollows are present. Where the platform surface is located near beach level, it is overlain by stratified medium–coarse sand (element 3) that is arranged in dm-scale normally graded planar beds. Occasionally, these beds terminate in erosional reactivation surfaces that show thin (mm-scale) clay drapes (Fig. 5). The relationship between the massive gravel and stratified sand units is uncertain.

The central part of the sediment sequence is similar across the site (Fig. 6), and comprises (at base) angular and flat-lying boulders with a minor coarse sand and granule matrix. Clast size and frequency decrease upwards. The unit (1.2 m thick; element 2) is laterally continuous and overlain by planar-bedded sands that contain dispersed angular clasts (element 5 or 6). The sand unit (4 m high) comprises laterally continuous dm-scale beds of poorly sorted medium–coarse sand. Clasts (<15 cm diameter) are flat-lying and increase in frequency upwards. In places this unit is overlain by massive–laminated, well-sorted fine sand (1.5 m thick). The upper boundary to element 6 or 7 is sharply erosional; the lowermost boundary is not observed. The sediment sequence is capped throughout by a clast-dominated diamicton (1.5–3 m thick; element 7) containing subangular–subrounded cobbles and boulders (<35 cm diameter) that are sometimes imbricated and which decrease in size

Fig. 5. Planar-bedded sands at Ballintra West (element 3) terminate in erosional reactivation surfaces that are marked by coarser granules (layer above the trowel handle) or clay drapes (layer at the level of the trowel handle).

Fig. 6. View of the sediment stratigraphy at Ballintra West, with gravel-dominated basal sediments (element 2; where the trowel is located) grading up to planar stratified diamicton with sand interbeds (element 5 or 6). The sand component becomes more important upwards. Clast-dominant diamicton (element 7) caps the exposure.

and frequency upwards. Discontinuous sand stringers and some planar bedding is also present at the top of the unit.

Cnoc na nAcrai

This site (52°08′N, 10°05.5′W) is located in a disused quarry 12 km east of the town of Dingle (County Kerry), about 600 m inland from a steep boulder and cliff coastline, and at an elevation of about 70 m OD. In this area, hillslopes are steep and rectilinear with a strong structural alignment imparted by the underlying Devonian sandstone bedrock. Rockhead is generally smooth and sharp to diffuse; in some places it grades upwards into fractured bedrock fragments that have been displaced in a downslope direction, even on low-angle slopes (Fig. 7a). Elsewhere, the rock-head surface is intact, sharply defined and shows conjugate fractures. These fractures are infilled from above with wind-blown sand (Fig. 7b). The sand unit (element 3) is 0.4–1.0 m thick and

Fig. 7. (**a**) View of deformed bedrock at Cnoc na nAcrai (overturned to the south) and the sharp bedrock surface that is overlain by aeolian sand (element 3), pinching out to the right, and overlain by angular local boulders (elements 4 or 7). Maximum sediment thickness for scale is 1.5 m here. (**b**) View of the fractured bedrock surface at Cnoc na nAcrai and overlying aeolian sands (element 3). Note the sand infills within the fractures.

comprises vaguely planar-bedded fine sand with dispersed granules. The upper boundary of the unit is generally diffuse and marked by angular local boulders. These boulders (30–60 cm diameter) are present over a vertical height of 1–4 m (element 4 or 7), are variably clast- to matrix-supported by the subjacent sand, and show no internal organization. Elsewhere, the boulder unit overlies rockhead and thickens towards the slope foot. There is no marked change in sand sedimentology with height

Fig. 8. (**a**) View of the smoothed bedrock surface (element 1) and overlying breccia (element 4) at White Strand. (**b**) Cryoturbation structures at White Strand that are picked out by the intermixing of clast-supported gravels from above with matrix-dominant diamicton from below.

through elements 3 and 4 or 7. The fracture-fills at the base of element 3 are linear and pinch out to 1.2–1.5 m depth from the bedrock surface (Fig. 7b). No internal structures are observed within the fracture-fills.

White Strand

This site (51°56′N, 10°17′W), also called An Trá Bhán (Warren 1987), backs a west-facing sandy beach located within Doulus Bay (Valentia

Harbour), 3 km west of Cahersiveen, County Kerry. Here, the bedrock surface (element 1) is variable in morphology: low-relief bedrock surfaces are sharply overlain by a cemented breccia of angular–subangular local clasts (2–70 cm diameter) that show marked lateral size variations (element 4). The unit (< 1.5 m thick) is variously clast- to matrix-supported and does not show any clear internal organization (Fig. 8a); laterally, the unit becomes finer grained, and shows planar bedding (20 cm thick) with planar lenses of sorted sand. Areas of higher bedrock relief show fractured and displaced bedrock blocks that decrease in size upwards. In areas with strong slatey cleavage, fractured and angular blocks are sharply overturned and cemented.

This breccia unit is overlain with a variably diffuse–sharp deformed boundary by a sedimentary unit in which larger and more lithologically diverse subrounded boulders (consistently 20–40 cm diameter) are present (element 4 or 6). This boulder unit (1.0–1.6 m thick) is matrix-supported and fines upwards. Deep cryoturbation structures are developed in this unit (Warren 1987), identified by variability in long-axis alignment of the boulders and by the presence of matrix-dominant patches within the unit. In the uppermost part of the unit boulders have a vertical long-axis alignment, are wedge to pillar shape in section, and define the lateral margins of cryoturbation structures that are 1.8 m wide and extend 1.2 m depth from the top of the feature (Fig. 8b). Within the vertical pillars are located circular patches of matrix-supported diamicton that forms 'cores' defined by the vertically aligned boulders. In places, high-amplitude U-shaped structures (0.3 m wide, 1 m high) are developed within the matrix-supported diamicton by boulder interfingering from above.

Interpretation of field evidence

Regional-scale lithostratigraphies are, by definition, based on data from individual sites. Generalizing from the local to regional scale, therefore, necessarily downplays the lithological variability found within and between individual sites. A summary of the main lithostratigraphic elements observed at the five sites, and attribution of these to Wright & Muff's (1904) scheme, is shown in Figure 9a. Not all stratigraphic elements are observed at all sites, which has implications for the regional applicability of any one lithostratigraphic scheme, and for the interpretation of missing elements. Some issues involved in stratigraphic relationships between the sediments observed at the five field sites, and implications for the relative ordering and age of events and reconstruction of past climates, are now discussed with reference to Wright & Muff's (1904) scheme (Fig. 2).

The bedrock platform (element 1) is regionally extensive and as such has been used extensively as a stratigraphic marker (e.g. Bowen 1973; Synge 1981). Interpretation as an interglacial platform (Bowen 1973) is based largely on its elevation, discussed earlier, and on amino acid racemization of limpet (*Patella vulgata*) shells from within stratigraphically equivalent gravels from sites in SW England and South Wales (Bowen *et al.* 1985). McCabe & Ó Cofaigh (1996), however, argued that subglacial erosion of the Ballycroneen Strand platform took place during a glacial MIS. The platform could therefore be both polygenetic and diachronous, which makes its age and climatic interpretation difficult. Hollows on the bedrock platform surface have also acted as local sediment traps, helping preserve transgressive and downlapping gravels (element 2) at Knockadoon Head and Ballintra West.

Although the rounded and clast-supported nature of the gravel unit (element 2) has been used to support interpretation as an interglacial raised beach (Wright & Muff 1904), the intact but displaced bedrock raft within the gravels at Knockadoon Head suggests rockfall from cliffs behind. This process of rockfall and subjacent gravel deformation is common on periglacial beaches that are backed by steep cliffs (e.g. Knight 2005*a*). Other sediments suggest more complex coastal environments. Sand and clay drapes at Ballintra West were probably formed in a backbarrier estuarine or lagoonal setting; the gravels at Ballycroneen Strand are probably reworked glacial outwash.

Element 3 (aeolian sand) is most clearly present at Cnoc na nAcrai, where it overlies fractured bedrock. Aeolian sand has been recorded in an uppermost stratigraphic position at many other extra-glacial sites in southern Britain (e.g. Scourse 1996; Clarke *et al.* 2007). An aeolian origin is supported by the laminated and well-sorted sand, and sediment draping over larger, angular fragments that are probably derived by rockfall. There are a number of reasons why aeolian sand is absent at other described sites. Sediment supply is a likely issue. It is notable that Cnoc na nAcrai is the only site underlain by sandstone bedrock. Any blown sand deposited at the other sites was likely to have been incorporated within, and thus diluted by, sediments deposited by other processes. The absence of sand elsewhere indirectly supports the contention that glacial ice was present across southernmost Ireland, pushing wind-blown sand activity to outside of this ice margin. McCabe (1999) interpreted sand found at this stratigraphic level on the south coast of Ireland as marine-influenced, based on the presence of

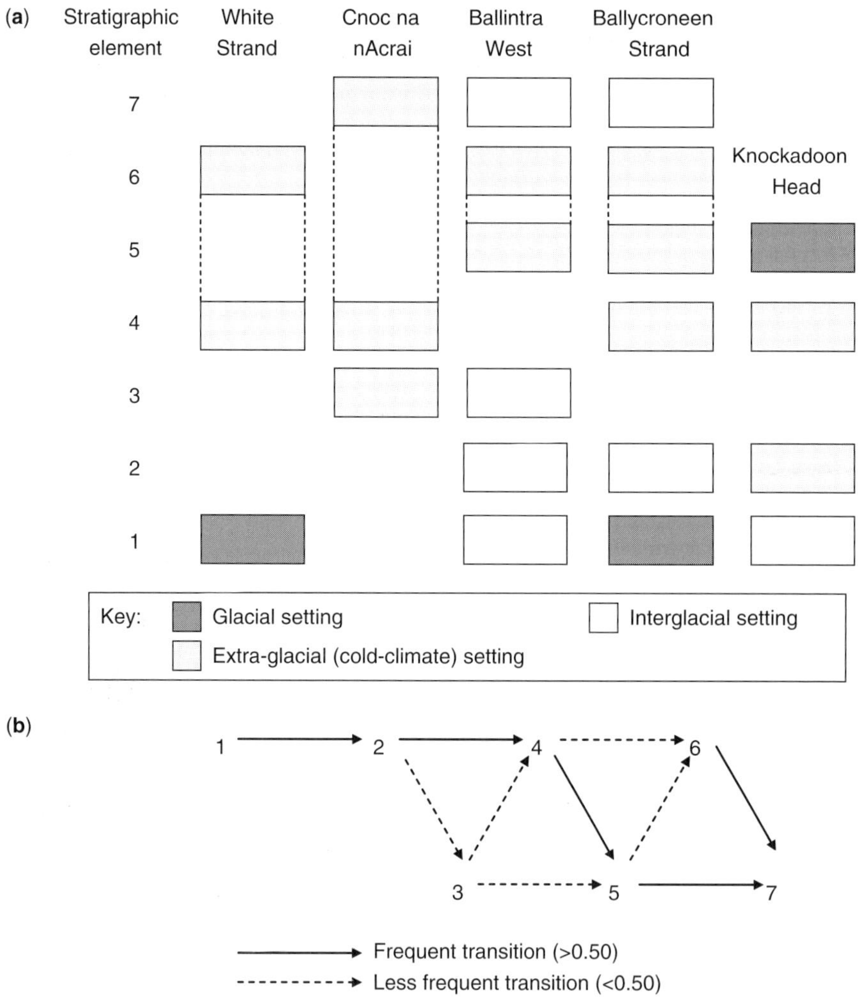

Fig. 9. (**a**) Interpretation of the climatic setting associated with the formation of stratigraphic elements at the described sites (shown from west to east). The shading of boxes is shown in the key. No box is shown where stratigraphic elements are absent at a site. Where sediment units could be assigned to several stratigraphic elements, the boxes are linked by dotted lines. (**b**) Illustration of the frequency of transitions (arrows) between the stratigraphic elements identified in Figure 2 (after Miall 1981; Dardis 1987). Data are shown in Table 1.

hummocky and swaley cross-bedding (McCabe & Ó Cofaigh 1996).

The diamictons and diamictic breccias (element 4) overlying the basal gravels (element 2) at most sites is best interpreted as loose materials of different origins that have been reworked downslope subsequent to primary deposition. In terms of their texture, clast composition and sedimentary structures, these are likely to have been glacial tills intermixed with frost-shattered local debris. The presence of these materials of different origins suggests sediment reworking by ice-margin oscillation in a proglacial environment associated with deposition of overlying tills (elements 5, 6). The bedded sediments at Ballycroneen Strand provide evidence for repeated episodes of sediment delivery, probably related to water availability. At Knockadoon Head and White Strand, sandy stringers and sand–silt couplets are also likely to have been deposited in a proglacial, waterlain setting.

The uppermost elements at all sites (elements 5–7) are most stratigraphically and lithologically variable, and thus probably represent more local-scale controls through sediment availability and accommodation space, rather than regional scale controls through climate forcing. This is particularly the case with these uppermost elements, which are stratigraphically distanced from those elements (1, 2) that are probably tied to sea-level position. Sea level therefore decreases in importance upwards as a control on regional sediment stratigraphic patterns. The uppermost elements (5–7) at Knockadoon Head, Ballycroneen Strand and Ballintra West may have a number of alternative origins, including subglacial or glacimarine tills, or proximal diamictons interbedded with outwash. Gravels associated with these units at Ballycroneen Strand and White Strand are rounded and lithologically diverse, and are best interpreted as proglacial outwash. These diverse origins and depositional settings can be reconciled when viewed as part of a continuum from ice-interior (subglacial) to ice-marginal (proglacial and glacimarine) and extra-glacial (periglacial), associated with advance and retreat of ice margins. The sedimentary units therefore can be considered as stratigraphic equivalents, irrespective of differences in depositional processes and setting.

The periglacial breccias and cryoturbation structures observed at White Strand are indicative of subaerial frost shattering of local bedrock outcrops, and sediment mixing within the active layer. These periglacial processes are linked by upslope sediment supply to downslope sediment transport and deposition on stable footslopes. The presence of clast-supported gravels around the margins of, and matrix-supported diamicton within, the cryoturbation structures also suggests density-driven soft-sediment deformation through the depth of the active layer. If so, seasonal ground saturation (and destabilization) associated with rapid summer thawing is a likely process (cf. McCarroll & Rijsdijk 2003). The capping sand unit at Ballycroneen Strand and Cnoc na nAcrai is considered to be the lateral equivalent of the breccia unit, and deposited by enhanced wind activity in an exposed extra-glacial (periglacial) environment (e.g. Clarke *et al.* 2007; Kolstrup 2007).

The limitations of Quaternary lithostratigraphy

Interpretation of the sediment units found at the five sites, described in the previous section, highlights uncertainty in the likely depositional processes and environmental setting of individual units, and shows that adjacent sediment units might have very different processes and settings. This poses challenges for both the correct genetic origin and interpretation of individual units, and for linking lithostratigraphic sequences from individual sites into a regional framework that is able to explain the spatial and temporal evolution of palaeoenvironments and sedimentary systems (Walker 2001). The key role of sedimentary evidence, therefore, in reconstructing Quaternary climates and environments is somewhat different to the requirement of standard lithostratigraphic procedure, which is focused on correlation of individual stratigraphic elements and sequences without regard to origin or interpretation (Salvador 1994). Bowen (1999) clearly identifies the tension between these two potential imperatives of Quaternary sediment successions, and argues that reliance on glacial–interglacial cycles as an interpretative framework for Quaternary sediments is very problematic.

This is shown very clearly by considering the age and palaeoclimatic significance of the bedrock platform (element 1), which is important because it has been used as a key chronostratigraphic marker across southernmost Ireland. Many previous workers (e.g. Synge 1981; Devoy 1983; Warren 1985; Hoare 1991; McCabe 1999) have adopted a strict lithostratigraphic approach by working upsequence from the platform and overlying beach (elements 1, 2), which are of (presumed) last-interglacial age (MIS 5e). Evidence from the five described sites, however, shows clearly that this assumption of platform age may be incorrect, and likewise the age-assignation of overlying

Table 1. *Transition count matrix (transition probability) shown for stratigraphic elements at sites discussed by Wright & Muff (1904)* (n = *21). The elements are coded according to Figure 2*

		Uppermost element						
		2	3	4	5	6	7	Total
Lowermost element	1	16 (0.80)	1 (0.05)	2 (0.10)	1 (0.05)	0	0	20
	2		3 (0.19)	9 (0.56)	2 (0.12)	0	2 (0.12)	16
	3			2 (0.50)	2 (0.50)	0	0	4
	4				9 (0.75)	2 (0.17)	1 (0.08)	12
	5					2 (0.33)	4 (0.66)	6
	6						1 (1.00)	1
Total		16	4	13	14	4	8	59

sediments. This evidence also shows that, upsequence, local-scale factors of sediment supply, accommodation space and slope aspect become more important, and that simple climate-forcing is not always appropriate. This is seen clearly in Figure 9a, which outlines the relationship between stratigraphic elements at the five sites and likely climatic conditions during sediment deposition. Sediments formed in an extra-glacial setting are most common, and reflect the interplay between climatic variables (e.g. temperature, precipitation), sediment supply and slope reworking.

The validity of this stratigraphic approach (Fig. 9a) can also be assessed by considering relationships between Wright & Muff's (1904) seven stratigraphic elements using embedded Markov chain analysis (Krumbein & Dacey 1969). This procedure involves counting up the number of transitions between different stratigraphic elements in order to assess whether some transitions are more commonly recorded than others (Miall 1981). This method has been used previously in the Irish Quaternary to examine stratigraphic relationships within drumlin sediments (Dardis 1987) and glacial lake deposits (Richards 2002). Results from 21 stratigraphic patterns at the 28 sites described by Wright & Muff (1904) are shown in Table 1. There are strong relationships between elements 1 and 2, 2 and 4, and 4 and 5, but elements 3 and 6 are poorly represented (Fig. 9b). This clearly shows the dominance of the bedrock platform as a morphological feature and stratigraphic marker; and that the most commonly occurring stratigraphic transitions take place lowermost in the profile. Further, extra-glacial sediments (elements 4, 7) are common irrespective of their position within the sediment sequence. This suggests regional-scale depositional models (e.g. Eyles & McCabe 1989; Ó Cofaigh & Evans 2001) overemphasize the role of glacial till (elements 5, 6) as chronostratigraphic units along this coast.

A more nuanced understanding of the principles of lithostratigraphy, as applied to Quaternary sediments, needs to consider the role of time transgression (e.g. ice-margin advance and retreat) and lateral equivalence (e.g. spatial relationships between glacial moraines and outwash, or energy gradient from coarse to fine sediments). These factors are important because they are related to processes of sediment deposition that are instrumental in forming the lithostratigraphic record. They are also important in an interpretive sense, because they inform understanding of how the lithostratigraphic elements found at individual sites can be correlated. In this context, wind-blown sediments, such as the sand at Cnoc na nAcrai, might be particularly useful as a stratigraphic and palaeoenvironmental marker.

References

BALLANTYNE, C. K. & HARRIS, C. 1994. *The Periglaciation of Great Britain*. Cambridge University Press, Cambridge.

BALLANTYNE, C. K., MCCARROLL, D. & STONE, J. O. 2006. Vertical dimensions and age of the Wicklow Mountains ice dome, Eastern Ireland, and implications for the extent of the last Irish Ice Sheet. *Quaternary Science Reviews*, **25**, 2048–2058.

BARTHOLDY, J. & AAGAARD, T. 2001. Storm surge effects on a back barrier tidal flat of the Danish Wadden Sea. *Geo-Marine Letters*, **20**, 133–141.

BATES, M. R., KEEN, D. H. & LAUTRIDOU, J.-P. 2003. Pleistocene marine and periglacial deposits of the English Channel. *Journal of Quaternary Science*, **18**, 319–337.

BOWEN, D. Q. 1973. The Pleistocene succession of the Irish Sea. *Proceedings of the Geologists' Association*, **84**, 249–272.

BOWEN, D. Q. (ed.) 1999. *A Revised Correlation of Quaternary Deposits in the British Isles*. Geological Society, London, Special Report, **23**.

BOWEN, D. Q., PHILLIPS, F. M., MCCABE, A. M., KNUTZ, P. C. & SYKES, G. A. 2002. New data for the Last Glacial Maximum in Great Britain and Ireland. *Quaternary Science Reviews*, **21**, 89–101.

BOWEN, D. Q., SYKES, G. A. *ET AL*. 1985. Amino acid geochronology of raised beaches in south west Britain. *Quaternary Science Reviews*, **4**, 279–318.

CHARLESWORTH, J. K. 1928. The glacial retreat from central and southern Ireland. *Quarterly Journal of the Geological Society, London*, **84**, 293–344.

CHARLESWORTH, J. K. 1963. Some observations on the Irish Pleistocene. *Proceedings of the Royal Irish Academy*, **62B**, 295–322.

CLARKE, M. L., MILODOWSKI, A. E., BOUCH, J. E., LENG, M. J. & NORTHMORE, K. J. 2007. New OSL dating of UK loess: indications of two phases of Late Glacial dust accretion in SE England and climate interpretation. *Journal of Quaternary Science*, **22**, 361–371.

DARDIS, G. F. 1987. Sedimentology of late-Pleistocene drumlins in south-central Ulster, Northern Ireland. *In*: MENZIES, J. & ROSE, J. (eds) *Drumlin Symposium*. Balkema, Rotterdam, 215–224.

DEVOY, R. J. 1983. Late Quaternary shorelines in Ireland: an assessment of their implications of isostatic land movement and relative sea-level changes. *In*: SMITH, D. E. & DAWSON, A. G. (eds) *Shorelines and Isostasy*. Institute of British Geographers, Special Publication, **16**, 227–254.

DEWOLF, Y. 1988. Stratified slope deposits. *In*: CLARK, M. J. (ed.) *Advances in Periglacial Geomorphology*. Wiley, Chichester, 91–110.

EYLES, N. & MCCABE, A. M. 1989. The Late Devensian (<22,000 BP) Irish Sea Basin: the sedimentary record of a collapsed ice-sheet margin. *Quaternary Science Reviews*, **8**, 307–351.

FARRINGTON, A. 1947. Unglaciated areas in southern Ireland. *Irish Geography*, **1**, 89–97.

FARRINGTON, A. 1965. A note on the correlation of some of the glacial drifts of the south of Ireland *Irish Naturalists' Journal*, **15**, 29–33.

FARRINGTON, A. 1966. The early-glacial raised beach in County Cork. *Scientific Proceedings of the Royal Dublin Society, Series A*, **2**, 197–219.

FINCH, T. & SYNGE, F. M. 1966. The drifts and soils of west Clare and the adjoining parts of Counties Kerry and Limerick. *Irish Geography*, **5**, 161–172.

GALLAGHER, C. & THORP, M. 1997. The age of the Pleistocene raised beach near Fethard, County Wexford, using infra red stimulated luminescence (IRSL). *Irish Geography*, **30**, 68–89.

HARRIS, C. 1998. The micromorphology of paraglacial and periglacial slope deposits: A study from case Morfa Bychan, west Wales, UK. *Journal of Quaternary Science*, **13**, 73–84.

HEIJNIS, H., RUDDOCK, J. & COXON, P. 1993. A uranium–thorium dated Late Eemian or Early Midlandian organic deposit from near Kilfenora between Spa and Fenit, Co. Kerry, Ireland. *Journal of Quaternary Science*, **8**, 31–43.

HIEMSTRA, J., EVANS, D. J. A., SCOURSE, J. D., MCCARROLL, D., FURZE, M. F. A. & RHODES, E. 2006. New evidence for a grounded Irish Sea glaciation on the Isles of Scilly, UK. *Quaternary Science Reviews*, **25**, 299–309.

HOARE, P. G. 1991. Pre-Midlandian glacial deposits in Ireland. *In*: EHLERS, J., GIBBARD, P. L. & ROSE, J. (eds) *Glacial Deposits in Great Britain and Ireland*. Balkema, Rotterdam, 37–45.

HOARE, P. G. & MCCABE, A. M. 1981. The periglacial record in east-central Ireland. *Biuletyn Peryglacjalny*, **28**, 57–78.

KARTE, J. & LIEDTKE, H. 1981. The theoretical and practical definition of the term 'periglacial' in its geographical and geological meaning. *Biuletyn Peryglacjalny*, **28**, 123–135.

KASSE, C. 2002. Sandy aeolian deposits and environments and their relation to climate during the Last Glacial Maximum and Lateglacial in northwest and central Europe. *Progress in Physical Geography*, **26**, 507–532.

KIRKBY, M. J. 1995. A model for variations in gelifluction rates with temperature and topography: implications for global change. *Geografiska Annaler*, **77A**, 269–278.

KOLSTRUP, E. 2007. Lateglacial older and younger coversand in northwest Europe: chronology and relation to climate and vegetation. *Boreas*, **36**, 65–75.

KNIGHT, J. 2005*a*. Regional climatic versus local controls on periglacial slope deposition: a case study from west Cornwall. *Geoscience in South-west England*, **11**, 151–157.

KNIGHT, J. 2005*b*. The Irish Sea Basin. *In*: LEWIS, C. A. & RICHARDS, A. E. (eds) *The Glaciations of Wales and Adjacent Areas*. Logaston Press, Almeley, Herefordshire, 177–188, 212–216.

KRUMBEIN, W. C. & DACEY, M. F. 1969. Markov chains and embedded chains in geology. *Journal of the International Association of Mathematical Geology*, **1**, 79–96.

LEWIS, C. A. 1978. Periglacial feature in Ireland: an assessment 1978. *Journal of Earth Science, Royal Society of Dublin*, **1**, 135–142.

MCCABE, A. M. 1998. Striae at St. Mullin's Cave, County Kilkenny, southern Ireland: their origin and chronological significance. *Geomorphology*, **23**, 91–96.

MCCABE, A. M. 1999. Ireland. *In*: BOWEN, D. Q. (ed.) *A Revised Correlation of Quaternary Deposits in the British Isles*. Geological Society, London, Special Report, **23**, 115–124.

MCCABE, A. M. 2008. *Glacial Geology and Geomorphology. The Landscapes of Ireland*. Dunedin Press, Edinburgh.

MCCABE, A. M. & Ó COFAIGH, C. 1996. Upper Pleistocene facies sequences and relative sea-level trends along the south coast of Ireland. *Journal of Sedimentary Research*, **66**, 376–390.

MCCARROLL, D. & RIJSDIJK, K. F. 2003. Deformation styles as a key for interpreting glacial depositional environments. *Journal of Quaternary Science*, **18**, 473–489.

MIALL, A. D. 1981 *Analysis of Fluvial Depositional Systems*. AAPG Educational Course Series, **20**.

MITCHELL, G. F. 1977. Periglacial Ireland. *Philosophical Transactions of the Royal Society of London, Series B*, **280**, 199–209.

MITCHELL, G. F., COLHOUN, E. A., STEPHENS, N. & SYNGE, F. M. 1973. Ireland. *In*: MITCHELL, G. F., PENNY, L. F., SHOTTON, F. W. & WEST, R. G. (eds) *A Correlation of Quaternary Deposits in the British Isles*. Geological Society, London, Special Publication, **4**, 67–80.

MURTON, J. B., BATEMAN, M. D., BAKER, C. A., KNOX, R. & WHITEMAN, C. A. 2003. The Devensian periglacial record on Thanet, Kent, UK. *Permafrost and Periglacial Processes*, **14**, 217–246.

Ó COFAIGH, C. & EVANS, D. J. A. 2001. Sedimentary evidence for deforming bed conditions associated with a grounded Irish Sea glacier, southern Ireland. *Journal of Quaternary Science*, **16**, 435–454.

Ó COFAIGH, C. & EVANS, D. J. A. 2007. Radiocarbon constraints on the age of the maximum advance of the British–Irish Ice Sheet in the Celtic Sea. *Quaternary Science Reviews*, **26**, 1197–1203.

RENSSEN, H., ISARIN, R. F. B., VANDENBERGHE, J., LAUTENSCHLAGER, M. & SCLESE, U. 2000. Permafrost as a critical factor in paleoclimate modelling: the Younger Dryas case in Europe. *Earth and Planetary Science Letters*, **176**, 1–5.

RICHARDS, A. E. 2002. Self-organization, fractal scaling and cyclicity in Late Midlandian glacio-deltaic sediments associated with Glacial Lake Blessington, Co. Wicklow. *Sedimentary Geology*, **149**, 127–143.

SALVADOR, A. (ed.) 1994. *International Stratigraphic Guide. A Guide to Stratigraphic Classification, Terminology, and Procedure*. IUGS/GSA, Boulder, Colorado.

SCOURSE, J. D. 1996. Late Pleistocene stratigraphy and palaeobotany of north and west Cornwall. *Transactions of the Royal Geological Society of Cornwall*, **22**, 2–56.

STEPHENS, N. 1957. Some observations on the 'interglacial' platform and the early post-glacial raised beach on the east coast of Ireland. *Proceedings of the Royal Irish Academy*, **58B**, 129–149.

STEPHENS, N. 1970. The west country and southern Ireland. *In*: LEWIS, C. A. (ed.) *The Glaciations of*

Wales and Adjoining Regions. Longman, London, 267–314.

STEPHENS, N., CREIGHTON, J. R. & HANNON, M. A. 1975. The late-Pleistocene period in north-eastern Ireland: an assessment 1975. *Irish Geography*, **8**, 1–23.

SYNGE, F. M. 1970. The Irish Quaternary: current views 1969. *In*: STEPHENS, N. & GLASSCOCK, R. E. (eds) *Irish Geographical Studies in Honour of E. Estyn Evans*. Queen's University, Belfast, 34–48.

SYNGE, F. M. 1979. Quaternary glaciations in Ireland. *Quaternary Newsletter*, **28**, 1–18.

SYNGE, F. M. 1981. Quaternary glaciation and changes of sea level in the south of Ireland. *Geologie en Mijnbouw*, **60**, 305–315.

WALKER, M. J. C. 2001. Rapid climate change during the last glacial–interglacial transition; implications for stratigraphic subdivision, correlation and dating. *Global and Planetary Change*, **30**, 59–72.

WARREN, W. P. 1985. Stratigraphy. *In*: EDWARDS, K. J. & WARREN, W. P. (eds) *The Quaternary History of Ireland*. Academic Press, London, 39–65.

WARREN, W. P. 1987. Periglacial periods in Ireland. *In*: BOARDMAN, J. (ed.) *Periglacial Processes and Landforms in Britain and Ireland*. Cambridge University Press, Cambridge, 101–111.

WILLIAMS, R. B. G. 1968. Some estimates of periglacial erosion in southern and eastern England. *Biuletyn Peryglacjalny*, **17**, 311–335.

WRIGHT, W. B. & MUFF, H. B. 1904. The pre-glacial raised beach of the south coast of Ireland. *Scientific Proceedings of the Royal Dublin Society*, **10**, 250–307.

Geotechnical controls on a steep lateral moraine undergoing paraglacial slope adjustment

ALASTAIR M. CURRY[1]*, TIM B. SANDS[2] & PHILIP R. PORTER[2]

[1]*Department of Geography and Land Management, Royal University of Phnom Penh, Confederation de la Russie Boulevard, Phnom Penh, Cambodia*

[2]*Division of Geography and Environmental Sciences, University of Hertfordshire, Hatfield AL10 9AB, UK*

**Corresponding author (email: currys@omf.net)*

Abstract: Sustained post 'Little Ice Age' retreat of the northern lobe of the Feegletscher, Valais, Switzerland, has exposed lateral moraines that show pronounced oversteepening on the upper proximal slopes, with upper-slope segments displaying angles of up to about 70°. Paraglacial processes have eroded gullies into the upper-slope segments, and associated debris-flow deposits result in lower angles of between 34° and 25° in the mid-slope and slope-foot zones, respectively. In order to assess the geotechnical properties of morainic sediments that permit development of quasi-stable, oversteepened slope segments, a standard suite of geotechnical measures was applied to samples of Feegletscher moraine sediments. Shear box testing yielded angles of friction ranging from 35° for loose samples to 52° for dense samples. Although the heterogeneous nature of moraine deposits makes laboratory testing of the whole size range of *in situ* sediments impractical, shear box test results imply that *in situ* upper-slope angles exceed the angles of friction of moraine sediments by 26°–40°. We are unable to replicate angles of friction in shear box tests that correspond to *in situ* angles of the upper-slope sections measured in the field. However, we suggest that distal dipping mica-schist clasts may play an important role in permitting high-angle slope stability. Quasi-stable storage of glacigenic sediments in high-angle moraine sequences over decadal timescales has implications for understanding the period following deglaciation over which paraglacial reworking and redistribution of sediments may operate.

Glacier retreat triggers the progressive modification of glacial sediments, landforms, landscapes and land systems through their exposure to non-glacial Earth-surface processes and conditions. In recently deglaciated mountain areas a wide range of these *paraglacial* processes is responsible for the release, reworking and redeposition of large quantities of unstable glacigenic sediment over a wide range of timescales (Ballantyne 2002*a*). Steep, sediment-mantled valley slopes, for example, are eroded by rapid mass-movement processes, often creating a land system of intersecting gullies, coalescing debris cones and valley-floor deposits of reworked sediment in a matter of decades (e.g. Ballantyne & Benn 1994; Ballantyne 1995; Curry 1999; Curry *et al.* 2005). These phenomena also pose a risk to inhabitants and infrastructure in deglaciated valleys, some of which are frequented by large numbers of visitors. While recent studies have focused attention on the processes, rates, and sedimentological and morphological consequences of this activity, few have specifically considered the geotechnical characteristics of sediments that may be susceptible to failure by paraglacial processes.

High-angle, gullied lateral-terminal moraines are a common feature of deglaciating land systems in the European Alps and elsewhere. Their morphology and steep gradient are strong conditioning factors for rapid and extensive paraglacial slope adjustment (e.g. Curry 2000; Curry *et al.* 2005). Although such lateral-terminal moraines are eroded rapidly following deglaciation, their upper slopes commonly retain a pronounced oversteepened form, at least during the early period of deglaciation and associated paraglacial modification. The detailed geotechnical characteristics of lateral-terminal moraines that permit them to stand in a quasi-stable state at high angles have yet to be adequately explained, however. This lack of detailed understanding arises partly due to the difficulties of accessing steep moraine slopes that are subject to regular paraglacial activity (such as debris flows and release of clastic debris from the fine-grained matrix) and the problems of replicating *in situ* conditions when undertaking geotechnical analysis in the laboratory. This research applies geotechnical methods to a lateral-frontal moraine sequence in an attempt to explain the geotechnical properties

From: KNIGHT, J. & HARRISON, S. (eds) *Periglacial and Paraglacial Processes and Environments.*
The Geological Society, London, Special Publications, **320**, 181–197.
DOI: 10.1144/SP320.12 0305-8719/09/$15.00

of glacigenic sediments that may allow them to stand at high angles in a quasi-stable form.

Lateral-frontal moraines are formed as debris falls, slumps, slides or flows down the ice surface and accumulates around the glacier margin (e.g. Small 1983; Owen & Derbyshire 1989; Owen 1994). The relative balance of debris supply from supraglacial and valley-side sources will affect the resultant detailed moraine morphology and sedimentary facies, while successive ice advances may result in superimposition of lateral-frontal moraine structures (Bennett & Glasser 1996). If the glacier remains in a stable position, the accumulation of dumped material often produces a wedge-shaped moraine with strong fabric and crude internal bedding dipping away from the glacier at angles of 10°–40° (Small 1983; Bennett & Glasser 1996). However, the ice-proximal parts of lateral-frontal moraines tend to be structurally complex, reflecting widespread collapse and reworking following removal of ice support, together with ice-marginal glacifluvial activity. Moreover, melt-out of buried ice and post-glacial gravitational reworking may destroy bedding. Sediment facies commonly consist of stacked diamictons with variable clast content intercalated with thin sand and gravel layers, reflecting intermittent glacifluvial deposition and reworking (Benn *et al.* 2005). Most diamicton facies are sandy boulder gravels containing predominantly angular debris from passive transport of rockfall material, although more rounded clasts may be found within lateral moraines where a higher proportion of basal-zone debris is delivered to the terminus (Matthews & Petch 1982), or where proglacial sediment is entrained during glacier advance (Slatt 1971). Where valley-constrained glacier termini repeatedly occupy similar positions, large, multicrested lateral moraines may form, reflecting episodic aggradation separated by periods of erosion or non-deposition. Complex depositional histories may be preserved in the internal moraine structure, with multiple depositional sequences bounded by erosion surfaces. Periods of non-deposition may be indicated by buried palaeosols (e.g. Röthlisberger *et al.* 1980).

The extent, conditioning factors, morphological consequences and rates of recent paraglacial reworking of sediment at this site are considered in detail elsewhere (Curry *et al.* 2005), while geotechnical investigation has previously been undertaken at this locality (Whalley 1975). Whalley applied shear test results from the nearby Allalingletscher lateral moraine to the similar form of moraine slopes at Feegletscher Nord in an attempt to explain the preservation of pronounced oversteepening on the upper moraine slopes. Here we report the results of geotechnical testing of samples drawn directly from the Feegletscher Nord moraine sequence.

Study area

Investigation of the paraglacial modification of valley-side sediment-mantled slopes was undertaken in August 2001 and September 2006 in the forefield of the Feegletscher Nord (46°06′N, 7°54′E), in the Saaser Valley, Valais, Switzerland (Fig. 1). The Feegletscher is a small (<10 km^2) ice-field outlet glacier that descends from an altitude of over 4000 m asl (m above sea level) on the eastern flank of the Mischabel to a terminus at approximately 2200 m asl (as of September 2006). The front of the Feegletscher divides into two distinct tongues; a heavily crevassed ice fall characterizes the top of the wider southern lobe snout area. The northern lobe (hereafter referred to as Feegletscher Nord) is also heavily crevassed in the upper reaches, but becomes increasingly constrained and narrowed by steep rock-walled topography in its lower reaches. Both lobes are orientated in a NE direction. The site is dominated by Bernard nappe mica-schists, with serpentinite, amphibolite and quartzite (Swiss Geological Commission 1980; Hsü 1995). It experiences the inner alpine, relatively dry climate of the Valais surrounded by high mountains, with an estimated 800–1200 mm annual precipitation and mean annual temperatures of approximately +1.5 °C (Swiss Meteorological Survey, Zürich). The lower limit of discontinuous permafrost in Valais is approximately 2350 m for north-facing slopes and 2650 m for south-facing slopes (Lambiel pers. comm. 2003).

The site was deglaciated by about 9 ka BP, but was repeatedly reoccupied by ice during the Holocene (e.g. Röthlisberger & Schneebeli 1979; Röthlisberger *et al.* 1980). Feegletscher Nord reached its 'Little Ice Age' maximum position at AD 1818, since when it has experienced overall retreat of *c.* 1200 m and lowered *c.* 90 m (Bircher 1982; GK/SCNAT & VAW/ETZH 2006). After a slight re-advance during the 1970s and 1980s, annual retreat of the Feegletscher Nord has been sustained since 1989 (Schnyder pers. comm. 2006).

The forefield area displays marked within-valley asymmetry, with much larger moraine volume on the northern side reflecting increased debris supply from extensive rockwalls on that side of the valley (Fig. 1). On the northern side of the forefield, the pattern of glacier thinning and retreat has exposed steep glacigenic deposits composed of a stacked, multicrested, lateral moraine, which has subsequently been locally reworked, with deep, intersecting gullies forming on most upper slopes at or just below the moraine crest, and cones or sheets of reworked sediment accumulating at the slope foot (Fig. 2).

The dominant agent of reworking of glacigenic sediment at this site is debris-flow activity and

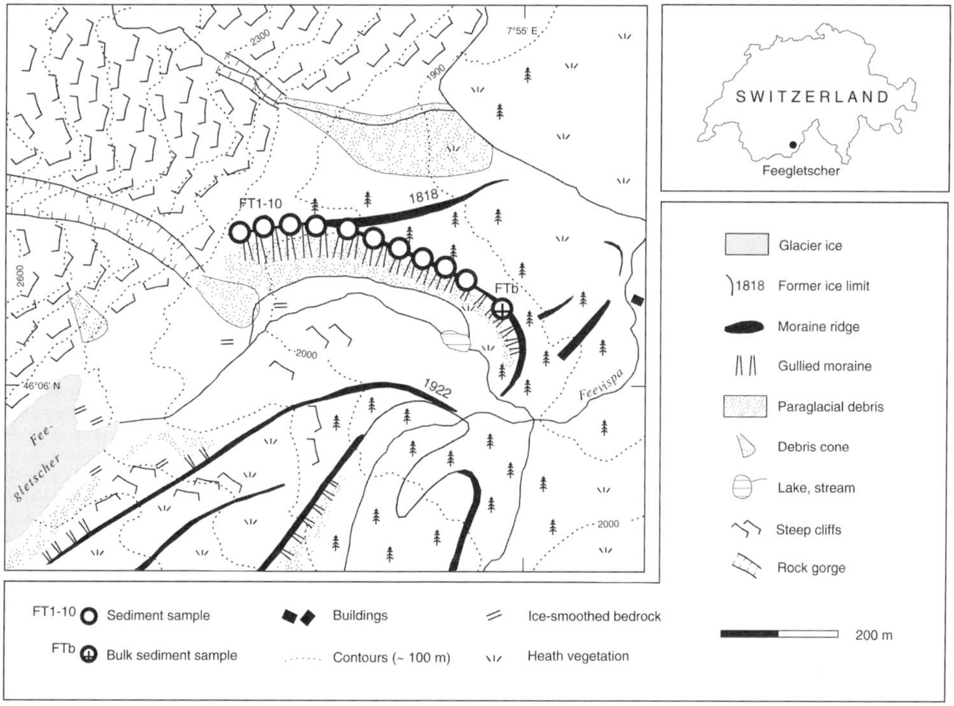

Fig. 1. Location and geomorphology of the Feegletscher Nord study site, Valais, Switzerland.

Fig. 2. Steep lateral moraine proximal slope incised by paraglacial processes operative since AD 1922 in the Feegletscher Nord forefield area. Debris slides and flows descend across the lower slope surfaces. Note crude, distal-dipping stratification exposed in the gully sidewalls.

translational sliding (Curry *et al.* 2005), triggered by rainstorms and snowmelt. There is further localized evidence of snow/slush avalanches, falls and surface wash. The Feevispa outlet stream has incised the lateral-terminal moraine in the SE, and glacifluvial gravel and sand has accumulated on the valley floor within the proximal proglacial zone.

Moraine morphology and evolution at Feegletscher

The stacked lateral moraine on the northern valley side is *c.* 60–120 m high, *c.* 700 m long (down-valley) and has a summit altitude that descends down-valley from 2060 m in the west to 1940 m in the east, considerably below the lower limit of discontinuous permafrost. A pronounced step is visible part-way along the moraine crest, at approximately 2020 m asl. At this point, a separate ridge diverges to the NE and represents the AD 1818 ice limit. Above (west of) this step, proximal slope height exceeds 100 m, but below it the height of the proximal slope is less than 70 m. Historical photographs suggest that the moraine may have been overtopped at this point during the 1890s.

The distal slope is fully vegetated, relatively stable (except for some surface creep as evidenced by curvature of tree trunks) and rests at a gradient of *c.* 33°. In contrast, the proximal slope is largely unvegetated, and steeper, although gradient is more variable, with the most recently deglaciated terrain generally steeper (<80°) than the older ground (<50°). On the oldest terrain, the most mature cones and sheets of reworked debris extend almost to the moraine crest and support a partial grass cover.

Previous investigations of the temporal pattern of moraine-slope adjustment at Feegletscher Nord indicate that following an initial period of gully incision, intervening gully divides are consumed through gully widening, resulting in progressive slope stabilization and levelling of inter-gully relief within 80 years (Curry *et al.* 2005). Thus, a dynamic land system of deep gullies, sharp arêtes and nascent debris cones is being replaced by one of increasingly vegetated coalescing cones and debris aprons, levées, lobes and low-relief scars. The resultant gullied moraine complex is morphologically similar to lateral moraine sequences observed elsewhere in the European Alps (e.g. Curry *et al.* 2005).

Field methods

Initial assessment of the extent of paraglacial reworking of glacigenic sediment comprised detailed geomorphological mapping on 1:5000 scale base maps, produced with the aid of ground and aerial photographs. Access to steep, potentially unstable moraine slopes is problematic. In particular, assessing the slope angles of upper-slope units (interfluves and gullies) that stand at angles of up to 80° is fraught with difficulty. Fifty upper-slope section angles were, therefore, measured by attaching a compass clinometer to a metre rule and placing this over the edge of the cliff at multiple locations along the moraine crest. Compass-clinometer measurements were also made at multiple locations along the moraine on the accumulated lower-angle debris cones. Although these lower-slope units were sufficiently accessible to allow the use of more accurate surveying techniques, in order to maintain a constant level of accuracy/error, compass-clinometer measurements were utilized throughout.

Diamicton facies exposed include sandy boulder gravels containing coarse, angular and subangular material. Distal-dipping of large boulders is clearly evident in the upper moraine slopes, exposed by recent erosion (Fig. 3), although for practical reasons of accessing steep, potentially unstable upper-slope units *in situ* fabric could not be quantitatively assessed. Clearly, the coarsest particles are impractical to sample and test using standard geotechnical laboratory techniques. The sedimentological size characteristics of 11 samples of *in situ* till deposit, each weighing 1 kg, were assessed in terms of fine-fraction (<8 mm, -3Φ) particle-size distributions representative of the finer matrix material within the moraine. Ten samples of massive, poorly sorted, compacted diamicton, spaced 50 m apart along the crest of the moraine ridge were extracted using a trowel, labelled 1–10 from the west to east (FT1-10, Fig. 1). Each of these samples was removed 0.2 m below the ground surface, *c.* 1 m from the proximal slope edge. Sample locations 1–4 are located above a clear step in the moraine crest at an altitude of 2020 m asl, while samples 5–10 are located east of this point. At each of these 10 sediment sampling locations, an *in situ* density test was carried out to BS1377: Part 9: 1990 using the 'sand replacement method'. These *in situ* measurements were carried out for comparison with the densities determined for samples used in the shear tests outlined below. The procedure involved excavating a 0.1 m-diameter hole to a depth of 0.1 m in a levelled area of moraine, using a metal tray with a hole in the middle, and weighing the excavated material. A preweighed pouring cylinder filled with sand of known density was then placed over the hole, and the sand released into the hole to fill it. The cylinder was weighed again and the mass of sand poured into the hole calculated. The volume of the hole was determined from the mass of sand poured, and the

Fig. 3. Flat faces of mica-schist boulders within the upper proximal slope dipping at approximately 33° towards the distal slope (left-hand side).

bulk density of the moraine calculated. The moisture content of a small sample of material taken from the hole was then determined in order to calculate the dry density of the moraine.

Finally, a bulk sample of moraine material weighing 18 kg was removed from a depth of 0.2 m on the moraine crest at a point 80 m SE of sample 10, for laboratory shear tests. An 11th particle-size sample weighing 1 kg was taken from this bulk sample, for granulometric analysis of material finer than 8 mm (-3Φ), as well as clast lithological and form analysis.

Laboratory procedures

To prepare samples for sieving, initial disaggregation was undertaken by prolonged (3 h) agitation in a 4% solution of sodium hexametaphosphate. The particle size distribution of fine gravel to very-fine sand-sized material was determined by wet sieving each of the 11 till samples, with weight per fraction calculated as a percentage of the total. Volume of material finer than 63 μm (4Φ) in each sample was quantified by a Malvern Instruments laser particle-size analyser, and the results converted into % weight data. A total of 167 clasts coarser than 10 mm were sampled from the bulk sediment, described, identified for lithology, and their three principal axes measured to calculate flatness and elongation ratios (Zingg 1935). Aggregate clast shape was also calculated in terms of the percentage of clasts with *c:a* axial ratios ≤ 0.4 (C_{40}) and ≤ 0.5 (C_{50}; Ballantyne 1982), and angularity expressed as the percentage of very angular plus angular clasts (RA; Benn & Ballantyne 1994).

The aim of the direct shear tests was to determine the effective angles of friction for the matrix material for a range of densities and fabric orientations. These values could then be compared with the variety of *in situ* angles of repose, as indicated by the proximal and distal slopes of the moraine, and the redeposited debris at the base of the proximal slope.

To prepare a sample with fabric at a high angle to the shear surface, either an undisturbed sample has to be obtained from the moraine or a reconstituted sample has to be manufactured. The former is virtually impossible because of the lack of cohesion, suction or cement, and the presence of large clasts. The latter method is also impractical because a 300 × 300 mm sample would have to be prepared using a procedure to place flat clasts horizontally, and the entire sample would then have to be rotated through 90° and placed into the shear box intact.

Clearly, a limitation of the shear box tests is that the material (and its fabric) has been disturbed, and then reconstituted within the shear box at densities that may differ from *in situ* densities. Furthermore, the range of matrix particle sizes does not include particles coarser than gravel size, which may affect the shear strength of the reconstituted moraine samples. The two halves of the shear box control the formation of the shear surface or zone. However, the use of a large shear box (300 × 300 mm) is the best available method for laboratory determination of the material shear-strength parameters.

A series of direct shear tests using a large shear box (300 × 300 mm) was carried out on an air-dried sample of the moraine matrix material, using the method in BS1377: Part 7, 1990. For the first series of three tests the lateral moraine sample was loosely placed into the shear box in three layers, to simulate the undisturbed distal and debris slopes. In a second series of three tests, an air-dried sample of material was compacted into the large shear box in three layers, each layer compacted for 60 s with a Kango vibrating hammer and plate rammer. The compaction of the sample is intended to simulate glacial compaction of the moraine during successive advances of Feegletscher Nord. A series of vibrating hammer compaction tests were carried out to BS 1377: Part 4: 1990, to determine the maximum dry density and optimum moisture content of the bulk sample, for comparison with the densities achieved in the shear box. For a third series of shear box tests, the air-dried sample was compacted into the large shear box in three layers and each layer compacted for 30 s with a Kango hammer and plate rammer. After compaction of

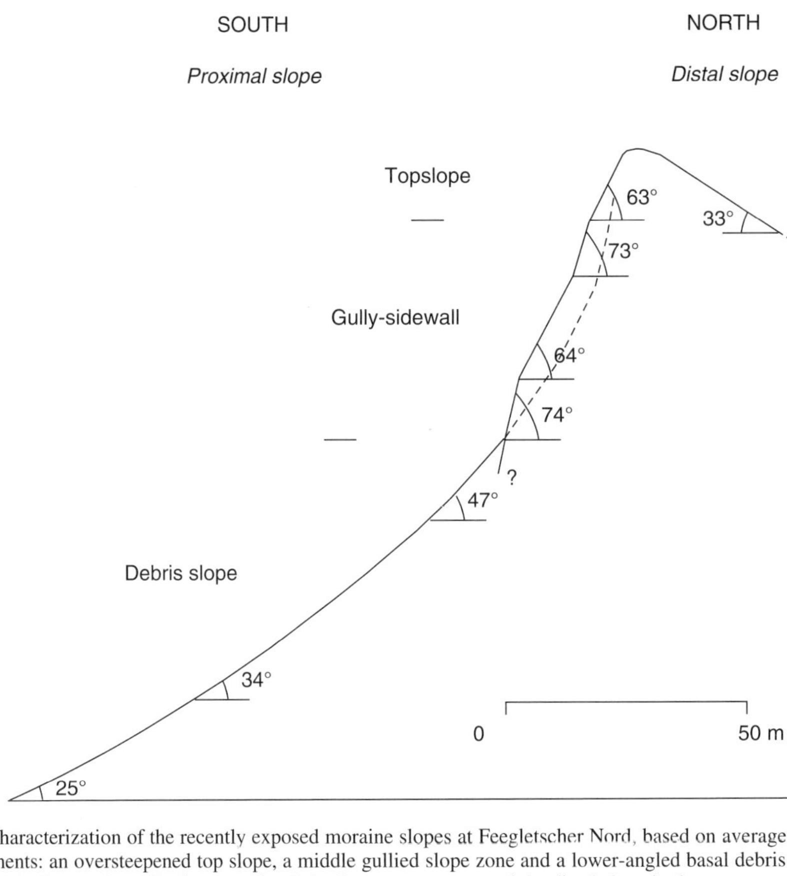

Fig. 4. Characterization of the recently exposed moraine slopes at Feegletscher Nord, based on average survey measurements: an oversteepened top slope, a middle gullied slope zone and a lower-angled basal debris sheet. The dashed line indicates the gully-floor profile. Only the uppermost part of the distal slope is shown.

the second layer 25 gravel-size flat clasts were placed in five rows with their *c*-axes parallel to the direction of shear, to partly simulate the distally dipping fabric of supraglacially derived material emplaced by dumping and gravity-flow deposition at the glacier margins (cf. Small 1983). In each series of tests the air-dried sample was subjected to three normal stresses equivalent to the likely *in situ* stresses within the near-surface sediments of the lateral moraine, based on the densities measured in the field, and then sheared at a rate of 1 mm min^{-1}. The lowest normal stress that could practically be applied by the shear box was used to simulate shear near the surface.

The effective angle of friction was determined from the Mohr–Coulomb failure criterion:

$$\tau_f = c' + \sigma_n' \tan \phi' \quad (1)$$

where τ_f is shear strength, c' is effective apparent cohesion, σ_n' is effective normal stress and ϕ' is effective angle of friction. Assuming that the

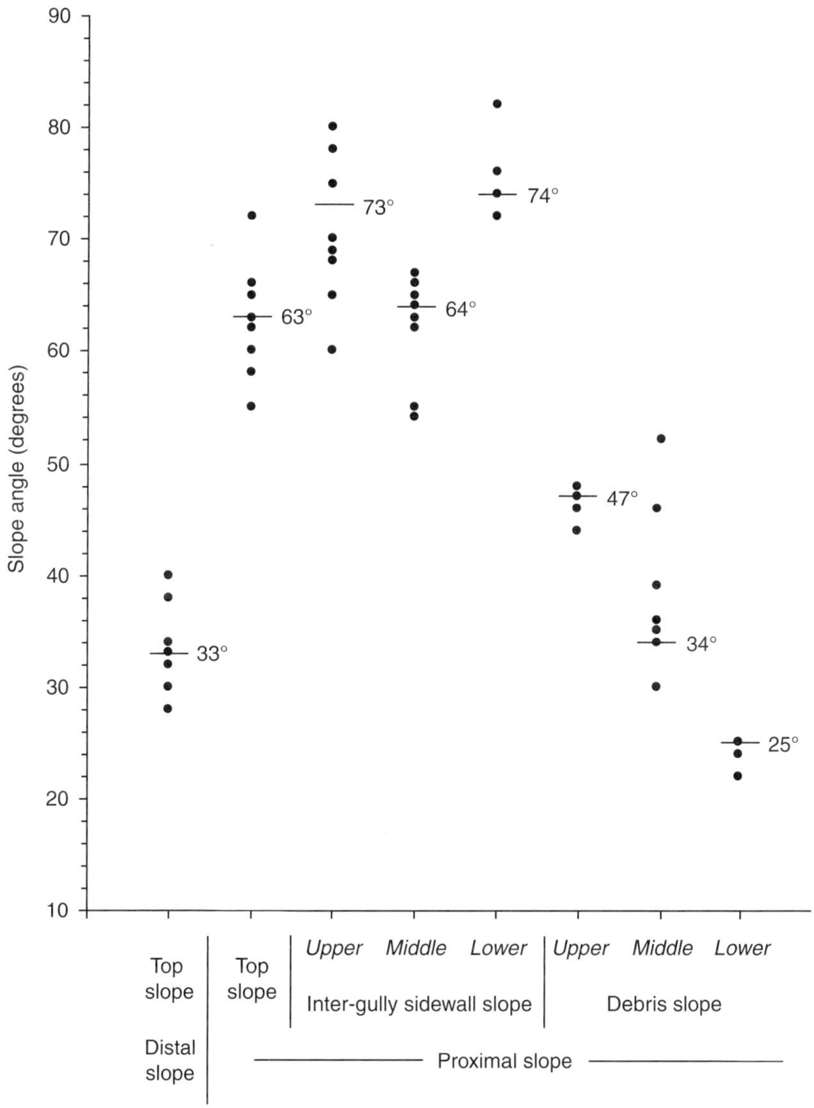

Fig. 5. Dispersion diagrams illustrating moraine slope gradient at Feegletscher Nord. Horizontal bars indicate median values. Each closed circle represents one surveyed slope.

Table 1. In situ *dry densities for samples taken along the crest of the Feegletscher Nord lateral moraine. Sample locations shown in Figure 1*

Sample	Dry density (Mg m^{-3})
1	1.29
2	0.84
3	n/a
4	0.90
5	2.34
6	2.20
7	1.96
8	2.04
9	n/a
10	2.07
Average	1.70

n/a = not available.

effective apparent cohesion of the material is zero, the effective angle of friction is therefore:

$$\phi' = \tan^{-1} \tau_f / \sigma'_n. \qquad (2)$$

Results

The proximal slope can be subdivided into three major slope elements, consisting of an oversteepened top slope, a middle gullied slope and a lower-angled debris sheet beneath (Fig. 4). The middle gullied slope section can be further subdivided into upper, middle and lower inter-gully sidewall slopes and gully floor slopes. Results of slope measurements made on the Feegletscher Nord moraine are shown in Figure 5. The median slope angle of the undisturbed and heavily vegetated distal slope is 33°. The median angle of the top proximal slope units was 63°. Median slope angles of the upper, middle and lower inter-gully sidewalls were 73°, 64° and 74°, respectively. The median measured proximal debris cone slopes were 47°, 34° and 25° for the upper, middle and lower sections, respectively.

Results of *in situ* dry density measurements are shown in Table 1. Values range from 0.84 to 2.34 Mg m^{-3} (Mg = 1 × 10^6 g) with an average of 1.70 Mg m^{-3}. The maximum dry density and optimum moisture content values of the bulk sample, obtained from the vibrating hammer compaction tests, were 2.25 Mg m^{-3} and 7.9%, respectively. *In situ* dry densities derived from the shear box samples range from 1.91 Mg m^{-3} for the loosely packed sample, 2.01–2.20 Mg m^{-3} for the densely packed sample and 2.12–2.16 Mg m^{-3} for the densely packed sample with partial perpendicular fabric (Table 2).

The particle-size distributions for the material finer than 8 mm (−3Φ) of the 10 *in situ* moraine samples and the single bulk moraine sample are shown in Figure 6. Particle-size analysis shows a generally poorly sorted, well-graded range of particle sizes, with clay-size particles making up less than 1% of material finer than 8 mm in all moraine samples. Ninety-four per cent of clasts sampled coarser than 10 mm were mica-schist, 5% were quartz, <1% were mylonite and <1% serpentinite. Clasts were found to be generally subangular (83%; average RA index = 84) and smooth (99%). Clast

Table 2. *Large shear box test results on the Feegletscher Nord moraine bulk sample*

Sample	Dry density (Mg m^{-3})	Normal stress (kPa)	Peak shear stress (kPa)	Critical shear stress (kPa)	Effective angle of friction (°)		Angle of dilation (°)
					Peak ϕ'	Critical ϕ'_{crit}	ψ
Loose	1.91	66.6	48	48	35.8	35.8	0.0
	n/a	133.1	94.2	94.2	35.3	35.3	0.0
	n/a	199.7	138.4	138.4	34.7	34.7	0.0
Average	1.91	–	–	–	35.3	35.3	0.0
Dense	2.09	66.6	51.8	36.7	37.9	28.9	9.0
	2.21	133.1	169.5	100.8	51.9	37.1	14.7
	2.21	199.7	190.2	139.4	43.6	34.9	8.7
Average	2.17	–	–	–	44.4	33.6	10.8
Dense fabric	2.12	66.6	72.5	62.2	47.4	43.0	4.4
	2.13	133.1	146.0	122.4	47.6	42.6	5.0
	2.16	199.7	228.8	168.6	48.9	40.2	8.7
Average	2.13	–	–	–	48.0	41.9	6.1

n/a = not available.

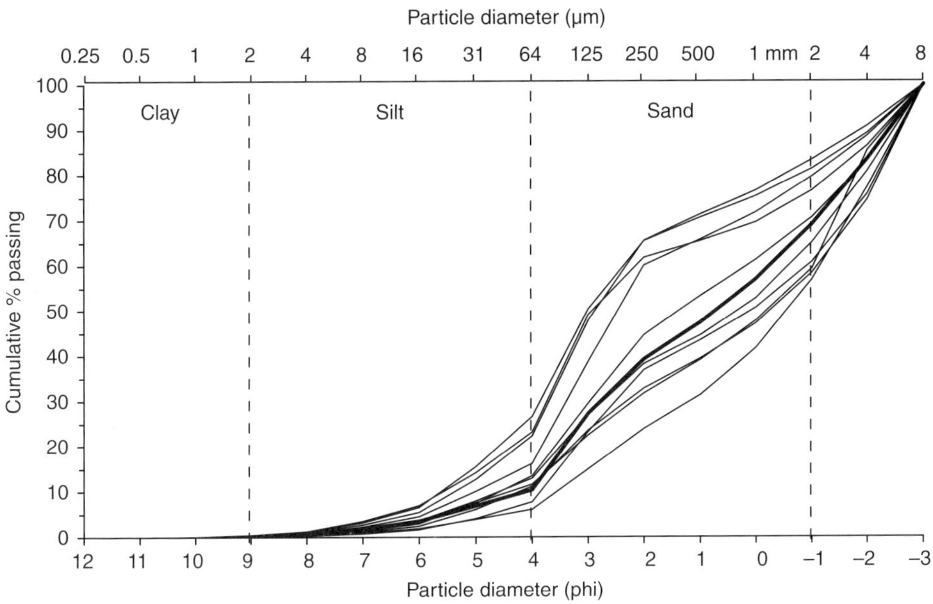

Fig. 6. Cumulative particle-size distributions (material finer than 8 mm/−3Φ) for 11 samples of Feegletscher Nord moraine. The bulk sample is indicated with a bold line.

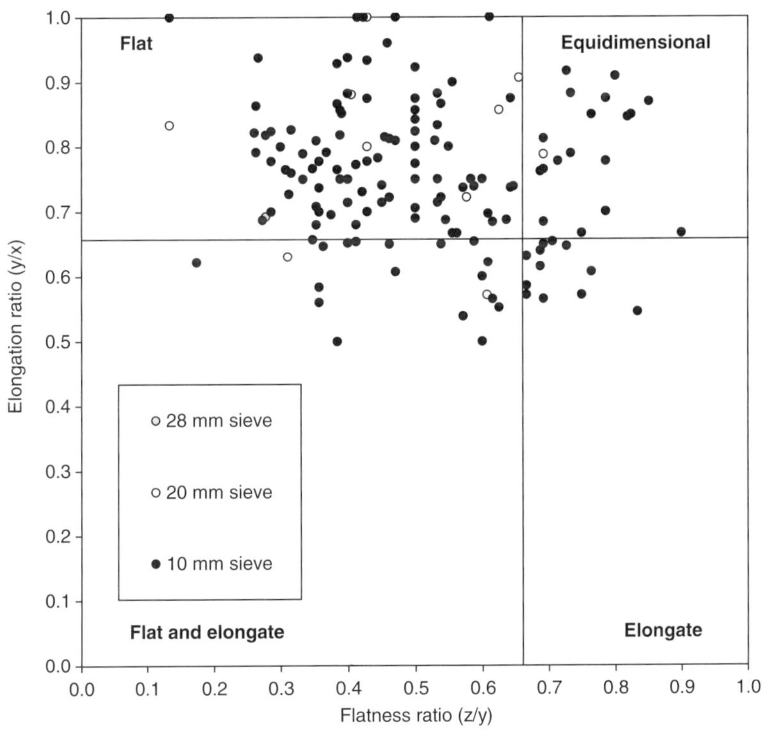

Fig. 7. Shape of 167 clasts taken from the Feegletscher Nord moraine bulk sample, summarized according to Zingg's (1935) flatness and elongation ratios.

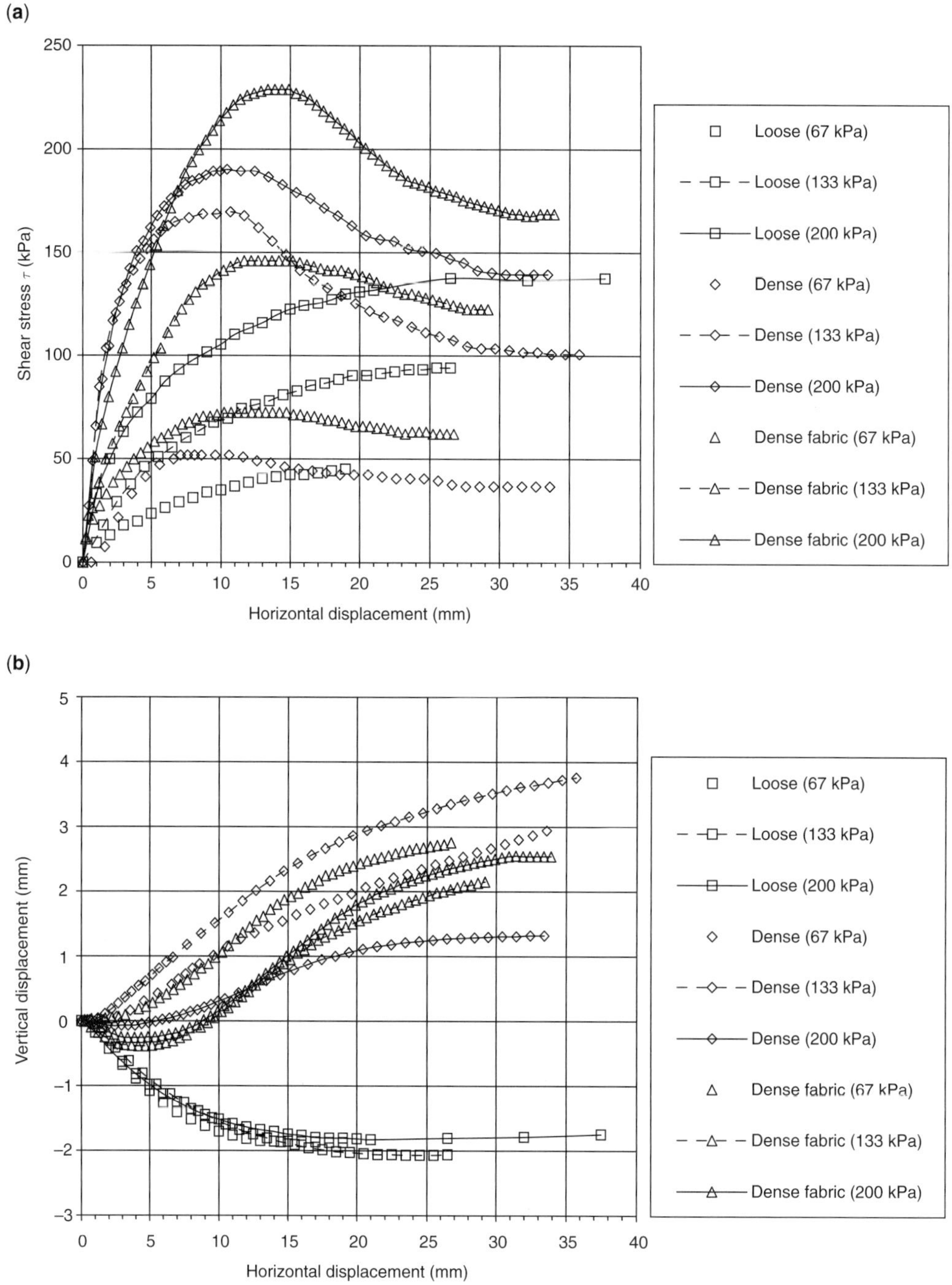

Fig. 8. Large shear box tests on the Feegletscher Nord moraine bulk sample tested 'loose', 'dense' and 'dense fabric' (the latter with a partial fabric perpendicular to the direction of shear). Normal stresses are shown in brackets. (**a**) Shear stress against horizontal displacement. (**b**) Vertical displacement against horizontal displacement. (**c**) Shear stress against effective normal stress.

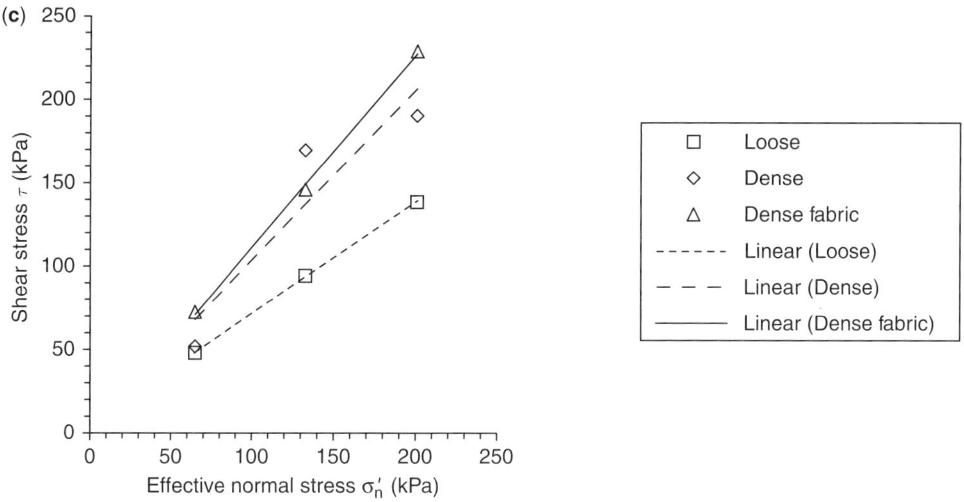

Fig. 8. (*Continued*).

form is described by C_{40} and C_{50} indices of 62 and 85%, respectively, indicating a slabby or elongate form typical of supraglacially transported clasts (Benn & Ballantyne 1994). Following Zingg's (1935) clast form method, 66% of the sampled particles were found to be flat (Fig. 7). Of the remainder, 14% were equidimensional, 13% flat and elongate, and 7% elongate. The average flatness and elongation ratios were 0.506 and 0.760, respectively, which indicates that the average clast form is flat (Zingg 1935). The material finer than 10 mm also contains mica flakes, which by their nature are flat in form.

The shear stress v. displacement and shear stress v. normal stress graphs for the well-graded muddy sandy gravel bulk moraine sample are shown in Figure 8. The effective angle of friction of the loosely packed sample (hereafter referred to as 'loose') ranges from 35° to 36°, with an average of 35° (Table 2). The peak effective angle of friction obtained for the compacted moraine sample (hereafter referred to as 'dense') ranged from 38° to 52°, with an average of 44°, while peak effective angle of friction obtained for the densely packed fabric samples (hereafter referred to as 'dense fabric') ranged from 47° to 49° with an average of 48°. Critical effective angles of friction were 35°, 34° and 42° for the loose, dense and dense fabric samples, respectively. The average dry densities of the loose, dense and dense fabric samples determined at the end of shear box testing were 1.91, 2.17 and 2.13 Mg m^{-3}, respectively. These are all within the range of the *in situ* densities shown in Table 1.

Discussion

This research tests the hypothesis that the angles of friction obtained from shear tests for samples of the lateral moraine prepared to different densities and fabrics should be similar to the natural angles of repose of the lateral moraine, assuming shallow translational failure. The grading of the moraine material supports field observations that suggest the type of failure is likely to be shallow translational sliding, flow and individual clast slide and/or fall, rather than deep-seated rotational slip, because of the very small percentage of clay-sized particles (cf. Selby 1993). Moreover, well-graded, imbricated and overlapping clasts prevent a discrete deep-seated rotational failure surface developing. Translational failure of an 'infinite' cohesionless soil slope is assumed to occur on a plane parallel to the ground surface, at shallow depth. The factor of safety, F, of the slope at the point of translational failure along a potential shear surface is commonly described as:

$$F = \tau_f/\tau = [1 - (u/\gamma z \cos^2 \beta)] \times [\tan \phi' / \tan \beta] = 1 \quad (3)$$

where τ is shear stress, u is pore-water pressure, γ is unit weight of soil, z the depth of failure surface and β is slope angle.

Assuming that the pore spaces within the moraine are unsaturated and that $u = 0$, then equation (3) yields:

$$\beta = \phi'. \quad (4)$$

The gradient of the ground surface is therefore equal to the effective angle of friction of the material. If there is a pre-existing slip surface, such as that which may exist between flat clasts with a distinct fabric parallel to the ground surface, then the mobilized effective angle of friction would be equal to the critical effective angle of friction, ϕ'_{crit}, and equation (4) becomes:

$$\beta = \phi'_{\mathrm{crit}}. \tag{5}$$

This appears to be true for the distal and middle proximal debris slopes, which, because of their general flat clast form, are orientated parallel to the slope and have undergone significant shearing during their formation, and so would be expected to be close to the ϕ'_{crit} for loose and dense samples. In fact, the median slope angles of 33° and 36° measured for the distal and middle proximal debris slopes, respectively, and those obtained by Whalley (1975) of 35°, compare well with the ϕ'_{crit} for loose and dense samples of 35° and 34°, respectively.

However, the average peak angle of friction obtained from shear tests on samples of the lateral moraine was 44° for the dense sample and 48° for the dense sample with partial perpendicular fabric. These angles are considerably lower than the maximum median slope angle of 74°, and those found at this site by Curry *et al.* (2005) and Whalley (1975) of 69° and 70°, respectively. This may reflect the fact that the majority of the flat mica-schist clasts in the reconstituted sample were likely to be orientated subhorizontally in the shear box and therefore their fabric was parallel or subparallel to the shear surface. The silt- and sand-sized material, made up of mainly mica particles, may also naturally lie flat when placed in the shear box, as do the gravel-sized particles that were mainly flat mica-schist clasts. In the upper proximal slopes of the lateral moraine, shear of the material near the surface would be at a high angle to the generally distal-dipping imbricated clasts. Therefore, for shallow translational shear to occur parallel with the proximal slope surface, shear would probably be between distally dipping flat clasts in addition to shear of some weak intact clasts. The shearing resistance along this potential shear surface would be significantly increased by the shear surfaces between clasts being at an oblique angle to the slope and by clast intact shear strength. For shear failure to occur parallel to the proximal slope, therefore, significant dilation of the dense distally dipping material is required. This dilation phenomenon is recognized in dense soils, which experience volumetric increase when sheared, until a peak shear stress is reached, followed by a reduction in the rate of dilation until a critical state is reached. At low normal effective stresses many soils exhibit 'dense' characteristics. In the shear box the material dilates and the platen rises at the angle of dilation, ψ. Although the shear surface is horizontal, it can be conceptualized that the microscopic intergranular shear planes are inclined at an angle of ψ. Therefore, the effective angle of friction being measured is:

$$\phi' = \phi'_{\mathrm{crit}} + \psi. \tag{6}$$

This is similar to shear between rock surfaces that are inclined at an angle, i, to the direction of shear (Patton 1966; Barton 1973). Given the presence of interlocking asperities along rough rock-mass discontinuity surfaces, shear stresses along these partings are often inclined to the overall applied shear stress direction. In this case the relationship between applied shear and normal stresses is:

$$\tau_{\mathrm{f}} = \sigma'_{\mathrm{n}} \tan(\phi'_{\mathrm{b}} + i) \tag{7}$$

In equation (7), ϕ'_{b} is the basic angle of friction for the rock, which is similar to the critical angle of friction in soils, since these are the angles measured when there is no volume change, and i is the roughness angle for the joint surface. It can be assumed that the moraine, a well-graded material of densely packed distally dipping clasts, from silt to boulders, behaves somewhere between a dense soil and a weak, highly jointed rock mass. The observed stratification and fabric within the undisturbed moraine might suggest that the geotechnical properties are more closely aligned with those of a highly jointed rock mass than a soil. If it is assumed that the stability of the upper proximal slope is governed by shallow translational failure and equation (4) is rewritten in terms of the angle of friction in equation (7), then:

$$\beta = \phi'_{\mathrm{b}} + i. \tag{8}$$

Barton (1973) presents values of $(\phi'_{\mathrm{b}} + i)$ from several authors for different rock types and rough discontinuity surfaces at low normal stresses, ranging from 66° to 80°, with an average of 72°. This value is close to the gradients of the proximal upper and lower inter-gully sidewall slopes, which have median angles of 73° and 74°, respectively.

Barton (1973) derived an empirical equation for curvilinear peak shear strength along rock joint surfaces:

$$\tau_{\mathrm{f}} = \sigma'_{\mathrm{n}} \tan(\phi'_{\mathrm{b}} + \mathrm{JRC}\ \log_{10}[\mathrm{JCS}/\sigma'_{\mathrm{n}}]) \tag{9}$$

where JRC is the Joint Roughness Coefficient (an empirical measure of joint roughness) and JCS is

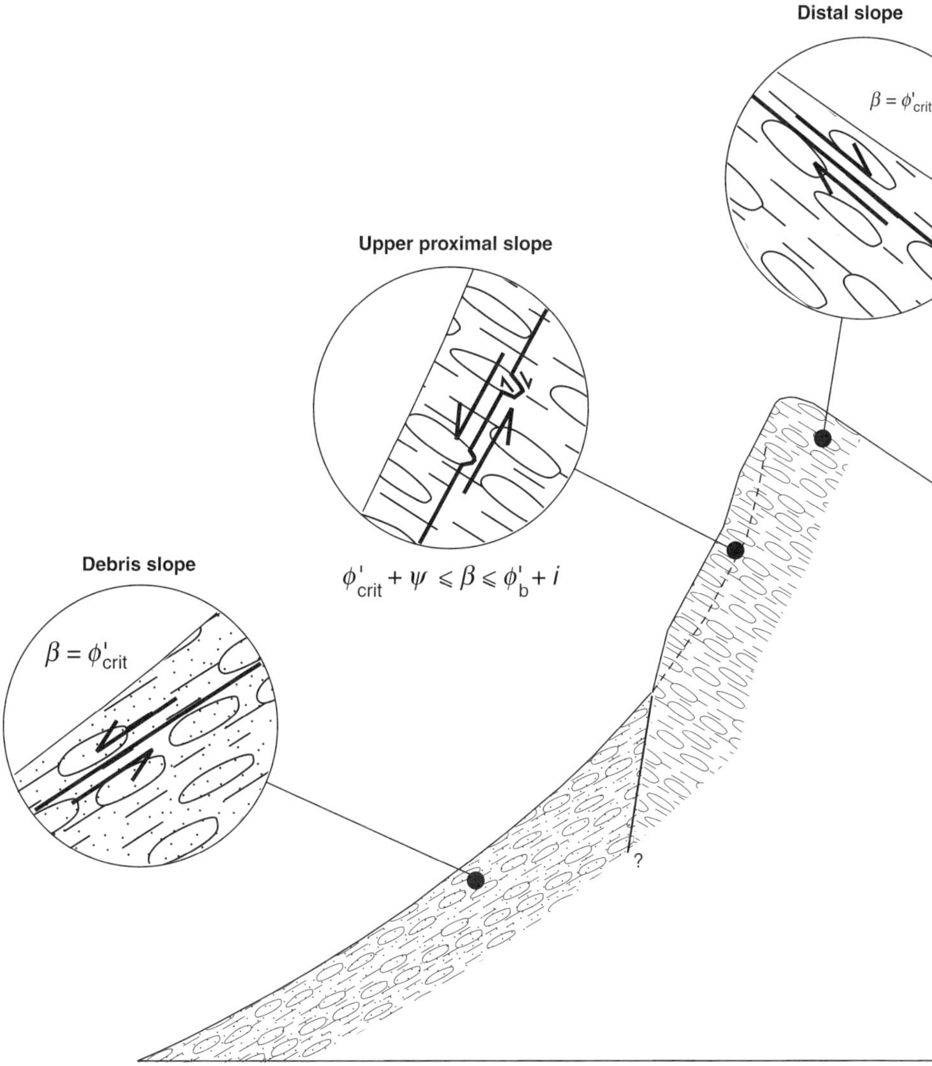

Fig. 9. Idealized section through steep moraine slopes showing the influence of clast shape and fabric on shearing resistance and slope angle. Insets show magnified fabric and potential shear surfaces.

the Joint wall Compressive Strength (compressive strength of weathered joint walls). At low normal stresses the value in parentheses in equation (9) becomes large for rough, undulating joints. Barton suggests that the maximum value for this section of equation (9) ($\phi'_b + \text{JRC} \log_{10}[\text{JCS}/\sigma'_n]$) should be 76° for $\text{JCS}/\sigma_n = 200$. The median angles of the upper proximal slopes lie just below this value.

Consequently, it seems plausible to suggest that the stability of steep, relatively undisturbed upper proximal slopes of the Feegletscher Nord moraine may be controlled by shallow translational failure, with inclined shear stresses and resulting dilation between the generally densely packed, distally dipping clasts (Figs 9 and 10). From the results of the shear box tests, the observations made in the field and the comparisons made with measurements of rock joint shear strengths carried out by others, the upper proximal-slope angles appear to lie between the values given by equations (6) and (8), and hence:

$$\phi'_{crit} + \psi \leq \beta \leq \phi'_b + i. \qquad (10)$$

Indeed, we consider that the major cause of the steep slopes at Feegletscher Nord probably reflects the distal-dipping fabric, whose stabilizing role is significantly enhanced by the generally flat form

Fig. 10. Naturally eroded, north–south gully section through the top of the middle part of the Feegletscher lateral moraine, showing a south-facing, steep, proximal slope on the left and densely packed, overlapping distally dipping flat clasts in the centre, following rough cleaning of the gully slope. The average dip and dip directions of the larger clasts are 31° and 008°, respectively.

of the mica-schist clasts. Imbrication of these flat-form clasts inhibits shallow translational shear on the proximal slopes and, hence, permits retention of steep slope angles. A further, secondary, temporary slope-stabilizing effect may be the natural 'buttressing' of the moraine material caused by the creation of inter-gully slopes through incision of adjacent gullies, prior to their erosion and removal, although this effect cannot be substantiated without further field investigation.

A possible alternative explanation for the difference between maximum observed slope angles and peak angles of friction concerns the role of cementing and suction (Whalley 1975). Coarse-grained particles, such as gravel and sand-sized material, may be held together by the intermolecular bonds of cementing precipitates such as silica, calcium carbonate and iron oxides, and by bridges of clays. The moraine material observed by the authors, however, was found to be friable and relatively easily broken up by finger pressure, suggesting a lack of suction and/or cementation. There is unlikely to be much cementing owing to the young age of the deposit and the lack of evidence for groundwater flow carrying minerals that could provide cement. There may be some temporary suction pressure holding the fine clasts together during dry periods following evaporation of infiltrated precipitation, because of their flat shape and the silt- and sand-sized matrix. This may explain the weak surface 'crust' that is sometimes observed, which holds finer particles together in larger agglomerates. However, any suction would probably be insufficient to provide a long-term stabilizing effect of the well-graded moraine. This suction is very difficult to measure *in situ* because of the well-graded nature of the sediment.

Two wider implications emerge from this study. The first concerns landform resilience to change and paraglacial sediment transfer rates. Without doubt, paraglacial reworking of glacigenic sediment stores has the potential to substantially modify land-surfaces and sediments during and following deglaciation. Minimum rates of ground-surface lowering on the Feegletscher Nord lateral moraine have averaged approximately 50–100 mm year^{-1}

since ice retreat in AD 1922 (Curry *et al.* 2005). These rates are similar to those calculated for sites in western Norway (Ballantyne & Benn 1994; Curry 1999), although greatly exceed 'normal' erosion rates in other settings (Young & Saunders 1986), emphasizing the extreme rapidity of geomorphic change and sediment transfer on steep, sediment-mantled slopes associated with paraglaciation (cf. Church & Ryder 1972; Church & Slaymaker 1989; Harbor & Warburton 1993). Yet, even within an active paraglacial setting, given favourable lithological conditions, moraine slopes, such as those observed in the lateral-terminus area of the Feegletscher Nord, are able to stand at steep angles in a quasi-stable form on a decadal timescale, resisting wholesale removal and reworking, and delaying the release of glacigenic sediment into the proglacial zone and paraglacial sediment cascade. Thus, while some models of primary paraglacial system behaviour indicate activity rates declining rapidly from a peak at the time of deglaciation (e.g. Matthews 1992), paraglacial sediment movement within the glacigenic sediment-mantled slope land system may in many cases peak shortly after, rather than immediately after, deglaciation.

Moreover, temporary, quasi-stable storage of glacigenic sediments in high-angle moraines, such as those observed in this study, is one of several processes likely to disrupt the simple, monotonic exponential decline of paraglacial sediment transfer rates and add a stochastic element to the timing of forefield sediment delivery by non-glacial processes. Significant levels of paraglacial reworking can take place in primary paraglacial systems some time after deglaciation in response to non-glacial extrinsic perturbations (Ballantyne 2002*b*) and, as this study demonstrates, where lithological conditions are favourable. Storage of sediments in moraine sequences can prevent release of glacigenic material by non-glacial processes for several decades following deglaciation. In general, the post-glacial disintegration processes of morainic deposits remain poorly understood (e.g. Sletten *et al.* 2001), and the ability of moraine slopes to resist wholesale disintegration immediately following deglaciation, as observed in this study, suggests that post-glacial moraine disintegration and associated sediment supply to forefield areas is not a simple linear process.

A second implication relates to the sensitivity of steep, recently deglaciated moraine slopes to future climate change. Clearly, the relevance of the paraglacial concept is particularly evident in the context of recent retreat of mountain glaciers and climatic amelioration. In contrast to the established use of geophysical and geotechnical approaches in the assessment of periglacial slope problems (e.g. Harris *et al.* 2001, 2003; Harris 2005), the application of engineering geology approaches and solutions to paraglacial slope problems is an underdeveloped research field, despite a growing awareness of non-glacial slope hazards associated with recent glacier retreat (e.g. Evans & Clague 1994; Haeberli *et al.* 1997; Holm *et al.* 2004; Huggel *et al.* 2004; Chiarle *et al.* 2007). In this study standard engineering techniques have shed light on controls on paraglacial modification of steep moraines. Further geotechnical work may facilitate improved spatial and temporal prediction of paraglacial mass movement on sediment-mantled slopes. Parameters commonly involved in modelling landslide susceptibility include climatic variables, slope morphology, land use and bedrock lithology. The research presented here highlights the potential role that clast form and fabric may play in enhancing the stability of steep moraine slopes, factors that should be included in landslide (especially debris flow) hazard assessment, monitoring and modelling on steep, glacigenic sediment mantles.

Conclusion

Although paraglacial reworking of sediments clearly has the capacity to induce slope instability and the movement of large volumes of sediment, given favourable lithological conditions, moraine slopes, such as those observed in the lateral-terminus area of the Feegletscher Nord, are able to stand at steep angles in a quasi-stable form on a decadal timescale. The current slope gradients of the lateral moraine reflect the mechanisms of original emplacement, post-glacial (paraglacial) modification, including translational failure, geotechnical properties, stratification and fabric, and high angles of dilation necessary for shallow translational shear to occur on the upper proximal slopes. Results from the shear box testing suggest: (i) that the distal and proximal debris slope angles are similar to the critical effective angles of friction for disturbed samples; and (ii) that a distal slope dip in fabric, throughout the undisturbed moraine, may assist in allowing proximal slopes to exceed peak effective angles of friction measured for disturbed samples in the laboratory. We propose a conceptual model, based on shallow translational shear failure, to explain the mechanisms that enable the preservation of distal, debris and very steep proximal slopes on the Feegletscher Nord and other similar moraines. However, detailed examination of *in situ* fabric samples from steep slope segments would be required to evaluate this model further, and it is difficult to envisage from a logistical point of view, how *in situ* fabric of these steeper slope segments could be quantitatively assessed. Detailed fabric analysis of actively forming lateral moraines at contemporary ice margins may provide additional information without the same level of logistical difficulty, although such studies would be unable to assess the role of factors such as glacifluvial reworking, removal of ice buttressing forces and enhanced paraglacial activity during deglaciation. The role of lithology and fabric in allowing steep slopes to exist in quasi-stable form requires further consideration from both a geomorphological point of view and by those considering the susceptibility of recently deglaciated terrain to mass movement hazards; wholesale slope instability is clearly not always an inevitable and immediate consequence of deglaciation and associated paraglacial activity.

This research was supported by a Nuffield Foundation Newly-Appointed Lecturer award to A. M. Curry. The authors thank J. Loring for field assistance, R. Delaloye and B. Schnyder for archive photography and site information, P. Coates for technical support and R. Langdon for assistance with laboratory testing.

References

BALLANTYNE, C. K. 1982. Aggregate clast form characteristics of deposits at the margins of four glaciers in the Jotunheimen Massif, Norway. *Norsk Geografisk Tidsskrift*, **36**, 103–113.

BALLANTYNE, C. K. 1995. Paraglacial debris cone formation on recently-deglaciated terrain. *The Holocene*, **5**, 25–33.

BALLANTYNE, C. K. 2002*a*. Paraglacial geomorphology. *Quaternary Science Reviews*, **21**, 1935–2017.

BALLANTYNE, C. K. 2002*b*. A general model of paraglacial landscape response. *The Holocene*, **12**, 371–376.

BALLANTYNE, C. K. & BENN, D. I. 1994. Paraglacial slope adjustment and resedimentation following recent glacier retreat, Fåbergstølsdalen, Norway. *Arctic and Alpine Research*, **26**, 255–269.

BARTON, N. R. 1973. Review of a new shear-strength criterion for rock joints. *Engineering Geology*, **7**, 287–332.

BENN, D. I. & BALLANTYNE, C. K. 1994. Reconstructing the transport history of glacigenic sediments: a new approach based on the co-variance of clast form indices. *Sedimentary Geology*, **91**, 215–227.

BENN, D. I., KIRKBRIDE, M. P., OWEN, L. A. & BRAZIER, V. 2005. Glaciated valley landsystems. *In*: EVANS, D. J. A. (ed.) *Glacial Landsystems*. Arnold, London, 372–406.

BENNETT, M. R. & GLASSER, N. F. 1996. *Glacial Geology: Ice Sheets and Landforms*. Wiley, London.

BIRCHER, W. 1982. *Zur Gletscher- und Klimageschichte des Saastales. Glazialmorphologische und dendroklimatologische Untersuchungen*. Phyische Geographie, **9**, Geographisches Institut der Universität, Zürich.

BS 1377: Part 4. 1990. *Methods of Test for Soils for Civil Engineering Purposes. Compaction-related Tests*. British Standards Institution, London.

BS 1377: Part 7. 1990. *Methods of Test for Soils for Civil Engineering Purposes. Shear Strength Tests (Total Stress)*. British Standards Institution, London.

BS 1377: Part 9. 1990. *Methods of Test for Soils for Civil Engineering Purposes. In-situ Tests*. British Standards Institution, London.

CHIARLE, M., IANNOTTI, S., MORTARA, G. & DELINE, P. 2007. Recent debris flow occurrences associated with glaciers in the Alps. *Global and Planetary Change*, **56**, 123–136.

CHURCH, M. & RYDER, J. M. 1972. Paraglacial sedimentation: a consideration of fluvial processes conditioned by glaciation. *Geological Society of America Bulletin*, **83**, 3059–3071.

CHURCH, M. & SLAYMAKER, O. 1989. Disequilibrium of Holocene sediment yield in glaciated British Columbia. *Nature*, **337**, 452–454.

CURRY, A. M. 1999. Paraglacial modification of slope form. *Earth Surface Processes and Landforms*, **24**, 1213–1228.

CURRY, A. M. 2000. Observations on the distribution of paraglacial reworking of glacigenic drift in western Norway. *Norsk Geografisk Tidsskrift*, **54**, 139–147.

CURRY, A. M., CLEASBY, V. & ZUKOWSKYJ, P. 2005. Paraglacial response of steep, sediment-mantled slopes to post-'Little Ice Age' glacier recession in the central Swiss Alps. *Journal of Quaternary Science*, **21**, 211–225.

EVANS, S. G. & CLAGUE, J. J. 1994. Recent climatic change and catastrophic geomorphic processes in mountain environments. *Geomorphology*, **10**, 107–128.

GK/SCNAT & VAW/ETZH. 2006. *The Swiss Glaciers, Yearbooks of the Glaciological Commission of the Swiss Academy of Science (SAS) Published by the Laboratory of Hydraulics, Hydrology and Glaciology (VAW) of ETH Zürich*, **1–122** (1881–2002). http://glaciology.ethz.ch/swiss-glaciers/.

HAEBERLI, W., WEGMANN, M. & VONDER MÜHLL, D. 1997. Slope stability problems related to glacier shrinkage and permafrost degradation in the Alps. *Eclogae Geologicae Helvetiae*, **90**, 407–414.

HARBOR, J. & WARBURTON, J. 1993. Relative rates of glacial and nonglacial erosion in alpine environments. *Arctic and Alpine Research*, **25**, 1–7.

HARRIS, C. 2005. Climate change, mountain permafrost degradation and geotechnical hazard. *In*: HUBER, U. M., BUGMANN, H. K. M. & REASONER, M. A. (eds) *Global Change and Mountain Regions*. Springer, Dordrecht, 215–224.

HARRIS, C., DAVIES, M. C. R. & ETZELMÜLLER, B. 2001. The assessment of potential geotechnical hazards associated with mountain permafrost in a warming global climate. *Permafrost and Periglacial Processes*, **12**, 145–156.

HARRIS, C., VONDER MÜHLL, D. *ET AL*. 2003. Warming permafrost in European mountains. *Global and Planetary Change*, **39**, 215–225.

HOLM, K., BOVIS, M. & JAKOB, M. 2004. The landslide response of alpine basins to post-Little Ice Age glacial thinning and retreat in southwestern British Columbia. *Geomorphology*, **57**, 201–216.

HSÜ, K. J. 1995. *The Geology of Switzerland: An Introduction to Tectonic Facies*. Princeton University Press, Princeton, NJ.

HUGGEL, C., HAEBERLI, W., KÄÄB, A., BIERI, D. & RICHARDSON, S. 2004. An assessment procedure for glacial hazards in the Swiss Alps. *Canadian Geotechnical Journal*, **41**, 1068–1083.

MATTHEWS, J. A. 1992. *The Ecology of Recently Deglaciated Terrain: A Geo-ecological Approach to Glacier Forelands and Primary Succession*. Cambridge University Press, Cambridge.

MATTHEWS, J. A. & PETCH, J. R. 1982. Within-valley asymmetry and related problems of Neoglacial lateral moraine development at certain Jotunheimen glaciers, southern Norway. *Boreas*, **11**, 225–247.

ORWIN, J. F. & SMART, C. C. 2004. The evidence for paraglacial sedimentation and its temporal scale in the deglacierizing basin of Small River Glacier, Canada. *Geomorphology*, **58**, 175–202.

OWEN, L. A. 1994. Glacial and non-glacial diamictons in the Karakoram Mountans and Western Himalayas. *In*: WARREN, W. P. & CROOT, D. G. (eds) *Formation and Deformation of Glacial Deposits*. Balkema, Rotterdam, 9–28.

OWEN, L. A. & DERBYSHIRE, E. 1989. The Karakoram glacial depositional system. *Geografiska Annaler*, **76**, 33–73.

PATTON, F. D. 1966. Multiple modes of shear failure in rock. *In*: *Proceedings of the 1st Congress of the International Society for Rock Mechanics, Lisbon*, Volume 1. 509–513.

RÖTHLISBERGER, F. & SCHNEEBELI, W. 1979. Genesis of lateral moraine complexes, demonstrated by fossil soils and trunks; indicators of post-glacial climatic fluctuations. *In*: SCHLÜCHTER, C. (ed.) *Moraines and Varves*. Balkema, Rotterdam, 387–419.

RÖTHLISBERGER, F., HAAS, P., HOLZHAUSER, H., KELLER, W., BIRCHER, W. & RENNER, F. 1980. Holocene climatic fluctuations – radiocarbon dating of fossil soils (fAh) and woods from moraines and glaciers in the Alps. *Geographica Helvetica*, **35**, 21–52.

SELBY, M. J. 1993. *Hillslope Materials and Processes*, 2nd edn. Oxford University Press, Oxford.

SLATT, R. M. 1971. Texture of ice-cored deposits from ten Alaskan valley glaciers. *Journal of Sedimentary Petrology*, **41**, 828–834.

SLETTEN, K., LYSÅ, A. & LØNNE, I. 2001. Formation and disintegration of a high-arctic ice-cored moraine complex, Scott Turnerbreen, Svalbard. *Boreas*, **30**, 272–284.

SMALL, R. J. 1983. Lateral moraines of Glacier De Tsidjiore Nouve: form, development and implications. *Journal of Glaciology*, **29**, 250–159.

SWISS GEOLOGICAL COMMISSION. 1980. *Carte géologique de la Suisse, 1:500,000*, 2nd edn. Swiss Geological Commission, Bern.

WHALLEY, W. B. 1975. Abnormally steep slopes on moraines constructed by valley glaciers. *In*: *Proceedings of the Midland Soil Mechanics and Foundation Engineering Society Symposium*: *The Engineering Behaviour of Glacial Materials, University of Birmingham*, 21–23 April, 1975, 60–66.

YOUNG, A. & SAUNDERS, I. 1986. Rates of surface processes and denudation. *In*: ABRAHAMS, A. D. (ed.) *Hillslope Processes*. Allen & Unwin, Boston, MA, 3–27.

ZINGG, T. 1935. Beiträge zur Schotteranalyse: Die Schotteranalyse und ihre Anwendung auf die Glattalschotter. *Schweizerische Mineralogische und Petrographische Mitteilungen*, **15**, 39–140.

Fluvial response to Holocene glacier fluctuations in the Nostetuko River valley, southern Coast Mountains, British Columbia

KENNA WILKIE & JOHN J. CLAGUE*

Department of Earth Sciences, Simon Fraser University, 8888 University Drive, Burnaby, British Columbia, Canada V5A 1S6

**Corresponding author (e-mail: jclague@sfu.ca)*

Abstract: Mountain rivers, like alpine glaciers, are sensitive indicators of climate change. Some rivers may provide a more complete record of Holocene climate change than the glaciers in their headwaters. We illustrate these points by examining the record preserved in the upper part of the alluvial fill in the Nostetuko River valley in the southern Coast Mountains, British Columbia (Canada). Glacier advances in the upper part of the watershed triggered valley-wide aggradation and complex changes in river planform. Periods when glaciers were restricted in extent coincide with periods of incision of the valley fill and floodplain stability. As many as 10 overbank aggradation units are separated by peat layers containing tree roots and stems in growth position. Twenty-five radiocarbon ages on roots, tree stems and woody plant detritus in several of the peat layers closely delimit periods of aggradation. The oldest phase of aggradation occurred about 6500 years BP and coincides with the Garibaldi Advance documented elsewhere in the southern Coast Mountains. A second phase of aggradation, recorded by several units of clastic sediment, dates to about 2500 years BP, near the peak of the middle Neoglacial Tiedemann Advance. The third phase occurred shortly after 1400 years BP during or shortly after the First Millennium Advance, which has been recently documented in coastal British Columbia and Alaska. The most recent phase of aggradation began about 800 years BP and continued until recently. It coincides with the Little Ice Age, when glaciers in the Nostetuko River basin and elsewhere in the southern Coast Mountains attained their greatest Holocene size. Several periods of peat deposition during the Little Ice Age indicate periods of floodplain stability separated by brief intervals of floodplain aggradation that coincide with Little Ice Age glacier advances in western Canada. The results imply that the west fork of Nostetuko River is sensitive to upvalley glacier fluctuations and, indirectly, to relatively minor changes in climate.

The proglacial fluvial archive is a largely unexploited source of information on upvalley glacier fluctuations. Streams may respond to fluctuations of glaciers in their headwaters by aggrading up or incising their floodplains. The resulting changes in local base level can be preserved in the valley-fill stratigraphy. Although potentially difficult to decipher, valley-fill stratigraphies may be more complete than the record of glacier fluctuations derived from landforms and sediments the forefields themselves. At the very least, they complement and strengthen the glacier forefield evidence.

This paper documents the response of the west fork of the Nostetuko River valley, located in the southern Coast Mountains of British Columbia (Canada), to changes in sediment supply during Neoglaciation – the last half of the Holocene. We have two objectives: first, to add to the knowledge of Holocene glacier fluctuations in British Columbia; and second, and more generally, to demonstrate the potential of fluvial archives for deciphering past alpine glacier activity. Field inspection of the upper part of the sediment fill revealed a series of clastic sediment units interstratified with peats containing rooted stumps that were subsequently radiocarbon dated. We show that sediment supply is intimately linked to fluctuations of glaciers at the head of the valley. Radiocarbon ages on stumps at the tops of the peat layers closely constrain times of glacier advances within the watershed. These times agree with those determined independently by other researchers working elsewhere in western North America.

Study area

The study area is the valley of the west fork of Nostetuko River in the southern Coast Mountains of British Columbia, 220 km north of Vancouver (Fig. 1). The west fork flows 11 km north and east from its source to the main stem of Nostetuko River (Fig. 2). It is fed mainly by meltwater from valley glaciers at the edge of Homathko Icefield. A major tributary of the west fork flows from Queen Bess Lake, a moraine-dammed lake that partially drained during an outburst flood in August 1997 (Kershaw 2002; Kershaw *et al.* 2004).

Queen Bess Lake is impounded by a large composite moraine produced by at least two advances of Diadem Glacier (Kershaw 2002). The lake formed

From: KNIGHT, J. & HARRISON, S. (eds) *Periglacial and Paraglacial Processes and Environments.*
The Geological Society, London, Special Publications, **320**, 199–218.
DOI: 10.1144/SP320.13 0305-8719/09/$15.00

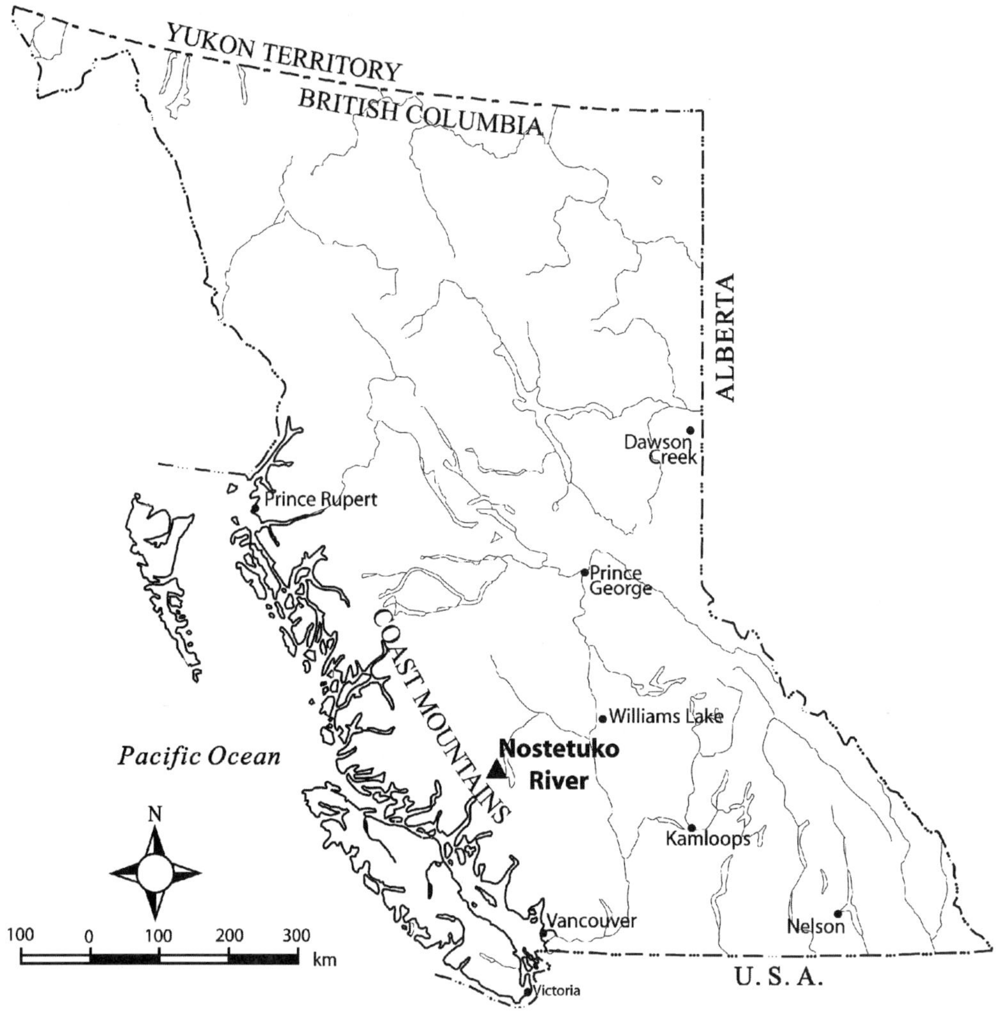

Fig. 1. Location of the study area in the southern Coast Mountains of British Columbia (modified from BMGS data; reproduced with permission of the Province of British Columbia).

behind the composite moraine during glacier retreat in the late 1800s and early 1900s. In 1997, a large ice avalanche fell from the toe of Diadem Glacier into Queen Bess Lake, generating displacement waves that overtopped and incised the moraine. The resulting flood eroded sediments in the valley below the dam, causing aggradation upstream and downstream of channel constrictions (Fig. 2). It also created exposures of the upper part of the valley fill, which enabled this study.

Methods

The aggradation history of the west fork of Nostetuko River was determined through stratigraphic, sedimentological and geochronological analyses of sections. Fieldwork was conducted during the summer of 2004. Detailed topographic maps (1:5000 scale), constructed from aerial photographs flown in 1998 – one year after the outburst flood – were used to map deposits and landforms. Locations of sections, terraces, trimmed colluvial fans and tree stumps exhumed by river incision were located (± 10 m) using a hand-held GPS unit and cross-referenced with the topographic maps. These data were subsequently entered into a Geographic Information System (GIS).

Detailed sedimentological and stratigraphic logs were made of exposed valley-fill sediments at seven sites (Fig. 2), and additional observations of sediments and landforms were made at many other locations. Sections were logged using a metric

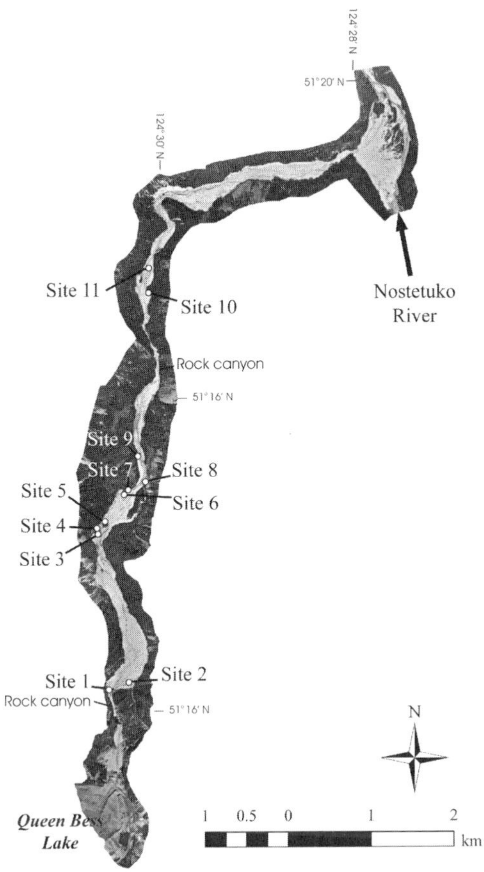

Fig. 2. Aerial photomosaic of the valley of the west fork of Nostetuko River, showing locations of studied sections. The aerial photographs were flown in 1998, 1 year after the outburst flood.

tape and a barometric altimeter (elevation accuracy of ±5 m). Recorded data included grain size (field estimates), sorting, sedimentary structures, Munsell colour, unit thickness, the nature of unit contacts and fossil plant remains.

Samples of *in situ* tree stems, fossil roots and detrital plant fossils were collected from peat layers and rooting horizons, and submitted to Beta Analytic for conventional (radiometric) ^{14}C analysis. Radiocarbon ages were calibrated using the software OxCal v. 4.0 (Bronk Ramsey 1995, 2001), which is based on the decadal data of Stuiver *et al.* (1998). The radiocarbon ages provide chronological control on periods of stability and aggradation in the valley.

Disks of radiocarbon-dated *in situ* fossil conifer stumps and logs in exposed peat layers were collected for tree-ring analysis. Samples were air-dried and sanded several times with progressively finer sand paper. Annual tree-ring widths were measured to the nearest 0.001 mm along up to four radii for each tree sample using a Velmex-type measuring stage, a Leitz stereomicroscope and the Measure J2X measuring program. Samples were cross-dated to establish floating chronologies by visually comparing marker rings and by employing the statistical correlation and verification procedures within the ITRDBL (International Tree-Ring Data Bank Library) tree-ring dating software program COFECHA (Holmes 1999; Grissino-Mayer 2001). Segments that were not significantly correlated were re-measured and corrected to account for radial growth anomalies and missing or false rings.

The age of the oldest living tree on a surface provides a minimum age for that surface, after corrections for local ecesis and sampling height have been applied (McCarthy *et al.* 1991; Wiles *et al.* 1999). Ecesis, defined as the time between surface stabilization and germination of the first seedling, has been shown to range from 1 to 100 years in the Pacific Northwest (Sigafoos & Hendricks 1969; Desloges & Ryder 1990; McCarthy *et al.* 1991; Smith *et al.* 1995; Wiles *et al.* 1999; Luckman 2000; Lewis & Smith 2004). Ecesis intervals of 1–4 years have been documented in the Coast Mountains at Tiedemann Glacier (Larocque & Smith 2003), and on Vancouver Island at Colonel Foster and Septimus glaciers (Lewis & Smith 2004). Seedlings growing on the floodplain scoured by the 1997 Queen Bess outburst flood were no more than 5 years old when we conducted fieldwork in 2004, suggesting that ecesis in the west fork valley is 1–2 years. Two years were therefore added to the outer ring ages of *in situ* stumps to correct for ecesis.

Sampling height errors occur when annual growth rings are lost due to sampling above the root crown (McCarthy *et al.* 1991). Larocque & Smith (2003) proposed a regional correction factor of 1.35 cm $year^{-1}$ for subalpine fir seedlings on valley floors in the Mount Waddington area. This correction was applied to all *in situ* stumps. Sampling height corrections cannot be applied to detrital logs.

Results

Geomorphology

The west fork of Nostetuko River flows through rugged terrain with local relief of up to 2000 m. Alluvial reaches are separated by short rock canyons located approximately 1 and 6 km north of Queen Bess Lake (Fig. 2). The canyons control valley gradients and local base levels. The average gradient of the west fork of the river is 3.8°, but it ranges from a maximum of 14° in the upper rock canyon to a minimum of about 0.5° along broad alluvial reaches (Kershaw 2002).

The river flows over sediments deposited by the 1997 outburst flood (Kershaw 2002; Kershaw *et al.* 2004). The flood sediments, which are nested within the fluvial sediments and peats that are the focus of this study, are poorly sorted, cobble–boulder gravel. Sand and silty sand were deposited locally along the margins of the flood path and in areas upstream of constrictions, where hydraulic ponding occurred during the outburst flood.

The flood altered the planform of the west fork of Nostetuko River. Prior to the flood, the west fork had a single dominant channel with local low-gradient, multi-channel reaches. The flood changed the position of the main channel and temporarily imposed a braided planform on the active floodplain. Over the past 10 years the west fork has incised its flood deposits and partly re-established its pre-flood form.

Lateral migration of the west fork is constrained by the steep valley slopes and colluvial fans and aprons draping the valley walls. The fans and aprons are eroded at times during floods and, thus, are an important source of sediment to Nostetuko River. A large moraine fan is located on the west side of the valley 1.5 km downvalley of Queen Bess Lake (site 1 in Fig. 2; see also Fig. 3). The fan comprises bouldery colluvium deposited on the distal side of a terminal moraine constructed during middle and late Neoglacial time. It is an important source of sediment to the west fork of Nostetuko River.

Four terraces are inset into a large gravel fan directly below the upper bedrock canyon and adjacent to the moraine fan mentioned above (site 2, Fig. 2; see also Fig. 4). The upper two terraces (T-1 and T-2) support sharp-crested boulder levees composed of crudely bedded boulder and cobble gravel. The levees are bordered by better-sorted cobble gravel associated with relict channels. The uppermost terrace (T-1) is partly vegetated, and it and T-2 support lichens (*Rhizocarpon* spp) up to 3 cm in diameter. The lower two terraces (T-3 and T-4) were swept by the 1997 outburst flood and, thus, lack lichens.

Aggradation of the outwash fan to the T-1 level probably occurred during the Little Ice Age when the outer, sharp-crested terminal moraine at the east end of Queen Bess Lake was constructed by Diadem Glacier (Kershaw 2002). T-2 is inset into, and therefore younger than, T-1. It probably dates to the late 1800s or early 1900s (Kershaw 2002). T-3 records the upper limit of aggradation of the 1997 flood. T-4 formed during the waning stage of the flood or soon thereafter.

A 30 cm-thick, buried soil exposed in the scarp of T-2 contains *in situ* stumps and abundant detrital plant material (Fig. 5). Kershaw (2002) reported a radiocarbon age of 370 $\pm$ 50 ^{14}C years BP (564–372 cal years BP; Table 1) on a root in the soil. The soil is underlain and overlain by cobble–boulder gravel. The stratigraphy at this site demonstrates that T-2 formed some time after AD 1400 (Kershaw 2002). The buried soil rests on a floodplain that records a bed elevation 8 m higher than present.

Study sites 3–11 are located on the floor of the west fork of Nostetuko River 3.4–6.6 km downvalley of Queen Bess Lake (Fig. 2). These study sites include riverbank exposures (sites 5, 7, 8, 9 and 10), the near-vertical wall of a channel eroded by the 1997 flood (sites 3 and 4), and localities where rooted stumps on the valley floor were exhumed by the flood (sites 6 and 11).

Terraces are uncommon along the west fork, but notable exceptions occur 3.5 and 7 km north of Queen Bess Lake (Fig. 2). Both terraces are 1–2 m above present river level, support forest at least 100 years old and were not inundated by the 1997 flood. They record a higher bed elevation prior to the twentieth century and may correlate with terrace T-2 further upvalley.

Fig. 3. Photomosaic of the eroded distal face of a large moraine fan located 1.5 km north of Queen Bess Lake (site 1, Fig. 2). The moraine was built during middle and late Neoglacial time.

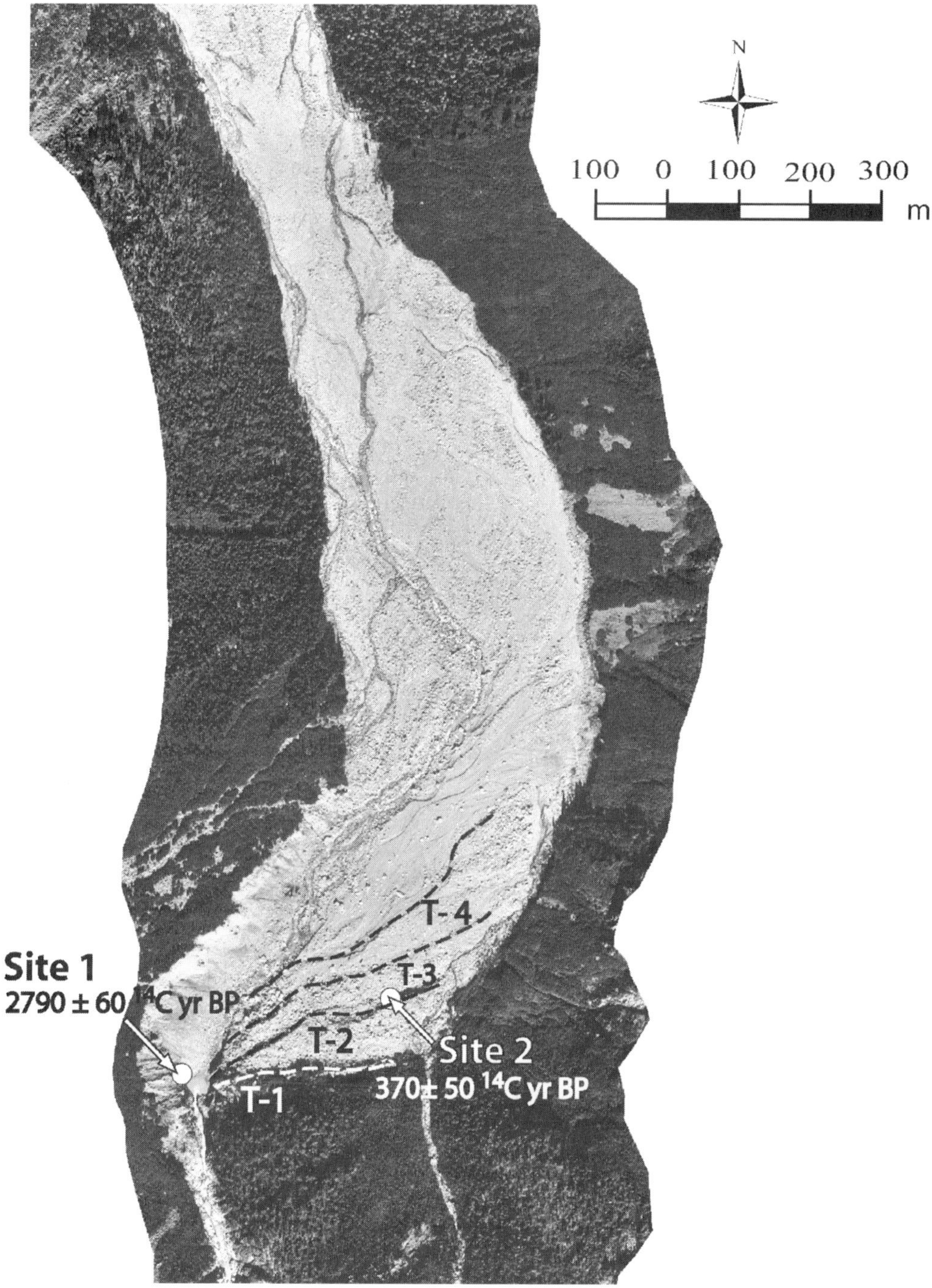

Fig. 4. Four terraces (T-1, T-2, T-3 and T-4) inset into a gravel fan below the upper bedrock gorge (site 2, Fig. 2). An *in situ* root in the scarp below T-2 yielded a radiocarbon age of 370 $\pm$ 50 ^{14}C years BP (Kershaw 2002).

Fig. 5. Scarp between terraces T-2 and T-3, showing organic soil overlain and underlain by cobble–boulder gravel. Dashed lines bracket the dated soil.

Sedimentology

In general, only the uppermost several metres of the Nostetuko valley sediment fill are exposed. Although this part of the fill varies both laterally and vertically, careful lithostratigraphic logging of sections revealed four dominant lithofacies, which are briefly described and interpreted below. The description excludes the capping 1997 outburst flood deposits because they have been described elsewhere (Kershaw 2002; Kershaw *et al.* 2004).

Gravel facies. Gravel units are present at six of the seven valley-floor sections (Figs 6 and 7). They are pebble-cobble in size, dominantly clast-supported, horizontally bedded, and locally imbricated and iron-stained. Clasts are subangular–well rounded, and the matrix comprises sand and granules. Gravel occurs at the base of four sections (e.g. Fig. 8) and as discontinuous lenses up to 15 cm thick within finer grained sediments at six sections (Fig. 6).

The gravel facies records deposition in high-energy channels of braided or wandering rivers. Horizontal stratification and clast imbrication suggest deposition on near-horizontal surfaces such as braid bars, medial and lateral bar complexes, and channel floors. These environments are common in braided and wandering gravel-bed rivers (Bluck 1979; Church 1983; Desloges & Church 1987; Brierley 1996; Sambrook Smith 2000; Lewin *et al.* 2005).

Sand facies. Massive and stratified sand is the dominant sediment of the uppermost part of the valley fill (Figs 6–8). Tabular and lenticular beds of mottled and oxidized, massive, well-sorted, very fine–medium sand occur at all sites. Beds of planar cross-stratified and ripple-stratified, medium–coarse sand are also common. Horizontally laminated fine–very fine sand is interstratified with the coarser sand. Laminae are typically flat to undulating. Small-scale, trough cross-stratified, fine–medium sand occurs in the uppermost 0.5 m of the sequence at three sites (Fig. 6). Sand beds commonly have sharp lower contacts and sharp–gradational upper contacts (Fig. 9).

The sand facies records deposition in channels, bars and levees. Rippled and horizontally bedded sand may have been deposited under a range of flow conditions, from lower-flow regimes in back

Table 1. *Radiocarbon ages from the west fork of Nostetuko River valley*

Radiocarbon age (^{14}C years BP)*	Calibrated age (cal years BP)†	Laboratory No.‡	Site No.	Location		Elevation (m)	Dated material	Unit
				Lat. (N)	Long. (W)			
110 ± 60	340–64	TO-8935	7	51° 17′ 24″	124° 30′ 18″	1378	*in situ* root	
130 ± 50	340–64	Beta-200730	7	51° 17′ 24″	124° 30′ 18″	1378	*in situ* root	
150 ± 60	346–57	TO-8932	3	51° 17′ 10″	124° 30′ 31″	1386	outer rings of *in situ* stump	7
270 ± 50	538–55	Beta-200723		51° 16′ 45″	124° 30′ 18″	1408	outer rings of *in situ* stump	
370 ± 50	564–372	TO-8923	2	51° 16.3′	124° 30.1′	1493	*in situ* root	
470 ± 60	696–378	TO-8942	10	51° 18′ 40″	124° 30′ 05″	1294	*in situ* root	
520 ± 50	705–556	Beta-200727	4	51° 17′ 10″	124° 30′ 32″	1388	outer rings of *in situ* stump	
530 ± 60	714–555	TO-8933	7	51° 17′ 24″	124° 30′ 18″	1373	outer rings of *in situ* stump	
580 ± 50	714–583	Beta-200726	3	51° 17′ 10″	124° 30′ 31″	1384	herbaceous plant tissue	6
600 ± 60	721–592	Beta-200733	5	51° 17′ 13″	124° 30′ 29″	1384	root	
620 ± 50	725–599	Beta-200725	6	51° 17′ 18″	124° 30′ 18″	1377	outer rings of *in situ* stump	
700 ± 60	791–610	Beta-200729	7	51° 17′ 24″	124° 30′ 18″	1376	*in situ* stump	
710 ± 60	797–609	Beta-200734	4	51° 17′ 10″	124° 30′ 32″	1387	*c.* 10 outer rings of log	
940 ± 50	992–798	Beta-200728	4	51° 17′ 10″	124° 30′ 32″	1388	outer rings of *in situ* stump	
990 ± 50	1108–839	TO-8931	3	51° 17′ 10″	124° 30′ 31″	1385	root	5
1030 ± 50	1117–856	Beta-200731	7	51° 17′ 24″	124° 30′ 18″	1375	outer rings of *in situ* stump	
1160 ± 50	1293–1019	Beta-200732	9	51° 17′ 37″	124° 30′ 09″	1395	outer rings of *in situ* stump	
1280 ± 60	1357–1127	Beta-200736	10	51° 18′ 40″	124° 30′ 05″	1293	outer rings of *in situ* stump	4
1300 ± 70	1389–1122	TO-8941	10	51° 18′ 40″	124° 30′ 05″	1293	outer rings of *in situ* stump	
2340 ± 60	2758–2216	Beta-200737	10	51° 18′ 40″	124° 30′ 05″	1291	outer rings of *in situ* stump	
2390 ± 70	2776–2380	TO-8943	10	51° 18′ 40″	124° 30′ 05″	1292	*in situ* root	3
2450 ± 70	2776–2413	TO-8939	10	51° 18′ 40″	124° 30′ 05″	1291	twig	
2490 ± 70	2796–2423	TO-8940	10	51° 18′ 40″	124° 30′ 05″	1291	*in situ* root	
2790 ± 4940	3123–2826	Beta-200721	1	51° 16′ 10″	124° 30′ 24″	1494	branch	2
5810 ± 70	6840–6506	Beta-200735	11	51° 18′ 53″	124° 30′ 04″	1285	outer rings of *in situ* stump	1

*Ages have been corrected for natural and sputtering fractionation to a base of $\delta^{13}C = 25.0‰$.
†Determined from atmospheric decadal data set of Stuiver *et al.* (1998) using the program OxCal v.4. The range represents the 95.4% confidence limits. The datum is AD 2009.
‡Laboratories: Beta – Beta Analytic Inc.; TO – IsoTrace Laboratory (University of Toronto).

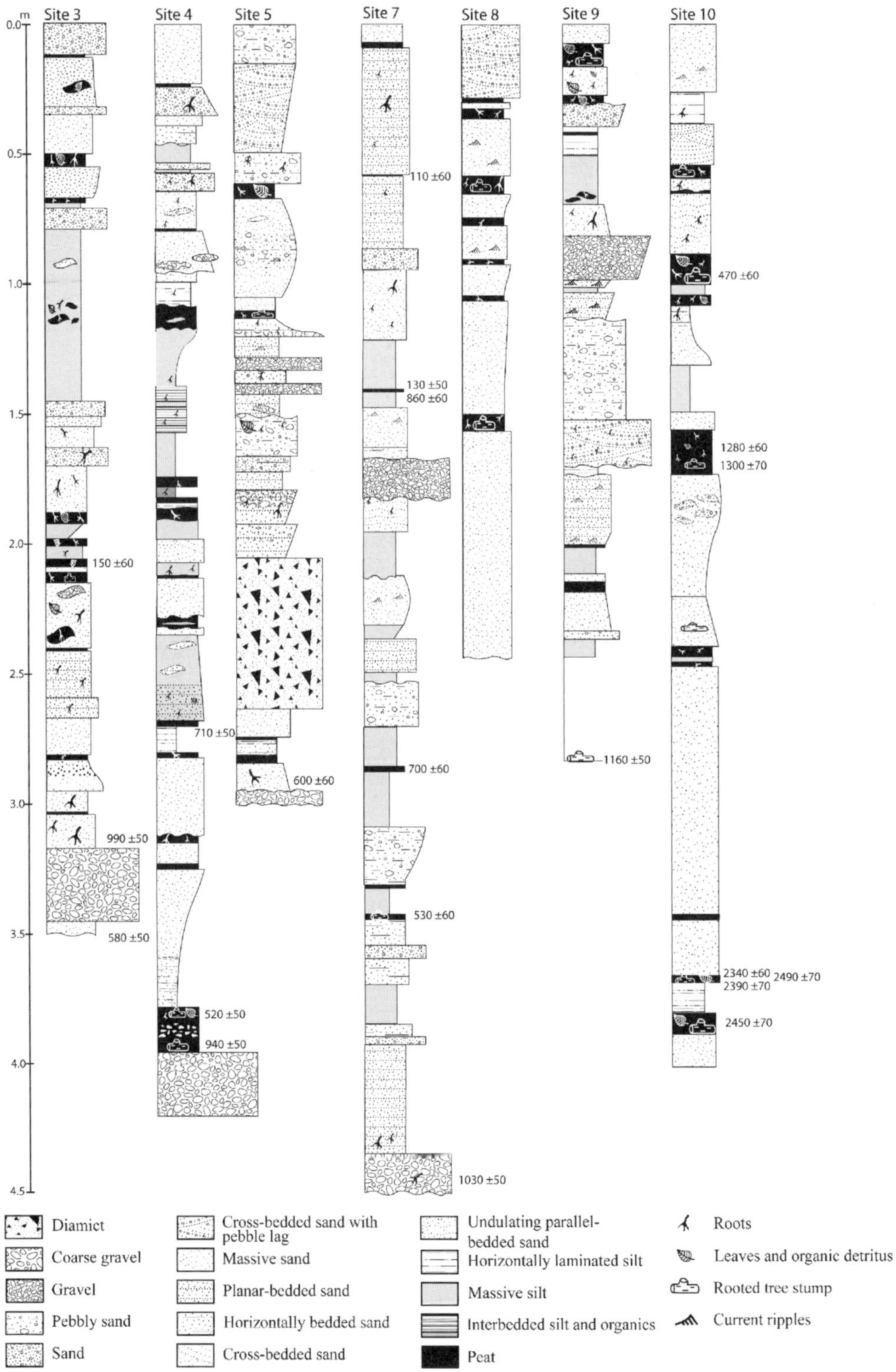

Fig. 6. Lithostratigraphy of late Holocene sediments exposed at seven sites in the study area. Peat layers are shown in black. See Figure 2 for site locations and Table 1 for details on radiocarbon ages.

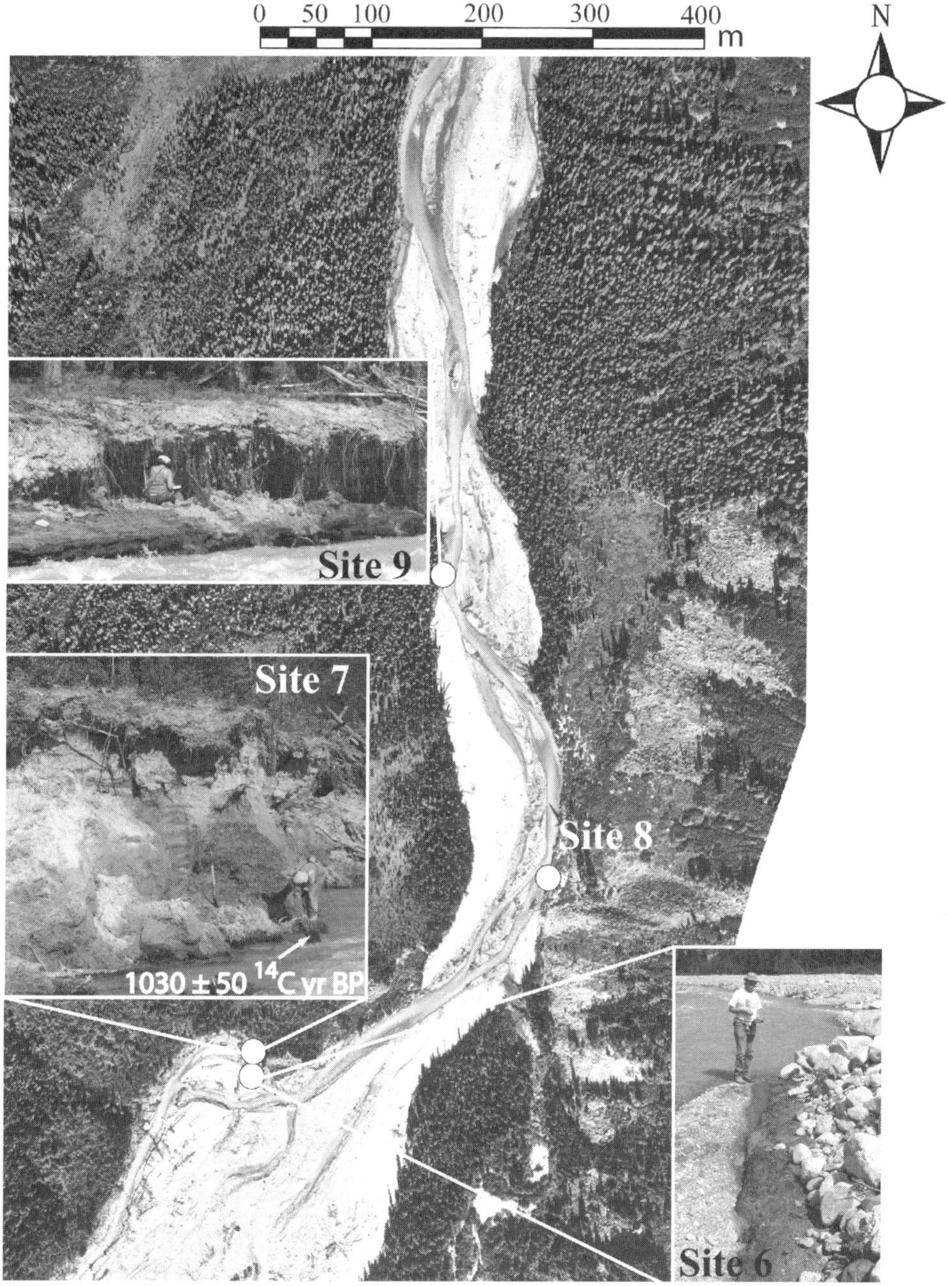

Fig. 7. Examples of sections documented in this study.

channels to upper-flow regimes in the main active channels (Desloges & Church 1987). Discontinuous, lenticular beds of structureless to cross-stratified sand were deposited in channels immediately before they were abandoned. Trough cross-stratified sand records in-channel migration of large lunate ripples and bars. These bedforms are typically transitory, as variable flow depths and velocities impede preservation (Desloges & Church 1987).

Fine facies. The fine facies consists of massive, bedded, and laminated very fine sand, silt and

Fig. 8. The lower part of the section at site 4. A basal gravel unit is sharply overlain by silt, sand and peat beds. The two arrowed stumps are rooted in two peat beds separated by about 10 cm. Dashed lines delineate the peat beds. Radiocarbon ages of 520 ± 50 and 940 ± 50 ^{14}C years BP (Table 1) were obtained on the outer rings of the two stumps.

minor clay (Fig. 10). Strata range from laminae a few millimetres thick to beds up 17 cm thick. Fine sediments dominate sections of valley fill up to 1 m thick, but isolated, lenticular strata are also common. Sediments are mainly olive grey, but locally are oxidized and mottled. Interbeds of coarser sand, fibrous peat and plant detritus occur within the fine facies.

The laminated and bedded fine sediments were deposited in an overbank depositional environment. Sediments commonly fine upwards from an erosive base, reflecting scour during the rising stage of a flood, followed by deposition during the waning stage (Brierley 1996). Massive, very fine sand and silt record either rapid deposition during floods or bioturbation (Collinson 1996).

Organic facies. Beds and laminae of brown peat and silty peat are abundant at all sites (e.g. Figs 6, 8 and 10). The strata range from a few millimetres to 18 cm

Fig. 9. Upper 2 m of the section at site 5, showing a prominent peat layer, an *in situ* stump and an oxidized horizon (directly above the tip of the trowel). The heavy dashed lines delineate the peat layer, and the dotted lines mark contacts between sand beds. A radiocarbon age of 600 ± 60 ^{14}C years BP was obtained on a root near the base of this section (not shown in the photograph).

thick. Thicker layers comprise woody and herbaceous peat mats with tree stumps in growth position (Figs 8–10). Roots of herbaceous plants and trees extend downwards from the organic horizons into underlying silt and sand. Contacts with overlying sediment are typically sharp, whereas basal contacts are either sharp or gradational.

The organic facies records soil development and peat accumulation on poorly drained, stable floodplain surfaces. River-bed elevation was either stable or dropping at times of peat deposition and soil development. Some peat layers are unconformably overlain by coarse sand or gravel, indicating that they were eroded and buried during aggradation following the stable floodplain phase.

Stratigraphy

Correlation of strata in high-energy proglacial fluvial systems based solely on lithofacies is difficult. Thicknesses and lithologies of units differ markedly over short distances, and peaty soils present at one site may be missing at others owing to erosion.

Fig. 10. Upper 2 m of the section at site 10, showing peat layers, laminated fine-grained sediments and a stump in growth position. Dashed lines delineate peat layers. A radiocarbon age of 470 $\pm$ 60 ^{14}C years BP was obtained on an *in situ* root (not shown in the photograph).

Sequences of gravel and sand facies alternate with sequences dominated by fine sediment at all of the studied sections in the west fork Nostetuko valley (Figs 6 and 9). Coarse gravel is exposed at the base of four sections in sharp contact with overlying fine sand. Gravel also overlies massive and laminated silt higher in the sequence at sites 7 and 9. Contacts between fine-grained sediments and overlying coarser sand and gravel are sharp and erosive. Coarse sand and granule-rich sand dominate the uppermost 1.5 m of the valley-fill sequence at all sites. These sediments overlie massive and laminated silt with layers of peat and plant detritus.

All sections have multiple peat layers that record episodic floodplain stability. Units of sand and silt with abundant plant matter alternate with the peat layers. Fine-grained beds commonly have gradational contacts with overlying peat layers, whereas the latter are typically sharply overlain by silt or sand.

The vertical succession of sediments, with numerous erosion surfaces and abrupt changes in facies, indicates frequent changes in river stage and channel position (Miall 1977, 1978). Laterally discontinuous units of sand and gravel within fine-grained sediment are typical of near-channel floodplain environments (Smith 1983; Marren 2005). Sequences of silt suggest periods of overbank deposition and floodplain accretion, whereas coarser sediment probably records deposition in channels.

Chronology

Age constraints provided by radiocarbon dating and stratigraphic relations allow tentative correlation of some peat layers and documentation of periods of floodplain stability. Dark brown, herbaceous peat mats at sites 6 and 9 contain discontinuous lenses of massive fine sand and silt, and may correlate (Fig. 6). Two thin, dark brown peat layers at sites 3 and 4 occur at depths of 2.3 and 2.1 m, respectively. At both sites they are separated by about 2 cm of fine sand and silt. Peat layers just above the basal gravel units at these sites and also at site 7 are correlated on the basis of radiocarbon ages obtained on *in situ* roots. Radiocarbon ages on tree stumps allow correlation of a thick, herbaceous peat at site 10 with the submerged peat bed at site 9.

Twenty-five samples of wood were radiocarbon dated, all but one for this study (Table 1). Samples were chosen to date peat beds that had been tentatively correlated based on stratigraphic relations. Nineteen of the 25 samples are outer rings of fossil trees in growth positions at or near the tops of peat layers. Their ages are interpreted to be the time of death, and, presumably, burial of the trees, and thus closely limit times of aggradation. The other six samples are fragile branches and twigs in peat that are unlikely to have been reworked. The ages of these six samples, nevertheless, must be considered maxima for the age of the sediments from which they were collected.

The oldest radiocarbon age (5810 ± 70 ^{14}C years BP) is from the outer 10 rings of a stump rooted in a peat below modern river level at the downvalley end of the study area (site 11, Fig. 2). A branch at the base of the moraine fan below the upper bedrock canyon (site 1) gave an age of 2790 ± 60 ^{14}C years BP. The two youngest radiocarbon ages, 110 ± 60 and 130 ± 50 ^{14}C years BP, are from silty and sandy peat layers within the uppermost 1.5 m of sediment at site 7.

Cluster analysis was performed on the suite of 25 radiocarbon ages to determine whether the ages are randomly distributed or grouped. Two methods (average linkage between clusters or Euclidean distances; and Ward's method with squared Euclidean distances) gave the same age groups when seven clusters were specified (Fig. 11). The age groupings are consistent with the provisional correlations of peat layers based on field observations and stratigraphic relations at measured sections.

The oldest 'group' is actually a single radiocarbon age of 5810 ± 70 ^{14}C years BP, obtained

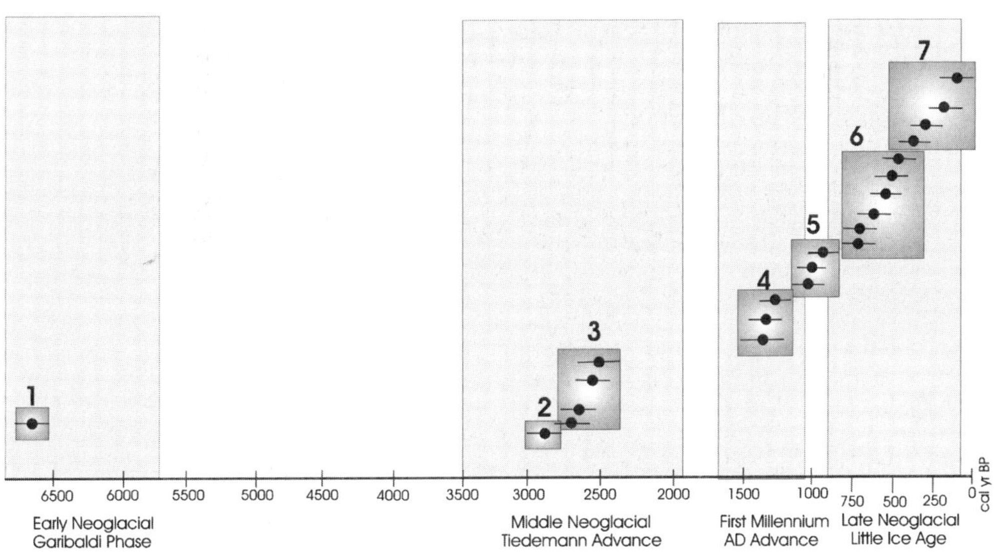

Fig. 11. Plot of radiocarbon ages obtained for this study and their relation to independently dated Neoglacial glacier advances in western Canada. Cluster analysis placed the radiocarbon ages into seven groups.

from the outermost rings of a rooted stump at site 11. The second group is a radiocarbon age of 2740 ± 60 ^{14}C years BP at site 1. The third group comprises four ages at site 10, centred on 2400 ^{14}C years BP. The fourth group consists of three radiocarbon ages ranging from 1300 ± 70 to 1160 ± 50 ^{14}C years BP at sites 9 and 10. The three samples that yielded these ages are growth-position fossils in a thick peat. The fifth group includes three ages ranging from 1030 ± 50 to 940 ± 50 ^{14}C years BP at sites 3, 4 and 7. The sixth group comprises eight ages ranging from 710 ± 60 to 470 ± 60 ^{14}C years BP at sites 3–7 and 10. The seventh group includes five ages ranging from 370 ± 50 to 110 ± 50 ^{14}C years BP at sites 2, 3 and 7. The greater abundance of ages in groups 6 and 7 reflects the better preservation of the youngest sediments and, thus, some bias in sampling.

Discussion

Ages of peat layers and aggradation episodes

The oldest stable floodplain recorded in this study is the surface at the north end of the study area (site 11), dated at 5810 ± 70 ^{14}C years BP (6840–6506 calendar years (cal years) BP). Bed elevation at this time was lower than today because the dated tree and others in the same area are rooted below present river level.

A radiocarbon age of 2790 ± 60 ^{14}C years BP (3123–2826 cal years BP) on a branch at the base of the moraine fan at site 1 indicates that moraine construction began at or shortly after this time. Cluster analysis separates this age from a group of four, statistically equivalent, radiocarbon ages, which mark the beginning of a period of significant aggradation about 2500 cal years ago at site 10.

The fourth and fifth groups of radiocarbon ages record floodplain stability at, respectively, about 1300–1200 (1400–1000 cal years BP) and 1000 ^{14}C years BP (1100–800 cal years BP). Both periods of floodplain stability were followed by aggradation. The ages are youngest at upvalley sites, which are nearer sediment sources, and oldest at downvalley, more distal sites.

Cluster analysis separates the youngest radiocarbon ages into two groups. The older group represents at least two peat beds that range in age from 710 ± 60 (797–609 cal years BP) to 470 ± 50 ^{14}C years BP (696–378 cal years BP) and occur primarily in fine-grained sediments. The younger group of radiocarbon ages is derived from peat layers in coarse sandy sediments. The ages range from 370 ± 50 (564–372 cal years BP) to 110 ± 60 ^{14}C years BP (340–64 cal year BP). Considering all the ages in groups 6 and 7, and their stratigraphic context, we suggest that there were several phases of aggradation during the Little Ice Age – an early phase about 600 cal years BP, one or more subsequent phases after 600 cal years BP, but before 370 cal years BP, and one or more phases late during the Little Ice Age, after 340 cal years BP, but before the beginning of the twentieth century.

We cross-dated the ring series of stumps rooted in Little Ice Age peat layers in order to better delimit durations of periods of floodplain stability. The length of time spanned by each peat layer, which must equal or exceed the lifespan of its associated trees, provides a constraint on the ages of stable and aggradation intervals. The length of time between the death of trees on a floodplain surface and the inception of tree growth on the next younger surface is a maximum for the period of intervening aggradation.

Three floating ring series were established for Little Ice Age peat layers containing rooted stumps (Fig. 12) (Wilkie 2006). One floating ring series is anchored by the radiocarbon age of 620 ± 50 ^{14}C years BP on outer rings of an *in situ* stump in peat. Cross-dating of this sample with ring series of other stumps in the same peat bed ($r = 0.434$, significant at the 99% confidence level) produced an uncorrected floating chronology of 176 years. Corrections for ecesis and sampling height extended the interval by 5 years to 181 years. Individual tree series end within 14 and 19 years of one another. The outer surfaces of the analysed samples are weathered, consequently an unknown number of rings have been lost and precise kill dates are unknown.

A second floating ring series is associated with a radiocarbon age of 370 ± 50 ^{14}C years BP on an *in situ* stump, reported by Kershaw (2002). A disk from an adjacent stump on the same surface was cross-dated with two other samples 2 km downvalley ($r = 0.420$, significant at the 99% confidence level). The uncorrected chronology records surface stability for a minimum of 237 years. This interval was extended to 254 years by applying corrections for ecesis and sampling height (Fig. 12). The minimum kill dates are within 12 and 34 years of each other.

A stump with 127 annual rings in the youngest peat layer cross-dated into a living subalpine fir master chronology of Larocque & Smith (2003) ($r = 0.438$, significant at the 99% confidence level). Assuming the cross-date is correct, the ring series dates to 168–41 cal years BP. With the addition of ecesis and sampling height corrections, the stump has a lifespan of about 142 years and dates to 183–41 cal years BP (Fig. 12).

In summary, three intervals of surface stability and peat deposition during the Little Ice Age span, from oldest to youngest, more than 181, 254 and

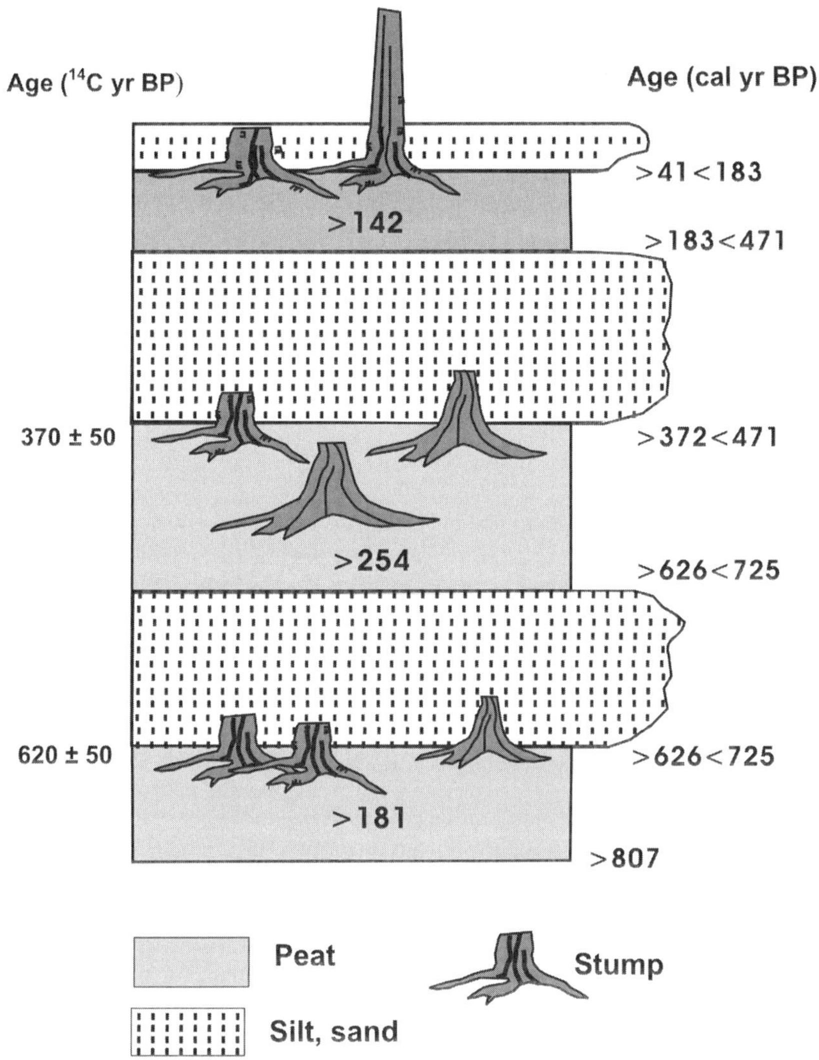

Fig. 12. Schematic diagram showing chronological framework of late Holocene peat and clastic units in the west fork of Nostetuko River valley. Unit thicknesses are representative of those at measured sections. Datum for calendar year ages is AD 2009.

142 years (i.e. the number of years in the floating tree-ring series). The three stable intervals were followed by three periods of aggradation, during which the peat layers were buried.

Times of Little Ice Age floodplain stability and aggradation can be further constrained by considering: (1) the extreme limits of the calendric age ranges; and (2) the minimum duration of deposition of the peat layers (Fig. 12). The youngest of the three periods of aggradation began some time between 183 and 41 cal years BP, probably in the nineteenth century. Therefore, the most recent interval of peat deposition must have begun before 183 cal years BP. It was preceded by a period of aggradation that began some time between 372 and 471 years BP (based on constraints imposed by the radiocarbon ages of 370 ± 50 and 620 ± 50 ^{14}C years BP; Fig. 12). The oldest aggradation interval began some time between 626 and 725 cal years BP (based on constraints imposed by the radiocarbon age of 620 ± 50 ^{14}C years BP and a minimum of 254 years of surface stability that followed; Fig. 12). The oldest of the three intervals of surface stability began more than 807 cal years BP (constraints imposed radiocarbon ages and accumulated minimum durations of subsequent stable intervals).

This simple analysis shows that the three intervals of floodplain aggradation during the Little Ice Age date, from youngest to oldest, to less than 183 cal years BP, between 471 and 183 cal years BP, and between 725 and 626 cal years BP. It must be emphasized that these dates are only limits on times of aggradation. Because peat layers without *in situ* stumps have not been included in the analysis, aggradation probably occurred over a shorter period within each of these intervals. In addition, thin peat layers, without stumps, have not been included in this analysis; the sequence of aggradation and incision thus is more complex than suggested here.

Relation between aggradation and glacier activity

The generally accepted paraglacial paradigm is that sediment yield increases during glacier retreat (Church & Ryder 1972). The rate of paraglacial sedimentation is initially highest during retreat and declines as sediment sources are exhausted or stabilized (Church & Ryder 1972; Church & Slaymaker 1989). The paraglacial concept, however, was originally developed for regional-scale, ice-sheet deglaciation, and its applicability to small alpine catchments with much smaller glaciers is uncertain.

During alpine glacier advance, initial incision due to increased competence of meltwater streams is quickly followed by aggradation as sediment supply increases (Maizels 1979). Sediment stored within and beneath glaciers is delivered at an increasing rate to fluvial systems as glaciers advance (Karlén 1976; Maizels 1979; Leonard 1986, 1997; Karlén & Matthews 1992; Lamoureux 2000). Similarly, subglacial erosion increases during glacier advance, and meltwater may carry more sediment into river valleys than at times when glaciers are more restricted (Clague 1986, 2000). Paraglacial sediment pulses may propagate rapidly downstream in narrow mountain valleys when glaciers advance to maximum positions and, subsequently, as they begin to retreat. Glacier retreat typically exposes large areas of unstable, poorly vegetated sediment that is easily transferred to the fluvial system, causing valley-wide aggradation and complex changes in channel planform (Church 1983; Desloges & Church 1987; Gottesfeld & Johnson-Gottesfeld 1990; Brooks 1994; Ashmore & Church 2001; Clague *et al.* 2003).

Many researchers have linked increased sediment yield and aggradation to periods of more extensive ice cover and glacier advances (Karlén 1976; Karlén & Matthews 1992; Leonard 1986; 1997; Nesje *et al.* 2000; Menounos 2002; Davies *et al.* 2003; Menounos *et al.* 2004). Increased glacial erosion and sediment production during glacier advance, coupled with climatically induced changes in discharge and sediment yield, can cause rivers to aggrade their beds (Knighton 1998). Sediment delivery to streams in the Coast Mountains, for example, increased during the Little Ice Age (Church 1983; Gottesfeld & Johnson-Gottesfeld 1990).

Times of forest death and sediment burial in the Nostetuko valley (Fig. 11) are similar to ages of previously documented Holocene glacier advances in the Coast and Rocky Mountains and parts of Alaska. This temporal association thus implies that sediment delivery to the fluvial system increased at times of glacier advance. Evidence for middle Holocene advances in western Canada is sparse because landforms and associated sediments were overridden, eroded and buried during later glacier expansion (Mathews 1951; Luckman 1986; Osborn 1986; Ryder & Thomson 1986; Ryder 1987; Desloges & Ryder 1990). However, dating of fossil stumps in a few glacier forefields and sediment records from proglacial lakes show that glaciers advanced 6900–5700 cal years BP (the Garibaldi Advance: Ryder & Thomson 1986; Koch *et al.* 2003, 2004; Smith 2003; Menounos *et al.* 2004). Advances of this age are also recognized in interior and coastal Alaska (Calkin 1988). The oldest phase of fluvial aggradation in the west fork of Nostetuko valley coincides with this event.

Another period of aggradation coincides with the Tiedemann Advance. At its type locality, 30 km NW of the study area, the Tiedemann Advance has been dated to 3600–1900 cal years BP, with a culmination around 2400 cal years BP (Ryder & Thomson 1986; Arsenault *et al.* 2007). The radiocarbon age of 2790 ± 60 ^{14}C years BP (3123–2826 cal years BP) from sediments just above bedrock in the moraine fan at site 1 records a local advance of glaciers to near their Neoglacial limit. The oldest fluvial aggradation age, several kilometres downvalley, is slightly younger, 2490 ± 70 ^{14}C years BP (2796–2423 cal years BP), suggesting a lag in response at that site.

Reyes *et al.* (2006) provide evidence from many mountain ranges in western North America for an advance of glaciers beginning about 1700 cal years BP and culminating after 1400–1300 cal years BP. Lichenometric studies record at least two advances of Tiedemann Glacier during this interval, one about 1380 cal years BP and the other around 1070 cal years BP (Larocque & Smith 2003). Aggradation recorded by group 5 radiocarbon ages may be a fluvial response to an advance late in the first millennium AD.

Aggradation shortly after 940 ± 50 ^{14}C years BP (992–798 cal years BP) coincides with the onset of earliest Little Ice Age activity in the Coast and Rocky Mountains. Lichenometric analysis places the earliest Little Ice Age advance of Tiedemann Glacier at about 900 cal years BP (Larocque & Smith 2003). One or more glacier advances in the southern Rocky Mountains at the same time have been documented by Osborn & Luckman (1988), Leonard & Reasoner (1999) and Luckman (2000). Koch *et al.* (2003) have identified a Little Ice Age advance of the same age in Garibaldi Park.

Floodplain stability commencing between 725 and 626 cal years BP, and ending between 471 and 372 cal years BP, lasted at least 254 years and probably corresponds to a documented warm interval within the Little Ice Age, separating early and late Little Ice Age glacier advances (Ryder & Thomson 1986; Ryder 1987; Desloges & Ryder 1990; Clague & Mathewes 1996; Calkin *et al.* 1998; Wiles *et al.* 1999; Luckman 2000; Smith & Desloges 2000; Koch *et al.* 2003; Larocque & Smith 2003). Berendon Glacier in the northern Coast Mountains was less extensive than during the historic period between 570 and 350 cal years BP (Clague *et al.* 2004). Temperatures in the Canadian Rocky Mountains during the sixteenth century and early and middle seventeenth century were also above average (Luckman 2000; Luckman & Wilson 2005).

The period of valley-wide aggradation delimited by ages of 471 and 183 cal years BP coincides with the climatic Little Ice Age advance of the Diadem Glacier and construction of its outermost moraine. Radiocarbon ages of 490 ± 60 ^{14}C years BP (708–391 cal years BP) from a peat clast within till of the outer moraine at Queen Bess Lake and 370 ± 50 ^{14}C years BP (564–372 cal years BP) from a stump in growth position below the T-2 terrace (Kershaw 2002) are maxima for the time that the Diadem Glacier achieved its greatest Holocene extent. A radiocarbon age of 180 ± 50 ^{14}C years BP (362–55 cal years BP) from fine-grained fluvial sediments overlying ice-proximal outwash is a minimum for the retreat of the glacier from the moraine (Kershaw 2002). Lichen measurements suggest that the outer Diadem Glacier moraine was abandoned in the middle or late nineteenth century (Kershaw 2002).

Glacier mass balance and dendroclimatic reconstructions for the Mount Waddington area indicate cool–wet conditions throughout the eighteenth and early nineteenth centuries (Larocque & Smith 2005*a*, *b*). Summer temperatures in the Rocky Mountains during the late seventeenth century were the lowest of the past 1000 years (Luckman & Wilson 2005). A major moraine-building episode in the late seventeenth and early eighteenth centuries, documented in Alaska, and the Coast and Rocky Mountains (Desloges & Ryder 1990; Clague & Mathewes 1996; Calkin *et al.* 1998; Wiles *et al.* 1999; Luckman 2000; Smith & Desloges 2000; Koch *et al.* 2003; Larocque & Smith 2003; Lewis & Smith 2004) probably coincides with construction of the outer moraine of the Diadem Glacier and the youngest aggradation period. Leonard (1997) documented high sedimentation rates at Hector Lake from the early eighteenth century until the mid-nineteenth century, with a peak in the first two decades of the eighteenth century.

Completeness and sensitivity of record

Moraine records have long been used to reconstruct glacial histories. However, most moraine systems record only recent glacier fluctuations because the major advances of the Little Ice Age destroyed or buried much of the evidence of earlier glacier activity. Researchers have partially overcome this problem by examining more complete, although indirect, proxy records, including lacustrine varves and sheared and detrital logs in glacier forefields (Karlén 1976; Leonard 1986, 1997; Karlén & Matthews 1992; Souch 1994; Leonard & Reasoner 1999; Luckman 2000; Nesje *et al.* 2000; Menounos 2002; Koch *et al.* 2003, 2004; Smith 2003; Menounos *et al.* 2004). Another, largely unexploited, archive of proxy information is fluvial deposits within glacierized basins. The sensitivity of the fluvial system to climate change has long been acknowledged, but fluvial deposits have not been widely used to reconstruct glacial histories. Our research shows that, in favourable settings, the fluvial system is extremely sensitive to low-magnitude climate change on decadal and centennial timescales. Increases in sediment supply, and attendant aggradation, in the west fork of Nostetuko River valley coincide with independently dated, late Holocene glacier advances. Fluvial sequences may also provide a more complete record of Holocene glacier and climate change than deposits and landforms in glacier forefields, which are strongly biased toward the late Little Ice Age.

Conclusion

Evidence of Holocene glacier fluctuations at the head of the west fork of Nostetuko River is preserved in the downvalley sedimentary sequence. The upper part of the valley fill records most of the independently documented, late Holocene glacier events in western Canada, including the Garibaldi Phase, Tiedemann Advance, First Millennium AD advance and the Little Ice Age. Much of the detail of glacier activity in the Nostetuko River watershed during the Little Ice Age appears to be archived in the valley fill. Specifically, periods of floodplain stability, recorded by peat beds and forest growth, coincide with times when glaciers were restricted. Major periods of aggradation coincide with times when glaciers were more extensive than today. The results of this study demonstrate that at least some mountain rivers are sensitive indicators of glacier fluctuations on decadal and centennial timescales. Fluvial archives in mountain valley provide a useful complement to evidence of glacier fluctuations preserved in glacier forefields.

We thank R. McKillop and M. Hanson for assistance in the field, and Dr B. Ward for assistance and support. Critical reviews by journal referees D. Swift and F. Tweed greatly improved the paper. The maps used in this project were produced by McElhanney Consulting Services (Vancouver, BC) from aerial photographs taken by Selkirk Remote Sensing (Richmond, BC). M. King (White Saddle Air Services) provided helicopter transport to and from the study area. NSERC (Natural Sciences and Engineering Research Council of Canada) and Simon Fraser University funded the project.

References

ARSENAULT, T. A., CLAGUE, J. J. & MATHEWES, R. W. 2007. Late Holocene vegetation and climate change at Moraine Bog, Tiedemann Glacier, southern Coast Mountains, British Columbia. *Canadian Journal of Earth Sciences*, **44**, 797–719.

ASHMORE, P. & CHURCH, M. 2001. *The Impact of Climate Change on Rivers and River Processes in Canada*. Geological Survey of Canada, Bulletin, **555**.

BLUCK, B. J. 1979. Structure of coarse grained braided alluvium. *Transactions of the Royal Society of Edinburgh: Earth Science*, **70**, 181–221.

BRIERLEY, G. J. 1996. Channel morphology and element assemblages: A constructivist approach to facies modeling. *In*: CARLING, P. A. & DAWSON, M. R. (eds) *Advances in Fluvial Dynamics and Stratigraphy*. Wiley, Chichester, 263–298.

BRONK RAMSEY, C. 1995. Radiocarbon calibration and analysis of stratigraphy: The OxCal program. *Radiocarbon*, **37**, 425–430.

BRONK RAMSEY, C. 2001. Development of the radiocarbon program OxCal. *Radiocarbon*, **43**, 355–363.

BROOKS, G. R. 1994. The fluvial reworking of late Pleistocene drift, Squamish River drainage basin, southwest British Columbia. *Géographie physique et Quaternaire*, **48**, 51–68.

CALKIN, P. E. 1988. Holocene glaciation of Alaska (and adjoining Yukon Territory, Canada). *Quaternary Science Reviews*, **7**, 159–184.

CALKIN, P. E., KAUFMAN, D. S., PRZYBYL, B. J., WHITFORD, W. B. & PECK, B. J. 1998. Glacier regimes, periglacial landforms, and Holocene climate change, Seward Peninsula, Alaska. *Arctic and Alpine Research*, **30**, 154–165.

CHURCH, M. 1983. Pattern of instability in a wandering gravel bed channel. *In*: COLLINSON, J. D. & LEWIN, J. (eds) *Modern and Ancient Fluvial Systems*. International Association of Sedimentologists, Special Publications, **6**, 169–180.

CHURCH, M. & RYDER, J. M. 1972. Paraglacial sedimentation: A consideration of fluvial processes conditioned by glaciation. *Geological Society of America Bulletin*, **83**, 3059–3072.

CHURCH, M. & SLAYMAKER, O. 1989. Disequilibrium of Holocene sediment yield in glaciated British Columbia. *Nature*, **337**, 452–454.

CLAGUE, J. J. 1986. The Quaternary stratigraphic record of British Columbia – evidence of episodic sedimentation and erosion controlled by glaciation. *Canadian Journal of Earth Sciences*, **23**, 885–894.

CLAGUE, J. J. 2000. Recognizing order in chaotic sequences of Quaternary sediments in the Canadian Cordillera. *Quaternary International*, **68–71**, 29–38.

CLAGUE, J. J. & MATHEWES, R. W. 1996. Neoglaciation, glacier-dammed lakes and vegetation change in northwestern British Columbia. *Arctic and Alpine Research*, **28**, 10–24.

CLAGUE, J. J., TURNER, R. J. W. & REYES, A. 2003. Record of recent river channel instability, Cheakamus Valley, British Columbia. *Geomorphology*, **53**, 317–332.

CLAGUE, J. J., WOHLFARTH, B. *ET AL*. 2004. Late Holocene environmental change at treeline in the northern Coast Mountains, British Columbia, Canada. *Quaternary Science Reviews*, **23**, 2413–2431.

COLLINSON, J. D. 1996. Alluvial sediments. *In*: READING, H. G. (ed.) *Sedimentary Environments: Processes, Facies and Stratigraphy*, 3rd edn. Blackwell, Oxford, 37–82.

DAVIES, T. R. H., SMART, C. C. & TURNBULL, J. M. 2003. Water and sediment outbursts from advanced Franz Josef Glacier, New Zealand. *Earth Surface Processes and Landforms*, **23**, 1081–1096.

DESLOGES, J. R. & CHURCH, M. 1987. Channel and floodplain facies in a wandering gravel-bed river. *In*: ETHRIDGE, F. G. & FLORES, R. M. (eds) *Recent Developments in Fluvial Sedimentology*. SEPM, Special Publications, **39**, 99–109.

DESLOGES, J. R. & RYDER, J. M. 1990. Neoglacial history of the Coast Mountains near Bella Coola, British Columbia. *Canadian Journal of Earth Sciences*, **27**, 281–290.

GOTTESFELD, A. S. & JOHNSON-GOTTESFELD, L. M. 1990. Floodplain dynamics of a wandering river, dendrochronology of the Morice River, British Columbia, Canada. *Geomorphology*, **3**, 159–179.

GRISSINO-MAYER, H. D. 2001. Evaluating cross-dating accuracy: A manual and tutorial for the computer program COFECHA. *Tree-Ring Research*, **57**, 205–221.

HOLMES, R. L. 1999. *Documentation for Programs in the Dendrochronology Program Library and the Dendroecology Program Library*. University of Arizona, Tucson, AZ.

KARLÉN, W. 1976. Lacustrine sediments and tree limit variations as evidence of Holocene climatic variations in Lappland, northern Sweden. *Geografiska Annaler*, **58A**, 1–34.

KARLÉN, W. & MATTHEWS, J. A. 1992. Reconstructing Holocene glacier variations from glacial lake sediments: Studies from Nordvestlandet and Jostedalsbreen–Jotunheimen, southern Norway. *Geografiska Annaler*, **74A**, 327–348.

KERSHAW, J. A. 2002. *Formation and Failure of Moraine-dammed Queen Bess Lake, Southern Coast Mountains, British Columbia*. MSc thesis, Simon Fraser University, Burnaby, BC.

KERSHAW, J. A., CLAGUE, J. J. & EVANS, S. G. 2004. Geomorphic and sedimentologic signature of a two-phase outburst flood from moraine-dammed Queen Bess Lake, British Columbia, Canada. *Earth Surface Processes and Landforms*, **30**, 1–25.

KNIGHTON, D. 1998. *Fluvial Forms and Processes: A New Perspective*. Arnold, London.

KOCH, J., MENOUNOS, B. P., CLAGUE, J. J. & OSBORN, G. D. 2004. Environmental change in Garibaldi Provincial Park, southern Coast Mountains, British Columbia. *Geoscience Canada*, **31**, 127–135.

KOCH, J., MENOUNOS, B. P., OSBORN, G. D. & CLAGUE, J. J. 2003. Holocene glacial history near Whistler, southern Coast Mountains, British Columbia. *In*: *XVI INQUA (International Union for Quaternary Research) Congress, Reno, NV, Program with Abstracts. 163.*

LAMOUREUX, S. 2000. Five centuries of interannual sediment yield and rainfall-induced erosion in the Canadian High Arctic recorded in lacustrine varves. *Water Resources Research*, **36**, 309–318.

LAROCQUE, S. J. & SMITH, D. J. 2003. Little Ice Age activity in the Mt. Waddington area, British Columbia Coast Mountains, Canada. *Canadian Journal of Earth Sciences*, **40**, 1413–1436.

LAROCQUE, S. J. & SMITH, D. J. 2005*a*. 'Little Ice Age' proxy glacier mass balance records reconstructed from tree rings in the Mt. Waddington area, British Columbia Coast Mountains, Canada. *The Holocene*, **15**, 748–757.

LAROCQUE, S. J. & SMITH, D. J. 2005*b*. A dendroclimatological reconstruction of climate since AD 1700 in the Mt. Waddington area, British Columbia Coast Mountains, Canada. *Dendrochronologia*, **22**, 93–106.

LEONARD, E. M. 1986. Varve studies at Hector Lake, Alberta, Canada, and the relationship between glacial activity and sedimentation. *Quaternary Research*, **25**, 199–214.

LEONARD, E. M. 1997. The relationship between glacial activity and sediment production: Evidence from a 4450-year varve record of neoglacial sedimentation in Hector Lake, Alberta, Canada. *Journal of Palaeolimnology*, **17**, 319–330.

LEONARD, E. M. & REASONER, M. A. 1999. A continuous Holocene glacial record inferred from proglacial lake sediments in Banff National Park, Alberta, Canada. *Quaternary Research*, **51**, 1–13.

LEWIN, J., MACKLIN, M. G. & JOHNSTONE, E. 2005. Interpreting alluvial archives: Sedimentological factors in the British Holocene fluvial record. *Quaternary Science Reviews*, **24**, 1873–1889.

LEWIS, D. H. & SMITH, D. J. 2004. Little Ice Age glacial activity in Strathcona Provincial Park, Vancouver Island, British Columbia, Canada. *Canadian Journal of Earth Sciences*, **41**, 285–297.

LUCKMAN, B. H. 1986. Reconstruction of Little Ice Age events in the Canadian Rocky Mountains. *Géographie physique et Quaternaire*, **40**, 17–28.

LUCKMAN, B. H. 2000. The Little Ice Age in the Canadian Rockies. *Geomorphology*, **32**, 357–384.

LUCKMAN, B. H. & WILSON, R. J. S. 2005. Summer temperatures in the Canadian Rockies during the last millennium: A revised record. *Climate Dynamics*, **24**, 131–144.

MAIZELS, J. K. 1979. Proglacial aggradation and changes in braided channel patterns during a period of glacier advance: An alpine example. *Geografiska Annaler*, **61**, 87–101.

MARREN, P. M. 2005. Magnitude and frequency in proglacial rivers: A geomorphological and sedimentological perspective. *Earth Science Reviews*, **70**, 203–251.

MATHEWS, W. H. 1951. Historic and prehistoric fluctuations of alpine glaciers in the Mount Garibaldi map-area, southwestern British Columbia. *Journal of Geology*, **59**, 357–380.

MCCARTHY, D. P., LUCKMAN, B. H. & KELLY, P. E. 1991. Sampling height–age error correction for spruce seedlings in glacial forefields, Canadian Cordillera. *Arctic and Alpine Research*, **23**, 451–455.

MENOUNOS, B. 2002. *Climate, Fine-sediment Transport Linkages, Coast Mountains, British Columbia, Canada*. PhD thesis, University of British Columbia, Vancouver, BC.

MENOUNOS, B., KOCH, J., OSBORN, G., CLAGUE, J. J. & MAZZUCCHI, D. 2004. Early Holocene glacier advance, southern Coast Mountains, British Columbia, Canada. *Quaternary Science Reviews*, **23**, 1543–1550.

MIALL, A. D. 1977. A review of the braided river depositional environment. *Earth Science Reviews*, **13**, 1–62.

MIALL, A. D. 1978. Lithofacies types and vertical profile models in braided river deposits: A summary. *In*: MIALL, A. D. (ed.) *Fluvial Sedimentology*. Canadian Society of Petroleum Geologists, Memoirs, **5**, 597–604.

NESJE, A., DAHL, S. O., ANDERSSON, C. & MATTHEWS, J. A. 2000. The lacustrine sedimentary sequence in Sygneskardvatnet, western Norway: A continuous, high-resolution record of the Jostedalsbreen ice cap during the Holocene. *Quaternary Science Reviews*, **19**, 1047–1065.

OSBORN, G. 1986. Lateral moraine stratigraphy and Neoglacial history of Bugaboo Glacier, British Columbia. *Quaternary Research*, **26**, 171–178.

OSBORN, G. & LUCKMAN, B. H. 1988. Holocene glacier fluctuations in the Canadian Cordillera (Alberta and British Columbia). *Quaternary Science Reviews*, **7**, 115–128.

REYES, A. V., WILES, G. C. *ET AL*. 2006. Expansion of alpine glaciers in Pacific North America in the first millennium AD. *Geology*, **34**, 57–60.

RYDER, J. M. 1987. Neoglacial history of the Stikine Iskut area, northern Coast Mountains, British Columbia. *Canadian Journal of Earth Sciences*, **24**, 1294–1301.

RYDER, J. M. & THOMSON, B. 1986. Neoglaciation in the southern Coast Mountains of British Columbia: chronology prior to the late Neoglacial maximum. *Canadian Journal of Earth Sciences*, **23**, 273–287.

SAMBROOK SMITH, G. H. 2000. Small-scale cyclicity in alpine proglacial fluvial sedimentation. *Sedimentary Geology*, **132**, 217–231.

SIGAFOOS, R. S. & HENDRICKS, E. L. 1969. *The Time Interval Between Stabilization of Alpine Glacial Deposits and Establishment of Tree Seedlings*. US Geological Survey, Professional Paper, **650-B**, B89–B93.

SMITH, D. G. 1983. Anastamosed fluvial deposits: modern examples from western Canada. *In*: COLLINSON, J. D. & LEWIN, J. (eds) *Modern and Ancient Fluvial Systems*. International Association of Sedimentologists, Special Publications, **6**, 155–168.

SMITH, D. J. 2003. The Garibaldi Phase revisited: Mid-Holocene glaciation in the British Columbia Coast Mountains. *Canadian Association of Geographers, Victoria, BC, Program with Abstracts*. 34.

SMITH, D. J. & DESLOGES, J. R. 2000. Little Ice Age history of Tzeetsaytsul Glacier, Tweedsmuir Provincial Park, British Columbia. *Géographie physique et Quaternaire*, **54**, 135–141.

SMITH, D. J., MCCARTHY, D. P. & COLENUTT, M. E. 1995. Little Ice Age glacial activity in Peter Lougheed and Elk Island provincial parks, Canadian Rocky Mountains. *Canadian Journal of Earth Sciences*, **32**, 579–589.

SOUCH, C. 1994. A methodology to interpret downvalley lake sediments as records of Neoglacial activity: Coast Mountains, British Columbia, Canada. *Geografiska Annaler*, **76A**, 169–185.

STUIVER, M., REIMER, P. J. *ET AL*. 1998. INTCAL98 radiocarbon age calibration, 24000–0 cal BP. *Radiocarbon*, **40**, 1041–1083.

WILES, G. C., POST, A., MULLER, E. H. & MOLNIA, B. F. 1999. Dendrochronology and late Holocene history of the Bering Piedmont Glacier, Alaska. *Quaternary Research*, **52**, 185–195.

WILKIE, K. M. K. 2006. *Fluvial Response to Late Holocene Glacier Fluctuations in the Nostetuko River Valley, Southern Coast Mountains, British Columbia*. MSc thesis, Simon Fraser University, Burnaby, BC.

Paraglacial geomorphology of Quaternary volcanic landscapes in the southern Coast Mountains, British Columbia

PIERRE A. FRIELE[1]* & JOHN J. CLAGUE[2]

[1]*Cordilleran Geoscience, P.O. Box 612, Squamish, British Columbia, Canada, V8B 0A5*

[2]*Centre for Natural Hazard Research, Simon Fraser University, Burnaby, British Columbia, Canada V5A 1S6 and Geological Survey of Canada, 101–605 Robson Street, Vancouver, British Columbia, Canada, V6B 5J3*

**Corresponding author (e-mail: pfriele@gmail.com)*

Abstract: An important paradigm in geomorphology is *paraglacial sedimentation*, a phrase first used almost 40 years ago to describe reworking of glacial sediment by mass wasting and streams during and after continental-scale deglaciation. The concept has been extended to include non-glacial landforms and landscapes conditioned by glaciation. In this paper we apply the paraglacial concept to volcanoes in southern British Columbia, Canada, that formed, in part, in contact with glacier ice. The Cheekye River basin, a small watershed on the flank of a volcano that erupted against the decaying Cordilleran ice sheet, has a Holocene history marked by an exponential decay in debris-flow activity and sediment yield. Its history is consistent with the primary exhaustion model of the paraglacial cycle. At larger spatial scales, this primary sediment is reworked by rivers and transported downstream and augmented by stochastic geomorphic events. Repeated large landslides from Mount Meager volcano in southern British Columbia have delivered a disproportionate volume of sediment to the fluvial system: although occupying only 2% of the watershed area, 25–75% of the 10 km^3 of sediment deposited in Lillooet River valley during the Holocene originated from the volcano. In these cases a significant overall reduction in sediment yield must await the removal, by erosion, of volcanic edifices, a process that could take up to millions of years. These examples of paraglacial activity on Quaternary volcanoes are end members in the spectrum of landscape response to Pleistocene deglaciation.

The phrase *paraglacial sedimentation* was coined by Church & Ryder (1972) to describe large-scale reworking of glacial sediment by colluvial and fluvial processes in SW British Columbia, Canada, during and following terminal Pleistocene deglaciation. Since then, the concept has been modified and extended to a wider range of landscapes and has become an important paradigm in geomorphology. Ballantyne (2002, p. 1938) defined paraglacial geomorphology as 'nonglacial earth-surface processes, sediment accumulations, landforms, landsystems and landscapes that are directly conditioned by glaciation and deglaciation'. Rates of landscape adjustment and the duration of the paraglacial period depend on a variety of factors, principally rock structure and lithology, the distribution and types of surficial materials, relief, storage, climate and spatial scale.

In small catchments (<100 km^2) paraglacial transfer of sediment is dominated by erosion and follows a 'primary exhaustion' model (Church & Ryder 1972; Ballantyne 2002). Here, sediment is transferred from slopes to fans as rapidly as the operative processes, including mass movement and fluvial erosion, allow. Eventually, sediment that can readily be moved to lower elevations is exhausted. As scale increases, sediment transferred to valley bottoms becomes available for fluvial reworking. As noted by Church & Slaymaker (1989), sediment yields in larger watersheds in British Columbia typically increase downstream (Fig. 1), reflecting migration of the paraglacial sediment pulse and reworking of sediment from Pleistocene glacial deposits (e.g. Brooks 1994).

Dadson & Church (2005) showed that, in order to reproduce the morphology and channel network of a typical post-glacial valley in SW British Columbia, it is necessary to combine: (1) stochastic landsliding, for example large rock avalanches; (2) non-linear diffusion processes, including slope-dependent rates of rockfall, soil creep and shallow slope failures; and (3) and fluvial transport. In their model, the pattern of Holocene fluvial sediment yield is complex, commonly multimodal, and varies strongly depending on the rate and timing of landslides and the diffusivity constant. Terrain with the highest landslide and diffusivity rates had the highest late Holocene fluvial sediment yields. We suggest that Quaternary volcanoes represent this end member.

Quaternary volcanoes in the area of the former Cordilleran ice sheet in SW British Columbia and

From: KNIGHT, J. & HARRISON, S. (eds) *Periglacial and Paraglacial Processes and Environments.*
The Geological Society, London, Special Publications, **320**, 219–233.
DOI: 10.1144/SP320.14 0305-8719/09/$15.00 © The Geological Society Publishing House 2009.

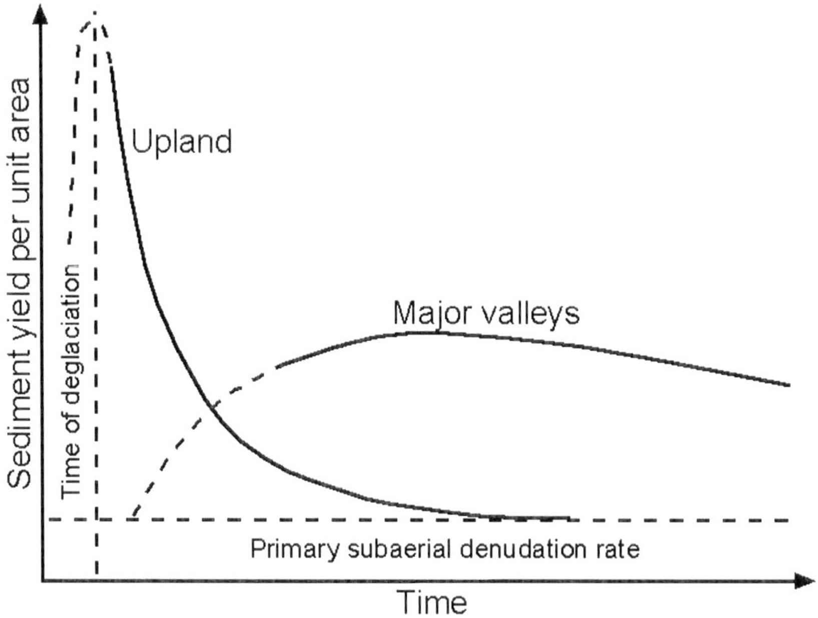

Fig. 1. The paraglacial cycle as formulated by Church & Slaymaker (1989).

NW Washington (Fig. 2) include Mount Garibaldi, Mount Cayley and Mount Meager in British Columbia, and Mount Baker and Glacier Peak in Washington. They differ from Cascade volcanoes south of the limit of the late Pleistocene ice sheet in that they have formed partly from eruptions into or against former glaciers, and they have a distinctive set of landforms reflecting this history (Mathews 1958; Hickson 2000), such as rapidly quenched, ice-marginal lava flows (tuyas) and subglacial flows similar in form to eskers (Mathews 1952). The volcanoes have been deeply dissected by valley glaciers (see Montgomery 2002) and rivers during the Holocene, leaving steep slopes and local relief in excess of 2000 m. They are the loci of frequent large Holocene landslides (Friele & Clague 2005; Friele *et al.* 2005, 2006; Simpson *et al.* 2006). Bedrock structure, relief and geomorphic processes active at these volcanoes have been conditioned by glaciation and, therefore, can be thought of as paraglacial.

This paper reviews the late-glacial and postglacial histories of three volcanoes that erupted against Pleistocene glaciers. Our objective is to explore these histories in the context of paraglacial sedimentation and related landscape evolution. The record of sediment delivery from one of the three volcanoes (Mount Garibaldi) broadly fits a sediment exhaustion model, which is the cornerstone of the paraglacial paradigm. However, sediment delivery from all three volcanoes is punctuated by periodic, very large landslides that strongly affect the style and rate of sedimentation far downstream.

Regional setting

Our study area is the southern Coast Mountains, a major NW-trending mountain belt bordered on the west by the Pacific Ocean and on the east by plateaus of the British Columbia interior (Fig. 2). High peaks in the southern Coast Mountains range from about 2000 to 4000 m asl (m above sea level), and local relief from valley floors to summits is commonly 1500–2000 m. The surface rocks are dominantly competent Cretaceous granodiorites, quartz diorites, metasedimentary rocks and metavolcanic rocks (Roddick *et al.* 1976). During the Pleistocene, glaciers repeatedly advanced and thickened to cover all but the highest peaks (Clague 1989, 1991). Ice-sheet glaciation ended about 12.8 cal ka BP (calendric ages hereafter designated cal ka BP), following a regional readvance during the Younger Dryas cooling event (Friele & Clague 2002*a*).

The Quaternary volcanic centres (Fig. 2) form a small, but important, part of this landscape. They have areas of up to 80 km^2, cover local basement rock across unconformable, high-relief surfaces, and are extremely unstable. Hydrothermal activity, which continues today at Mount Meager and Mount Cayley (Hickson 1994), has weakened the

Fig. 2. Locations of Quaternary volcanoes at the north end of the Cascade volcanic arc and place names mentioned in the text.

edifices (Finn *et al.* 2001). Mount Meager, which contains about 20 km^3 of Plio-Pleistocene volcanic rocks, is the largest of the volcanic complexes and has larger areas of hydrothermal alteration.

Climate in the area is maritime, with warm dry summers and cool wet winters. Mean annual precipitation is 1500–2200 mm, much of it as snow from October through to May. Late summer and fall rainstorms are common at high elevations, and rain can fall at any time of the year in low-lying valleys. Climate shifted from wet and cool at the end of the last glaciation to warm and dry during the early Holocene (Mathewes & Heusser 1981). Cooling and increased precipitation about 6000 years ago initiated Neoglaciation, which culminated in the climactic Holocene advances of the Little Ice Age (Clague 1989; Luckman 2000).

Valleys on the flanks of the Quaternary volcanoes are transport-limited (Jakob & Bovis 1996), with avalanching and rockfall providing a continuous supply of debris that can be mobilized into debris flows whenever threshold runoff conditions are exceeded. Spring and summer snow melt, and autumn and winter rains, are common triggering events. Rates of mass transfer of sediments have probably changed during the Holocene in response to centennial-scale climate fluctuations, not only on the volcanic massifs but elsewhere in southern British Columbia (Bovis & Jones 1992). Twentieth-century climate warming and debuttressing of slopes due to glacier retreat have increased the incidence of landslides in some areas (Bovis & Jakob 2000; Holm *et al.* 2004). However, Dadson & Church (2005) suggest that the geomorphic effects of Neoglacial climate change have been much less significant than those induced by Pleistocene deglaciation.

History of Mount Garibaldi and the Cheekye fan

Cheekye fan, on the west side of Mount Garibaldi (Fig. 3), provides an excellent example of primary paraglacial activity within Ballantyne's (2002) rockslope land system. In this context, 'primary' refers to the rapid transfer of sediment from a small catchment, in contrast to long-term storage and remobilization that characterize the classic paraglacial cycle of large fluvial (Church & Slaymaker 1989) and coastal (Shaw *et al.* 1990) systems.

Mount Garibaldi (2678 m asl) overlooks the town of Squamish at the head of Howe Sound (Fig. 3). Part of community is located on the distal margins of Cheekye fan, a large alluvial and colluvial fan derived from collapse and erosion of the west flank of Mount Garibaldi. The fan is a complex feature consisting of late Pleistocene ice-contact sediments and Holocene debris flow and fluvial deposits (Fig. 3). Cheekye basin terminates in steep (>45°) slopes directly west of Mount Garibaldi. Linear cracks on the ridge adjacent to these steep slopes (Fig. 3) are evidence of continuing gravitational instability. The basin has an area of 26 km^2, relief of 2200 m and an average basin slope of 25°. It is extensively gullied and supports a fourth-order channel network.

Growth of Garibaldi volcano

Mathews (1952) showed that the summit of Mount Garibaldi formed after the last glacial maximum, which locally dates to about 17 cal ka BP (Clague 1981; Porter & Swanson 1998). He suggested, on the basis of alteration characteristics of basement rocks in Cheekye basin, that the surface of the downwasting Cordilleran ice sheet stood at or below 1300 m asl when an explosive eruption built Mount Garibaldi. Pyroclastic flows and lahars from this eruption covered the glacier ice that filled Cheekye basin. Soon thereafter, the trunk glacier in Squamish valley readvanced and deposited granitic erratics up to 1660 m asl on the volcano's west flank.

Friele & Clague (2002*b*) documented a readvance of the Squamish valley glacier 13.5–12.9 ka BP to Porteau, 35 km south of Mount Garibaldi (Fig. 2 and Table 1). A line extending from the uppermost granitic erratics, noted by Mathews, to the Porteau end moraine has a slope of 3°–4°, similar to large valley glaciers in the Coast Mountains today. We infer that the summit cone of Mount Garibaldi formed shortly before 13.5 cal ka BP.

Deglaciation

Ice retreat from the Porteau end moraine debuttressed the west flank of Mount Garibaldi, causing it to collapse (Mathews 1952). By 12.8–12.5 cal ka BP, ice had thinned considerably and the ice surface in Cheekye basin stood at about 500 m asl (Fig. 4) (Friele & Clague 2002*a*). Debris derived from collapse of the west flank of Mount Garibaldi was carried down Cheekye River, but was deflected south down into Mashiter and Hop Ranch creeks (Fig. 3) along the decaying ice margin and into an ice-marginal lake. The lake overflowed to the south into Stawamus River valley (Fig. 3) and, from there, into Howe Sound (Mathews 1952; Friele & Clague 2002*b*). Radiocarbon ages from raised marine deltaic sediments at the mouth of Stawamus River (Friele *et al.* 1999) and from ice-contact sediments exposed at the Garibaldi Springs section (Fig. 3) indicate that ice persisted in the lower Squamish Valley until 11.3 cal ka BP (Table 1).

Three sections (CA, MC and GS in Fig. 3) at the margins of kame terraces on upper Cheekye fan provide 10–50 m-high exposures of sediments emplaced during the collapse of Mount Garibaldi (Mathews 1952). Aside from the lowermost unit at the Garibaldi Springs site (see later), the sediments are entirely of volcanic provenance and consist of beds ranging from several metres to 20 m thick (Fig. 5a, b). The sediments are massive–weakly stratified sandy gravel and diamicton with angular–subangular clasts up to 2 m across. Lower bed contacts are typically erosive and in some cases marked by channel cut-and-fill structures. The outer rings of a fragment of wood 7 m below

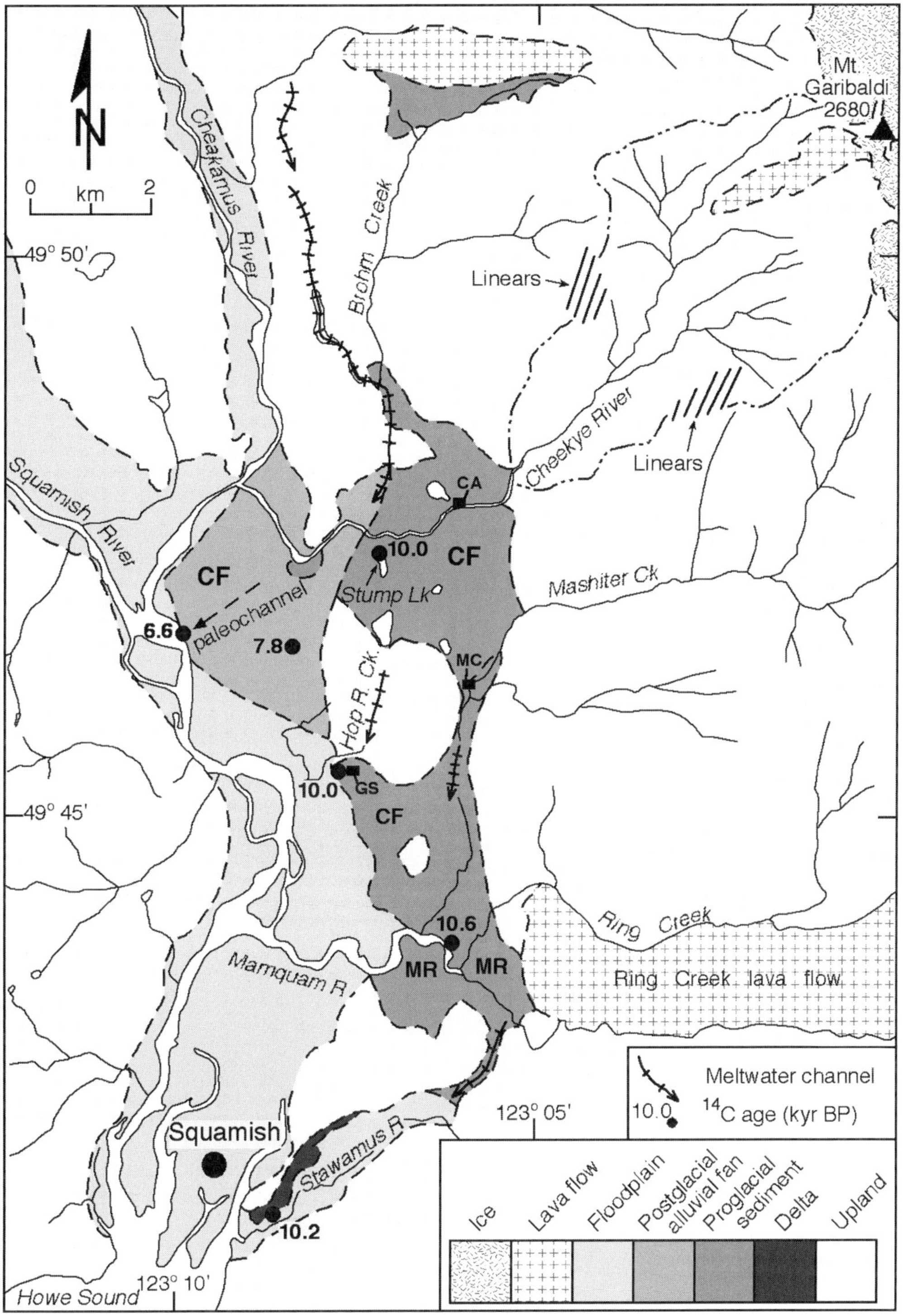

Fig. 3. Geology and landforms of the Cheekye basin and fan on the west slope of Mount Garibaldi. CF, Cheekye fan landforms; MR, Mamquam River landforms. Exposures in the upper Cheekye fan include Cat Lake (CA), Mashiter Creek (MC) and Garibaldi Springs (GS).

Table 1. *Selected radiocarbon ages relevant to the evolution of the Cheekye land system*

Age (^{14}C years BP)	Calibrated age* (cal years before AD 2000)	Laboratory number†	Dated material	Latitude (°N) longitude (°W)	Comment	Reference
810 ± 60	700–800	GSC-6639	Log	49°47.3′ 123°08.8′	Lower fan. Sanitary landfill. Age of surface debris-flow (2×10^6–3×10^6 m^3) unit	Clague *et al.* (2003)
4810 ± 80	5500–5700	GSC-6293	Sticks	49°47.1′ 123°08.4′	Lower fan. Sanitary landfill, 10 m below surface. Maximum age for four overlying debris-flow units	Clague *et al.* (2003)
6210 ± 60	7000–7300	TO-9228	Gyttja	49°47.1′ 123°08.4′	Stump Lake. Minimum age for largest debris flow (3×10^6–5×10^6 m^3) during Holocene	Clague *et al.* (2003)
6590 ± 130	7300–7700	TO-8275	Plant detritus	49°46.2′ 123°07.2′	Stump lake. Maximum age for largest debris flow (3×10^6–5×10^6 m^3) during Holocene	Clague *et al.* (2003)
6595 ± 90	7400–7700	GX-17894	Charcoal	49°46.2′ 123°10.0′	Lower fan. Minimum age for incision of palaeochannel in central sector	Thurber–Golder (1993)
7820 ± 95	8500–9000	GX-17397	Charcoal	49°46.0′ 123°08.3′	Lower fan. Pit, 0.8 m below surface. Minimum age for cessation of fan growth on southern sector	Thurber–Golder (1993)
10 020 ± 80	11 300–12 000	TO-9682	Twig	49°41.7′ 123°08.4′	Basal sediments from Stump Lake. Minimum age for complete deglaciation of Howe Sound	Clague *et al.* (2003)
10 090 ±70	12 130–11 280	Beta-203639	Wood fragment	49°45.4′ 123°07.8′	Garibaldi Springs section. Maximum age on ice-contact edifice collapse deposits	This paper
10 200 ± 100	11 600–12 400	GSC-6236	Wood fragment	49°41.7′ 123°08.4′	Stawamus River raised marine delta Minimum age for partial deglaciation of Howe Sound	Friele & Clague (2002*a*, *b*)
10 650 ± 70	12 400–13 000	Beta-43865	Stump, *in situ*	49°43.9′ 123°05.3′	Ring Creek section. Maximum age for Ring Creek lava flow and deposition of late-glacial deposits	Brooks & Friele (1992)

*Determined from dendrocalibrated data of Stuiver *et al.* (1998) using the program CALIB 4.2. The range represents the 95% confidence interval ($\pm 2\sigma$) calculated with an error multiplier of 1.0.
†Beta, Beta Analytic Inc; GSC, Geological Survey of Canada Radiocarbon Laboratory; GX, Geochron Laboratory; TO, Isotrace Laboratory.

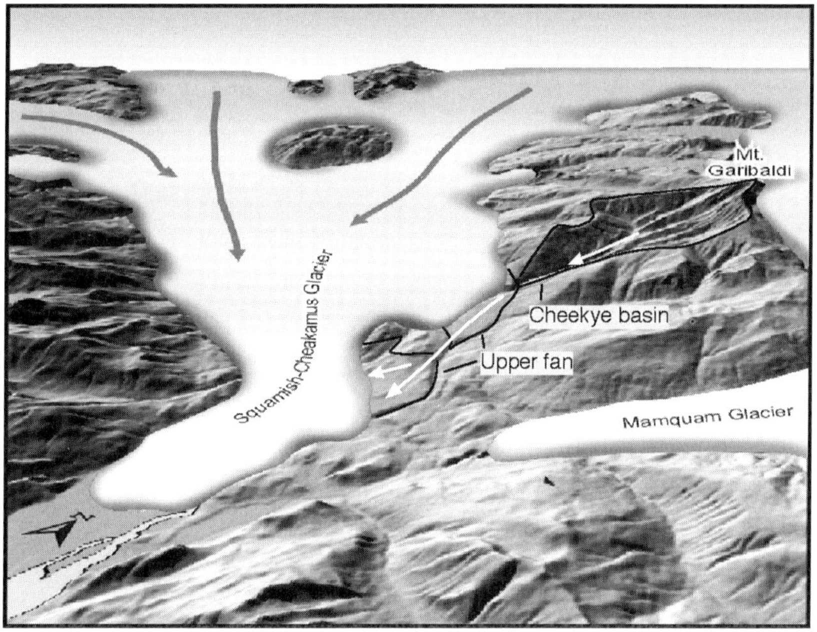

Fig. 4. Reconstruction of the terminus of the glacier at the head of Howe Sound 12.8 cal ka BP (Friele & Clague 2002*b*, fig. 4). The Cheekye River basin on the west flank of Mount Garibaldi is outlined. The Squamish Valley glacier backstopped collapsing debris to form the upper Cheekye fan deposits.

the ground surface at the Garibaldi Springs section yielded a radiocarbon age of 10 090 ± 70 ^{14}C years BP (12.4–11.3 cal ka BP; Table 1).

The sediments underlying the kame terraces were deposited by both rock avalanches and debris flows. Precursors of modern Cheekye and Mashiter rivers washed the sediments between mass-wasting events. Crude stratification suggests that destruction of Mount Garibaldi did not happen in one flank collapse, but rather in a sequence of smaller failures.

The basal sediment unit at Garibaldi Springs (Fig. 5c) lies on basement rock and consists of 2–5 m of massive, clast- to matrix-supported diamicton. Coarse fragments up to boulder size consist of about equal amounts of subangular–subrounded granitic rocks and angular–subangular volcanic rocks. Larger volcanic clasts display radial fractures, indicative of cooling *in situ* within the diamicton (Fig. 5d) and deposition associated with volcanism.

Continuing decay and final stagnation of the valley glacier are documented by a series of successively lower terraces and kettle lakes north and south of Cheekye River (Mathews 1952). A basal radiocarbon age of 10 020 ± 80 ^{14}C years BP (12.0–11.3 cal ka BP) from Stump Lake (Fig. 3 and Table 1) (Clague *et al.* 2003) is a minimum age for the deglaciation of middle Cheekye fan. Thus deglaciation of Cheekye basin lasted from about 13.0 to 11.3 cal ka BP.

Post-glacial phase

At the close of the Pleistocene, Howe Sound extended for several tens of kilometres north of its present head, and the lower Cheekye fan prograded laterally into this arm of the sea (Hickin 1989). Ground-penetrating radar (GPR) data indicate that most of lower Cheekye fan was deposited when sea level fell during local isostatic rebound between 12.0 and 10.0 ka BP (Friele *et al.* 1999).

Some time after 10.0 cal ka BP the fan extended across, and blocked, the fjord, impounding a lake in Squamish valley to the NW. The Squamish and Cheakamus rivers then crossed the toe of the fan, ultimately incising it and lowering local base level. Incision was complete by 7.5 cal ka BP; thus, most sediment transfer to the lower Cheekye fan took no more than 4000 years following complete deglaciation of Howe Sound (Friele & Clague 2005) (Table 1).

Paraglacial sediment budget for the Cheekye watershed

To determine a paraglacial sediment budget, we assume that Mount Garibaldi volcano erupted onto ice filling the Cheekye basin below 1300 m asl, as suggested by Mathews (1952). Comparison of the inferred form of the original volcano and the

Fig. 5. Sections of ice-contact sediments in the upper Cheekye fan. (**a**) Section along the Cheekye River near Cat Lake. (**b**) Section along the Mashiter Creek near Alice Lake Provincial Park. (**c**) Section at Garibaldi Springs near the mouth of Hop Ranch Creek. (**d**) Radially fractured volcanic boulder in the basal unit at Garibaldi Springs.

present topography above 1300 m asl on its west side suggests that about 7.3 km^3 of debris, or roughly half the original cone, were transferred to Cheekye fan. Friele *et al.* (1999) estimated that lower Cheekye fan contains 1.6 km^3 of debris, all but 0.2 km^3 of which was deposited before 7.5 cal ka BP. Accordingly, as much as 5.9 km^3, or 80% of the available debris, was deposited in ice-marginal positions on the upper fan during the late-glacial phase from 13.0 to 11.5 cal ka ago.

We convert volumes of transferred sediment to unit yield rates using the chronology of events established for the area (Table 2). A caveat in interpreting the yield rates is that sediment volumes are order-of-magnitude estimates. Acknowledging this caveat, we conclude that sediment yield decreased by two orders of magnitude, from 150 000 m^3 $year^{-1}$ km^{-2} in the late-glacial period to 1000 m^3 $year^{-1}$ km^{-2} after 7.5 cal ka BP (Fig. 6). Present-day sediment yields derived from data reported by

Table 2. *Paraglacial sediment budget*

Interval	Interval (cal years BP)	Duration (cal years)	Debris volume (km^3)	Sediment yield (m^3 $year^{-1}$)	Unit yield (m^3 km^{-1} $year^{-1}$)
Late Pleistocene	13 000–11 300	1700	5.7	3.4×10^6	130 000
Early Holocene	11 300–7500	3800	1.4	3.7×10^5	14 000
Middle and late Holocene	7500–present	7500	0.2	2.7×10^4	1000
Present day				1.8×10^4	700

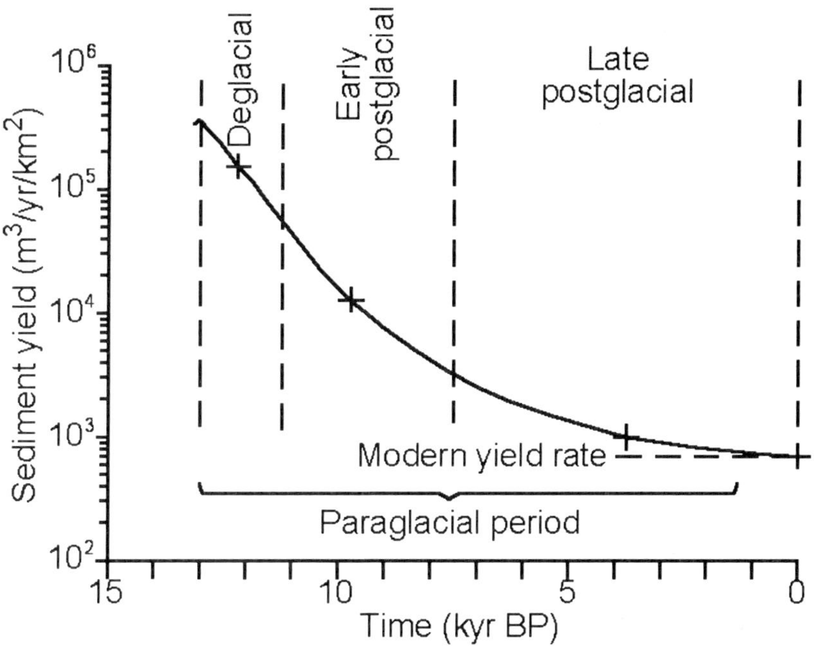

Fig. 6. Deglacial and post-glacial sediment yield for the Cheekye basin based on data presented in Table 2. The pattern of sediment yield follows the exponential decline first proposed by Church & Ryder (1972) and termed the primary exhaustion model by Ballantyne (2002).

Hickin (1989) are comparable to those calculated for the middle to late Holocene (Fig. 6) (Friele *et al.* 1999). The data from the Cheekye River basin thus support the primary exhaustion model.

Holocene sediment pulsing from Mount Meager

Data from the Lillooet River valley (Fig. 2), downstream of Mount Meager, provide a more complex picture of late Holocene sediment yield. Mount Meager (Fig. 7) is the largest of the Cascade volcanic massifs north of the International Boundary. It is located in the headwaters of Lillooet River, which has a watershed area upstream of Lillooet Lake of about 3150 km^2. The edifice is imposing and covers an area of about 80 km^2, but it represents only 2.5% of the area of the Lillooet River watershed.

Historic rates of advance of the Lillooet River delta into Lillooet Lake (Gilbert 1975), 75 km downstream from Mount Meager, have been used to calculate an average sediment yield of 417 $m^3\ km^{-2}\ year^{-1}$ (range 273–654 $m^3\ km^{-2}\ year^{-1}$) for the watershed (Jordan & Slaymaker 1991; Slaymaker 1993). Jordan & Slaymaker (1991) and Slaymaker (1993) attempted to construct a sediment budget for the Lillooet River basin, but were unable to reconcile the observed historic sediment yield at Lillooet Lake with a yield of 221 $m^3\ km^{-2}\ year^{-1}$ estimated from processes active over the majority of the contributing watershed, an area dominated by non-volcanic, basement rocks. They attributed the unaccounted yield to a number of factors, including landslides from Mount Meager, the paraglacial response to glacier advances during Neoglaciation, and historical river dyking and straightening. They proposed a modified version of the paraglacial model (*sensu* Church & Slaymaker 1989) to include large sediment pulses throughout the Holocene (Fig. 8).

Several large ($\sim 10^6\ m^3$) landslides have occurred in the historic period at Mount Meager, including: the 1931 Devastation Creek debris flow, which travelled the length of Meager Creek (Carter 1932) and a further 15 km along Lillooet River (Decker *et al.* 1977); the 1975 Devastation Creek landslide, which killed four exploration geologists (Mokievsky-Zubok 1977); and the 1998 Capricorn Creek event, which temporarily dammed Meager Creek (Bovis & Jakob 2000). However, inclusion of landslides of this size and frequency in the sediment flux to Lillooet River do not balance the sediment budget.

Recent work at Mount Meager has documented several large ($>10^8\ m^3$) prehistoric, non-eruptive

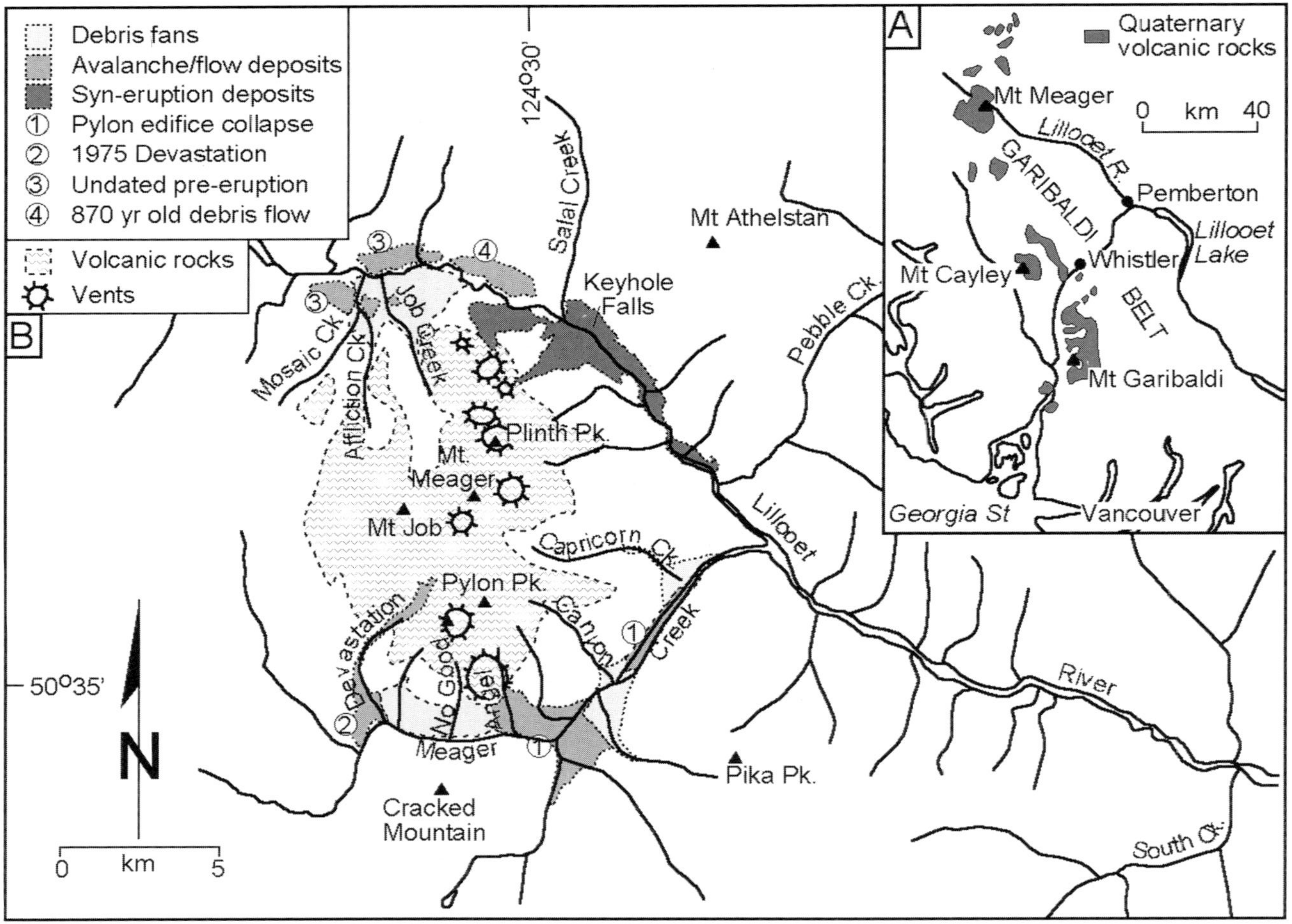

Fig. 7. Geology and landforms of the Mount Meager massif. The massif is drained on the north slopes by the upper Lillooet River and on the south by Meager Creek. Extensive hydrothermal alteration is associated with vents (from Friele *et al.* 2008).

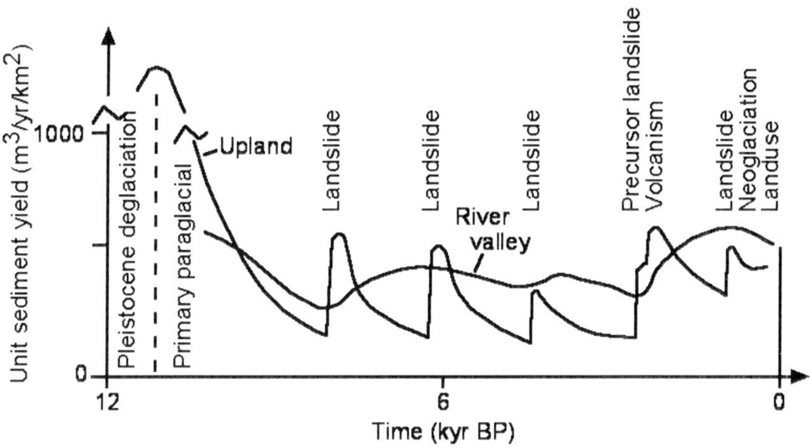

Fig. 8. Schematic sediment yield curve for the Lillooet River, modified from Jordan & Slaymaker (1991), showing pulsing attributed to post-glacial landslides, volcanism, Neoglaciation and land use. Numerical modelling incorporating large stochastic landslides (Dadson & Church 2005) can produce a high, multimodal Holocene sediment yield in formerley glaciated valleys.

landslides (Friele & Clague 2004). In addition, the last eruption at Mount Meager about 2400 years ago generated pyroclastic flows and a large (10^8 m^3) landslide that blocked the Lillooet River, causing an outburst flood (Stasiuk *et al.* 1996; Stewart 2002).

Friele *et al.* (2008) compiled all known landslides to prepare a frequency–magnitude model for Mount Meager (Fig. 9). The calculated denudation rate is about 3000 m^3 km^{-2} year^{-1}, or 3 mm year^{-1} over the entire 80 km^2 area. This estimate is 50 times higher than the average for hypermaritime areas of British Columbia (Martin *et al.* 2002; Guthrie & Evans 2004), suggesting that the massif may be the most active landslide region in Canada. Furthermore, it is two orders of magnitude larger than the value of 80 m^3 km^{-2} year^{-1} estimated for debris flows and landslides by Jordan & Slaymaker (1991) and Slaymaker (1993), and provides ample material with which to balance the sediment budget. Much of this material, however, is stored in the valleys as large landslide deposits and is gradually eroded.

The diffusion and fluvial transport of volcanigenic sediment is the crux of the unbalanced sediment budget problem. An exploratory drilling programme, conducted to document the stratigraphy of the Lillooet River valley fill 30–65 km downstream from Mount Meager, revealed a series of diamictic volcaniclastic sheets interbedded within Holocene fluvial deposits (Friele *et al.* 2005). The diamicton sheets have widths of several hundreds of metres up to 1.5 km and range from 1 to 8 m thick. They were deposited by very large, out-of-channel debris flows and hyperconcentrated flows from Mount Meager (Friele *et al.* 2005; Simpson *et al.* 2006). Metres of floodplain aggradation occurred suddenly during these events. A detailed radiocarbon chronology derived from drill core documents periods of rapid delta advance that are one or two orders of magnitude larger than the long-term average of 6 m year^{-1}. These pulses were coincident with large (10^8–10^9 m^3) landslides (Friele *et al.* 2005). In addition to large debris flows and hyperconcentrated flows, which consist of 75–100% sediment of volcanic provenance, about 25% of Lillooet River bedload is sourced from Mount Meager. In valleys on the flanks of the massif, eroding escarpments in volcanic landslide deposits are directly coupled to the river, providing large sediment loading over the full range of grain sizes.

These data suggest that the anomalously high historic sedimentation rates at Lillooet Lake may be explained by the high flux of sediments from Mount Meager. Contributions from the rest of the watershed are of secondary importance.

Other examples of landslides conditioned by glaciation

Large (10^6–10^8 m^3) Holocene landslides are also characteristic of Mount Garibaldi and Mount Cayley. Numerous landslides have occurred at The Barrier (Fig. 10), a 450 m-high ice-contact volcanic escarpment 20 km north of Cheekye River (Moore & Mathews 1978), with significant effects on Cheakamus River tens of kilometres from the source (Clague *et al.* 2002). This example illustrates how the presence of glacier ice can affect the shape and stability of a volcano. The situation at The

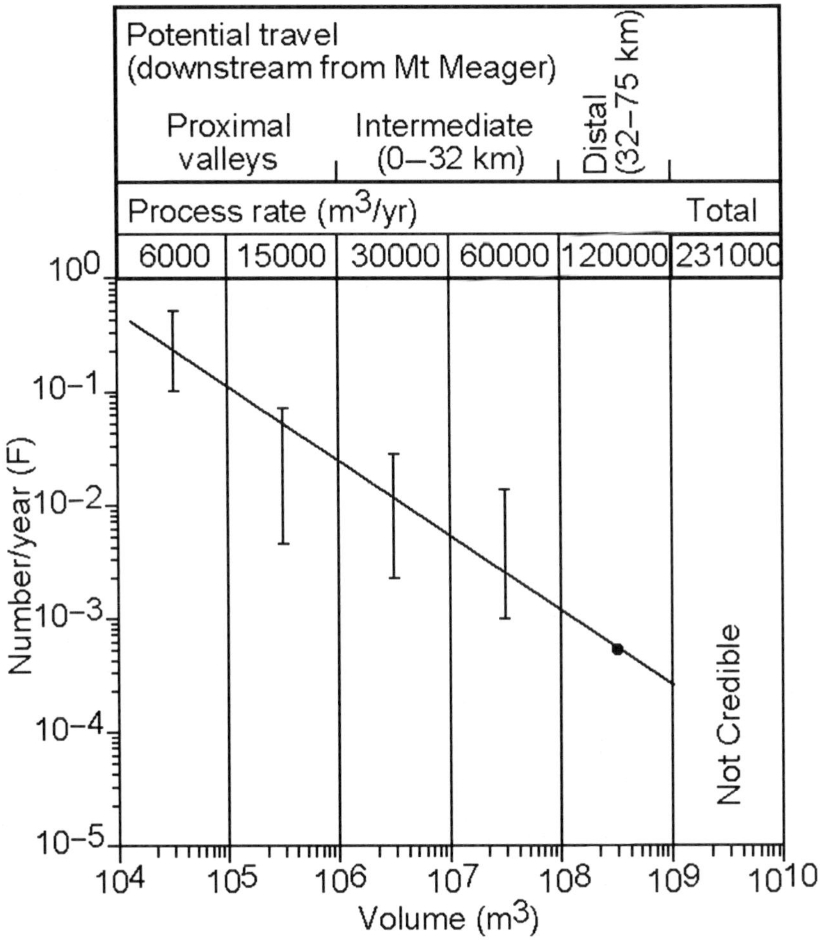

Fig. 9. Frequency–magnitude model for the Mount Meager volcanic complex derived from a compilation of all known landslides at Mount Meager (from Friele *et al.* 2008). The vertical bars represent frequency estimates derived from historic (upper-bound) and prehistoric (lower-bound) events; they provide an indication of the uncertainty associated with the estimates of the landslide process rate.

Barrier contrasts with that on the south flank of the Garibaldi massif, where the early post-glacial Ring Creek lava flow covered the Mamquam valley floor (Brooks & Friele 1992), sharply reducing sediment delivery to Mamquam River (Brooks 1994).

Although the bulk of sediment in Cheekye basin was transferred in late-glacial and early post-glacial time, sediment delivery continues to be punctuated by rare, but very large, out-of-channel debris flows. For example, a large (5.5×10^6 m^3) debris flow inundated the lower Cheekye fan some time between 7.5 and 7.1 cal ka BP, and at least four smaller debris flows were deposited sediment on the fan after 5.6 cal ka BP; the largest of the four (2×10^6–3×10^6 m^3) occurred about 800 years ago (Clague *et al.* 2003). Similarly, Evans & Brooks (1991) documented three large (10^6–10^8 m^3) rock avalanches from the Mount Cayley volcano into the Squamish valley, and dated them to about 5.5, 1.0 and 0.5 cal ka BP. Each of the three landslides successively blocked the Squamish River, creating temporary reservoirs that reached 6–8 km upstream of the dams. These and four or five additional landslide impoundments over the past 5000 years (Brooks & Hickin 1991) indicate an average recurrence of one in every 500 years.

Synthesis

The examples presented in this paper illustrate the tenets of the paraglacial paradigm (*sensu* Ballantyne 2002). The evolution of the Cheekye basin and fan illustrates the primary exhaustion model in a small

Fig. 10. An oblique aerial view of The Barrier, a 450 m-high escarpment formed by the quenching of the Clinker Peak lava flow (right foreground) against glacier ice filling the Cheakamus valley about 13.0 cal ka ago. The lava flows impounded the Garibaldi Lake (background). (Austin Post photograph.)

watershed (<100 km^2); an exponential decay resulted in 80% of the post-glacial sediment transfer happening within 4000 years of Pleistocene deglaciation (Fig. 6). The record from Mount Meager, a watershed two orders of magnitude larger than the Cheekye basin, is more complex. The high post-glacial landslide rate at Mount Meager (Friele *et al.* 2008), coupled with the disproportionately high transfer of volcaniclastic sediment downstream, can probably reconcile the unbalanced sediment budget in the Lillooet River basin.

The volcanoes and their proximal and distal deposits, including the Cheekye and Rubble Creek fans and the 50 km-long Lillooet River delta, are paraglacial landforms. They encompass the rockslope, alluvial, lacustrine and coastal landsystems of Ballantyne (2002), but the processes that have created them, including deep-seated creep, rockfalls, rock avalanches and debris flows, act primarily within the rockslope land system. The downstream land systems are temporary sediment-storage elements.

Ballantyne notes that sediment delivery by paraglacial activity is subject to extrinsic perturbation. The onset of increased precipitation and cooling in the middle Holocene (Mathewes & Heusser 1981) may have induced instability (see Bovis & Jones 1992; Blikra & Nemec 1998). Certainly, retreat from Neoglacial moraines has locally renewed paraglacial activity (Holm *et al.* 2004), but on a smaller scale than during Pleistocene deglaciation.

Finally, 'glacial inheritance in its *manifold forms* [italics added] emerges as the dominant control on the trajectory of subsequent landscape change and sediment flux, and paraglacial landscape modification' (Ballantyne 2002, p. 2008). Nothing illustrates this point better than comparison of sedimentation in the Lillooet and Chilko lakes, two large Coast Mountain fjord lakes in adjacent watersheds (Fig. 2). The two lakes are set in formerly glaciated and partly glacierized watersheds, thus their lacustrine deposits have been conditioned by glaciation. However, Chilko Lake lacks a large, rapidly prograding delta and sedimentation rates are low (Desloges & Gilbert 1998). In contrast, a 50 km-long Holocene delta has been built into Lillooet Lake (Friele *et al.* 2005), and sedimentation rates in the lake remain high (Desloges & Gilbert 1994; Gilbert *et al.* 2006). The difference is the existence of a paraglacial Quaternary volcano in the Lillooet River watershed.

In watersheds containing paraglacial stratovolcanoes, a significant overall reduction in sediment yield must await removal by erosion of volcanic edifices, a process that could take tens of thousands to many millions of years. The examples of paraglacial activity at Quaternary volcanoes that we have presented illustrate end members in the spectrum of landscape response to Pleistocene deglaciation.

We thank C. Ballantyne, J. Carrivick, J. Desloges and J. Knight for constructive reviews of earlier drafts of the paper. The research was supported by grants from the Natural Sciences and Engineering Research Council of Canada (NSERC) and Simon Fraser University.

References

BALLANTYNE, C. K. 2002. Paraglacial geomorphology. *Quaternary Science Reviews*, **21**, 1935–2017.

BLIKRA, L. H. & NEMEC, W. 1998. Post-glacial colluvium in western Norway: depositional processes, facies and paleoclimatic record. *Sedimentology*, **45**, 909–959.

BOVIS, M. & JAKOB, M. 2000. The July 29, 1998 debris flow and landslide dam at Capricorn Creek, Mount Meager Volcanic Complex, southern Coast Mountains, British Columbia. *Canadian Journal of Earth Sciences*, **37**, 1321–1334.

BOVIS, M. J. & JONES, P. 1992. Holocene history of earthflow mass movements in south-central British Columbia: the influence of hydroclimatic changes. *Canadian Journal of Earth Sciences*, **29**, 1746–1755.

BROOKS, G. 1994. The fluvial reworking of late-Pleistocene drift, Squamish River drainage basin, Southwestern British Columbia. *Géographie physique et Quaternaire*, **48**, 51–68.

BROOKS, G. & FRIELE, P. A. 1992. Bracketing ages for the formation of the Ring Creek lava flow, Mount Garibaldi Volcanic Field, Southwestern British Columbia. *Canadian Journal of Earth Sciences*, **29**, 2425–2428.

BROOKS, G. & HICKIN, E. J. 1991. Debris avalanche impoundments of Squamish River, Mt Cayley area, Southwestern British Columbia. *Canadian Journal of Earth Sciences*, **28**, 1375–1385.

CARTER, N. M. 1932. Exploration in the Lillooet River watershed. *Canadian Alpine Journal*, **21**, 8–18.

CHURCH, M. & RYDER, J. M. 1972. Paraglacial sedimentation: A consideration of fluvial processes conditioned by glaciation. *Geological Society of America Bulletin*, **83**, 3059–3072.

CHURCH, M. & SLAYMAKER, O. 1989. Disequilibrium of Holocene sediment yield in glaciated British Columbia. *Nature*, **337**, 452–454.

CLAGUE, J. J. 1981. *Late Quaternary Geology and Geochronology of British Columbia, Part 2: Summary and Discussion of Radiocarbon-dated Quaternary History*. Geological Survey of Canada, Paper, **80-35**.

CLAGUE, J. J. 1989. Quaternary geology of the Canadian Cordillera. *In*: FULTON, R. J. J. (ed.) *Quaternary Geology of Canada and Greenland*. Geological Survey of Canada, Geology of Canada, **1**, 15–96.

CLAGUE, J. J. 1991. Quaternary glaciation and sedimentation. *In*: GABRIELSE, H. & YORATH, C. J. (eds) *Geology of the Cordilleran Orogen in Canada*. Geological Survey of Canada, Geology of Canada, **4**, 419–434.

CLAGUE, J. J., FRIELE, P. A. & HUTCHINSON, I. 2003. Chronology and hazards of large debris flows in the Cheekye River basin, British Columbia, Canada. *Environmental and Engineering Geoscience*, **8**, 75–91.

CLAGUE, J. J., TURNER, R. & REYES, A. V. 2002. Record of recent river channel instability, Cheakamus River valley, British Columbia. *Geomorphology*, **53**, 317–332.

DADSON, S. J. & CHURCH, M. 2005. Post-glacial topographic evolution of glaciated valleys: a stochastic landscape evolution model. *Earth Surface Processes and Landforms*, **30**, 1387–1403.

DECKER, F., FOUGBERG, M. & RONAYNE, M. 1977. *Pemberton: The History of a Settlement*. Hemlock, Burnaby, BC.

DESLOGES, J. R. & GILBERT, R. 1994. Sediment source and hydroclimatic inferences from glacial lake sediments; the postglacial sedimentary record of Lillooet Lake, British Columbia. *Journal of Hydrology*, **159**, 375–393.

DESLOGES, J. R. & GILBERT, R. 1998. Sedimentation in Chilko Lake; a record of the geomorphic environment of the eastern Coast Mountains of British Columbia, Canada. *Geomorphology*, **25**, 75–91.

EVANS, S. & BROOKS, G. 1991. Prehistoric debris avalanches from Mt. Cayley volcano, British Columbia. *Canadian Journal of Earth Sciences*, **28**, 1365–1374.

FINN, C. A., SISSON, T. W. & DESZCZ-PAN, M. 2001. Aerogeophysical measurements of collapse prone hydrothermally altered zones at Mount Rainier volcano. *Nature*, **409**, 600–603.

FRIELE, P. A. & CLAGUE, J. J. 2002*a*. Readvance of glaciers in the British Columbia Coast Mountains at the end of the last glaciation. *Quaternary International*, **87**, 45–58.

FRIELE, P. A. & CLAGUE, J. J. 2002*b*. Younger Dryas readvance in Squamish River valley, southern Coast Mountains, British Columbia. *Quaternary Science Reviews*, **21**, 1925–1933.

FRIELE, P. A. & CLAGUE, J. J. 2004. Large Holocene landslides from Pylon Peak, Southwestern British Columbia. *Canadian Journal of Earth Sciences*, **41**, 165–182.

FRIELE, P. A. & CLAGUE, J. J. 2005. Multifaceted hazard assessment of Cheekye fan, a large debris flow fan in Southwestern British Columbia. *In*: JAKOB, M. & HUNGR, O. (eds) *Debris Flow Hazards and Related Phenomena*. Praxis-Springer, Berlin, 659–683.

FRIELE, P. A., CLAGUE, J. J., SIMPSON, K. & STASIUK, M. 2005. Impact of a Quaternary volcano on Holocene sedimentation in Lillooet River valley, British Columbia. *Sedimentary Geology*, **176**, 305–322.

FRIELE, P. A., EKES, C. & HICKIN, E. J. 1999. Evolution of Cheekye fan, Squamish, British Columbia: Holocene sedimentation and implications for hazard assessment. *Canadian Journal of Earth Sciences*, **36**, 2023–2031.

FRIELE, P. A., JAKOB, M. & CLAGUE, J. J. 2008. Hazard and risk from large landslides from Mount Meager volcano, British Columbia, Canada. *Georisk: Assessment and Management of Risk for Engineered Systems and Geohazards*, **2**, 48–64.

GILBERT, R. 1975. Sedimentation in Lillooet Lake, British Columbia. *Canadian Journal of Earth Sciences*, **12**, 1697–1711.

GILBERT, R., DESLOGES, J. R., LAMOUREUX, S. F., SERINK, A. & HODDER, K. R. 2006. The geomorphic and paleoenvironmental record in the sediments of Atlin Lake, northern British Columbia. *Geomorphology*, **79**, 130–142.

GUTHRIE, R. H. & EVANS, S. G. 2004. Analysis of landslide frequencies and characteristics in a natural system, coastal British Columbia. *Earth Surface Processes and Landforms*, **29**, 1321–1339.

HICKIN, E. J. 1989. Contemporary Squamish River sediment flux to Howe Sound, British Columbia. *Canadian Journal of Earth Sciences*, **26**, 1953–1963.

HICKSON, C. J. 1994. Character of volcanism, volcanic hazards, and risk, northern end of the Cascade magmatic arc, British Columbia and Washington State. *In*: MONGER, J. W. H. (ed.) *Geology and Geological Hazards of the Vancouver Region, Southwestern British Columbia*. Geological Survey of Canada, Bulletin, **481**, 231–250.

HICKSON, C. J. 2000. Physical controls and resulting morphological forms of Quaternary ice-contact volcanoes in western Canada. *Geomorphology*, **32**, 239–261.

HOLM, K., BOVIS, M. J. & JAKOB, M. 2004. The landslide response of alpine basins to post-Little Ice Age glacial thinning and retreat in Southwestern British Columbia. *Geomorphology*, **57**, 201–216.

JAKOB, M. & BOVIS, M. J. 1996. Morphometric and geotechnical controls of debris flow activity, southern Coast Mountains, B.C., Canada. *Zeitschrift für Geomorphologie*, **104**, 13–26.

JORDAN, P. & SLAYMAKER, O. 1991. Holocene sediment production in Lillooet River basin, British Columbia: A sediment budget approach. *Géographie physique et Quaternaire*, **45**, 45–57.

LUCKMAN, B. H. 2000. Little Ice Age in the Canadian Rockies. *Geomorphology*, **32**, 357–384.

MARTIN, Y., ROOD, K., SCHWAB, J. W. & CHURCH, M. 2002. Sediment transfer by shallow landsliding in the Queen Charlotte Islands, British Columbia. *Canadian Journal of Earth Sciences*, **29**, 189–205.

MATHEWES, R. W. & HEUSSER, L. E. 1981. A 12 000 year palynological record of temperature and precipitation trends in Southwestern British Columbia. *Canadian Journal of Botany*, **59**, 707–710.

MATHEWS, W. H. 1952. Mount Garibaldi, a supraglacial Pleistocene volcano in Southwestern British Columbia. *American Journal of Science*, **250**, 553–565.

MATHEWS, W. H. 1958. Geology of the Mount Garibaldi Map area, Southwestern British Columbia, Canada. *Geological Society of America Bulletin*, **69**, 179–198.

MONTGOMERY, D. R. 2002. Valley formation by fluvial and glacial erosion. *Geology*, **30**, 1047–1050.

MOKIEVSKY-ZUBOK, O. 1977. Glacier caused slide near Pylon Peak, British Columbia. *Canadian Journal of Earth Sciences*, **15**, 1039–1052.

MOORE, D. & MATHEWS, W. H. 1978. The Rubble Creek landslide, SWern British Columbia. *Canadian Journal of Earth Sciences*, **15**, 1039–1052.

PORTER, S. C. & SWANSON, T. W. 1998. Radiocarbon age constraints on rates of advance and retreat of the Puget Lobe of the Cordilleran ice sheet during the last glaciation. *Quaternary Research*, **50**, 205–213.

RODDICK, J. A., MULLER, J. E. & OKULITCH, A. V. 1976. *British Columbia–Washington Geological Atlas, Sheet 92, Fraser River*. Geological Survey of Canada, Map 1386A, scale 1:1 000 000.

SHAW, J., TAYLOR, R. B. & FORBES, D. L. 1990. Coarse clastic barriers in eastern Canada: patterns of glaciogenic sediment dispersal with rising seas. *Journal of Coastal Research*, **9**, 160–200.

SIMPSON, K., STASIUK, M., CLAGUE, J. J. & FRIELE, P. A. 2006. Evidence for catastrophic volcanic debris flows in Pemberton Valley, British Columbia. *Canadian Journal of Earth Sciences*, **43**, 679–689.

SLAYMAKER, O. 1993. The sediment budget of the Lillooet River basin, British Columbia. *Physical Geography*, **14**, 304–320.

STASIUK, M. V., RUSSELL, J. K. & HICKSON, C. J. 1996. Distribution, Nature, and Origins of the 2400 BP Eruption Products of Mount Meager, British Columbia: Linkages Between Magma Chemistry and Eruption Behavior. Geological Survey of Canada, Bulletin, **486**.

STEWART, M. L. 2002. *Dacite Block and Ash Avalanche Hazards in Mountainous Terrain: 2360 yr BP Eruption of Mount Meager, British Columbia*. MSc thesis, University of British Columbia, Vancouver, BC.

STUIVER, M., REIMER, P. J. *ET AL.* 1998. INTCAL98 radiocarbon age calibration, 24 000–0 cal BP. *Radiocarbon*, **40**, 1041–1083.

THURBER–GOLDER. 1993. *The Cheekye River Terrain Hazard and Land-use Study, Final Report.* British Columbia Ministry of Environment, Lands and Parks, Burnaby, BC. (Report compiled by Thurber Engineering Ltd and Goldes Associates Ltd.)

Glacially conditioned rock-slope failures and disturbance-regime landscapes, Upper Indus Basin, northern Pakistan

KENNETH HEWITT

Cold Regions Research Centre, Wilfrid Laurier University, Waterloo, Ontario, Canada N2L 3C5 (e-mail: khewitt@wlu.ca)

Abstract: The possible role of paraglacial adjustments in catastrophic rock-slope failures is investigated. Most of some 250 rockslide–rock-avalanche events identified in the Upper Indus Basin descended from slopes affected by Quaternary glaciations. Examples from the lower Gilgit Basin in the Karakoram Himalaya illustrate relations of rock-wall failure to former glacier activity and post-glacial responses, including *sackung* features. In many cases there is strong morphological evidence for paraglacial adjustments. However, other conditions must be considered as partial or alternative controls. They include a complex bedrock geology and seismo-tectonic conditions, and vertical shifts in periglacial and moisture climates. Given the enormous relief, the landslides themselves may involve a range of temporal and spatial distributions influenced by different topo-climates, geological terranes and stages of deglaciation. Meanwhile, post-glacial rock avalanche barriers have blocked and fragmented the river system. They exercise a major control over intermontane sedimentation and valley morphology. These impacts are treated as part of a broader 'disturbance-regime' geomorphology, where late Quaternary landscape evolution is driven by transient, but recurring, interactions among geomorphic process systems. Post-glacial conditions have also tended to partition the landscape into subunits with different mixes of geomorphic activity or rates of adjustment. The findings presented here depart from existing interpretations of Quaternary events in the region, and pose challenges for process geomorphology and the role of large landslides.

It has been suggested that the incidence or scale of mass movements in mountains can reflect former glacial activity (Ballantyne 2002). Rock slopes may be destabilized due to oversteepening by glacial erosion and by removal of support when ice retreats or debuttressing (Bovis 1990). In British Columbia more than half of some 30 historic landslides were identified with glacially debuttressed slopes (Evans & Clague 1994). A number of studies in other mountains identify large rock-slope failures with a glacial legacy (Ballantyne & Benn 1996; Wieczorek 2002; Geertsema *et al.* 2006). This suggests the mass movements would be 'directly conditioned by glaciation or deglaciation', the definition of *paraglacial* adjustments (Church & Ryder 1972, p. 3060). A paraglacial origin of landslides has been invoked by Dadson & Church (2005) in British Columbia, and by Fort (2005) for parts of the European Alps.

This study focuses on catastrophic rock-slope failures that generate rock avalanches. The events of interest are *catastrophic* in having sudden occurrence, great size ($>10^6$ m^3), exceptional rates of movement (100–250 km h^{-1}) and short duration (mins). The rock avalanche typically involves the rapid movement of thoroughly broken and crushed rock for distances of several kilometres, also referred to as a *sturzstrom* (Hsu 1975), or forms a 'long runout' landslide. Typically, the angle between the source or detachment zone and outermost reach of the debris is relatively small, often less than 12° (Heim 1932). The term 'rockslide' is introduced to indicate that there is an initial phase of movement before the mass is sufficiently crushed and pulverized to become a rock avalanche. Usually, the initial movement involves a sliding component along failure surfaces in bedrock, or down the pre-existing slope (Voight 1978). Some cases do appear more like vertical collapses or involve falls, toppling or air launch. There is disagreement about the exact nature of rock avalanches and their mechanics, and over terminology. Some view them as 'slides', perhaps travelling on a lubricated substrate (Legros 2006). Others argue they are flows (Hsu 1978). Attempts to include them among 'flowslides' have not been widely accepted (Rouse 1984), and some reject these terms altogether (Erismann & Abele 2001).

Of greater interest for mountain geomorphology is that hundreds of catastrophic rock-slope failures have been newly identified in the past few decades; initially, in most cases, from rock-avalanche deposits. They are reported throughout the mountains of the Alpine–Himalayan and Inner Asian ranges,

From: KNIGHT, J. & HARRISON, S. (eds) *Periglacial and Paraglacial Processes and Environments.*
The Geological Society, London, Special Publications, **320**, 235–255.
DOI: 10.1144/SP320.15 0305-8719/09/$15.00

where the trigger is known, it is most often a seismic or extreme weather event (Keefer 1984). However, triggers for the prehistoric events to be examined are not known.

A majority of the large landslides described here and elsewhere did not occur for centuries or millennia after deglaciation. An important question, therefore, is whether slopes merely await a sufficient triggering event or if there are intervening, destabilizing adjustments and whether these are paraglacial. Instrumentation of formerly glaciated rock walls has shown long-term and on-going gravitational creep (Bovis 1990; Blikra *et al.* 2006). An important indicator of this is called *sackung*, gravitational spread or 'sagging', widely reported from glacially sculpted rock slopes (Selby 1993). It generates features such as fractures and slide planes subparallel to slope geometry and, on interfluves above steep slopes, trenches, small graben or reverse scarps. While not confined to formerly glaciated mountain terrain, sackung features are common there and have been included among paraglacial slope adjustments (Ballantyne 2002).

The original concept also implies a temporal distribution of landslides diminishing over post-glacial time (Church 2002). Gardner (1983) suggested that high-magnitude rock falls in the Canadian Rockies were more frequent in the late-glacial, and immediate post-glacial, intervals than today. For the same region Cruden & Hu (1993) proposed an 'exhaustion' model for rockslide–rock-avalanche events in which unstable post-glacial sites are 'used up' when landslides occur, and diminish over time. Dadson & Church (2005) included paraglacial slope adjustments in their stochastic 'diffusion' model. In the absence of sufficiently large sets of dated events, however, there is an element of speculation in such models.

A further concern is that only a small fraction of steep, deglaciated rock walls have failed catastrophically, and only a few of the interfluves with sackung features. If post-glacial slope adjustments are important, and move slopes towards critical instability at different rates, this would affect the temporal and spatial distribution of events. It must also be considered whether just a few locations are sufficiently affected by glaciation, or do other conditions intervene?

Debuttressing is not the only condition affecting post-glacial slope stability. Deglaciation in high mountains is accompanied by shifts in temperature and moisture conditions, essentially topo-climatic responses. Periglacial conditions with permafrost may develop and follow the pattern of deglaciation, of rising snow lines or Equilibrium Line Altitudes (ELAs) on the glaciers. In the Karakoram, support for this includes 'fossil' periglacial features below today's alpine periglacial areas and extensive development of rock glaciers (Owen & England 1998).

Periglacial freeze–thaw has been invoked as a landslide-generating condition in mountain environments, as have 'active' layer processes over mountain permafrost (Whalley 1984; Lewkowitz 1988; Kaab *et al.* 2006). Conversely, it has been suggested that snow and ice or permafrost may serve to bind and stabilize mountain rock slopes. Warming or the degradation of permafrost are invoked as destabilizing factors (Harris *et al.* 2001; McSaveney 2002). Where a periglacial phase has intervened between deglaciation and catastrophic slope failures, it may complicate, mask or be mistaken for paraglacial adjustments. Studies of a glacially undercut and recently debuttressed slope in British Columbia showed that the deformation occurring was actually in response to cyclical fluctuations in groundwater (Bovis & Stewart 1998).

There is, of course, a need to bear in mind the full range of conditions affecting slope instability. Landslide studies recognize the roles of tectonics, structure and lithology, and of climatic controls on slope processes. Scheidegger (1998) argued for the 'predesign' of landslides by geotectonic history, suggesting their location and forms reflect crustal events and, especially in mountains, existing tectonic stresses. In the Karakoram, residual or on-going tectonic stresses cannot be ruled out anywhere, but there are no measurements of them that can be related specifically to landslides. Some events are associated with active faults, notably in the Nanga Parbat–Haramosh Massif (Shroder 1993). However, most have occurred where no active faults are known and rates of uplift believed to be less (Searle 1991). The immediate triggers are known for just two historical cases: the earthquake-generated 1841 event; and weather-related conditions for the Bualtar events. There are many examples in which geological structures or partings are associated with rock-slope failures, although the case studies given later include examples that are ambiguous in this regard. Meanwhile, in the Upper Indus Basin, catastrophic failures have been observed in all geological terranes and lithologies, and in a wide variety of structures (Hewitt 2002*a*). Also, in such high mountains climatic conditions and slope processes vary greatly with elevation zones or topo-climates (Hewitt 1993). As noted, they may also shift with climatic changes accompanying deglaciation.

In fact, it seems unlikely that purely paraglacial adjustments occur, with none of the other factors playing a role. Since it is hardly possible to adequately address all possibilities, investigations like those reported here must simply be alert to indicators of the influence of each set of factors. In a mountain landscape, all or most of which was under the ice, it seems necessary to work from actual catastrophic rock-wall failures to establish these relations.

Massive rock-slope failures of the lower Hunza and Gilgit valleys

In total, 28 late Quaternary rock avalanches have been identified in a small subregion of the Karakoram, within a radius of 30 km of Gilgit town (Fig. 2). Two are only known from small remnants, but dimensions for 26 events could be reconstructed from a combination of field observations and inspection of satellite imagery (Table 1). These transported a total volume of material of over 61 km^3. Subsequently, more than half (approximately 33 km^3) has been removed by the rivers. However, two-thirds of both volumes are accounted for by one event; the Nomal Complex (see below). The landslides have a distinctive relation to glacial activity and have profoundly influenced developments along the river valleys.

The subregion is chosen because most of the rock-slope failures occurred at or below ice levels of the local LGM and several detachment zones are relatively accessible. The bedrock geology, structures and tectonic history are relatively well known (Searle 1991), and the area has played a major role in reconstructions of Quaternary events (Paffen *et al.* 1956; Desio & Orombelli 1983; Li Jijun *et al.* 1984; Derbyshire & Owen 1990; Kalvoda 1992). However, until recently the large landslides were missed, their deposits and effects attributed to other processes. About half of the landslides were identified and described in Hewitt (2001). The remainder, and new material pertinent to a paraglacial interpretation, are described here for the first time.

Geological setting of the study area

The study area is in the 'Kohistan terrane' (Searle 1991), interpreted as a former island arc between the Indian and Karakoram plates (Fig. 1). The bedrock geology records multiple intrusive episodes and at least two major tectonic collisional events. The Kohistan Batholith comprises a series of magmatic episodes, thought to be associated with the main tectonic events and emplacing three sets of plutonics (Petterson *et al.* 1990).

- *Stage I deformed plutonics* of intermediate composition, emplaced between 102 and 85 Ma, and deformed by later tectonic and intrusive events (Petterson 1984).
- *Stage II 'undeformed' plutons* comprising three distinct intrusive episodes of basic–intermediate magmatism between 85 and 50 Ma.
- *Stage III intrusives* emplaced 8–34 Ma, and comprising leucogranite sheets, extremely dense near the Indus confluence, and later basic sills and dykes thought to reflect crustal melting as the Indian Plate thrust under Kohistan.

Sixteen of the landslides have detachment zones largely or wholly in exposed plutons. Five are wholly in metasedimentaries, the remainder mixed intrusives and either metasedimentaries or volcanics (Table 2). The 'deformed plutonics' cover less than one-fifth of the area, but feature in half the landslides. Six involve the largest of these plutons, 'Matum Das'. However, volcanics appear to comprise more than two-thirds of the Nomal Complex and so make up the larger fraction of all the landslide materials.

At the eastern edge of the subregion is the Rakiot Fault zone and earthquake activity identified with the Main Mantle Thrust (Fig. 1). The Shuta landslides on the Indus are closest to active faulting and to the most rapidly uplifting area in the whole region (Zeitler 1985). The 1841 earthquake-triggered event occurred just outside the study area, some 35 km south of Gilgit and in the Rakiot Fault zone (Butler 2000). Proximity to this zone suggests that some, if not all the prehistoric landslides, could have been seismically triggered.

In cases such as Dhak Chauki and Jut, the geometry of the detachment-zone scars suggests that structural planes of weakness played a role in the failures. Throughout the region planes of failure are often related to sheeting structure or foliation in plutons, and bedding or joint systems in metasedimentaries (Hewitt 2002*a*, 2006*a*). However, they were not found in detachment zones of the Upper Henzul, Nomal and Batkor events. Their failure surfaces follow the erosional topography, or appear as bowl-shaped exposures above deep-seated rotational slumping (Fig. 3). Perhaps the rock is relatively massive, lacks preferential planes of weakness or they are only present at depth.

It is evident that the massive rock-slope failures play an important role in unroofing plutons, as throughout the region's extensive batholithic terranes (Hewitt 2002*a*). The scale of post-glacial downslope removal of plutonic rock in the study area is enormous, as is that of overlying or flanking country rock.

Glacial–paraglacial relations

A few landslide detachment zones reach above LGM ice levels. Most lie below them. The location and geometry of some, at least, seem to reflect former ice-flow patterns (Fig. 4). Undercutting at a

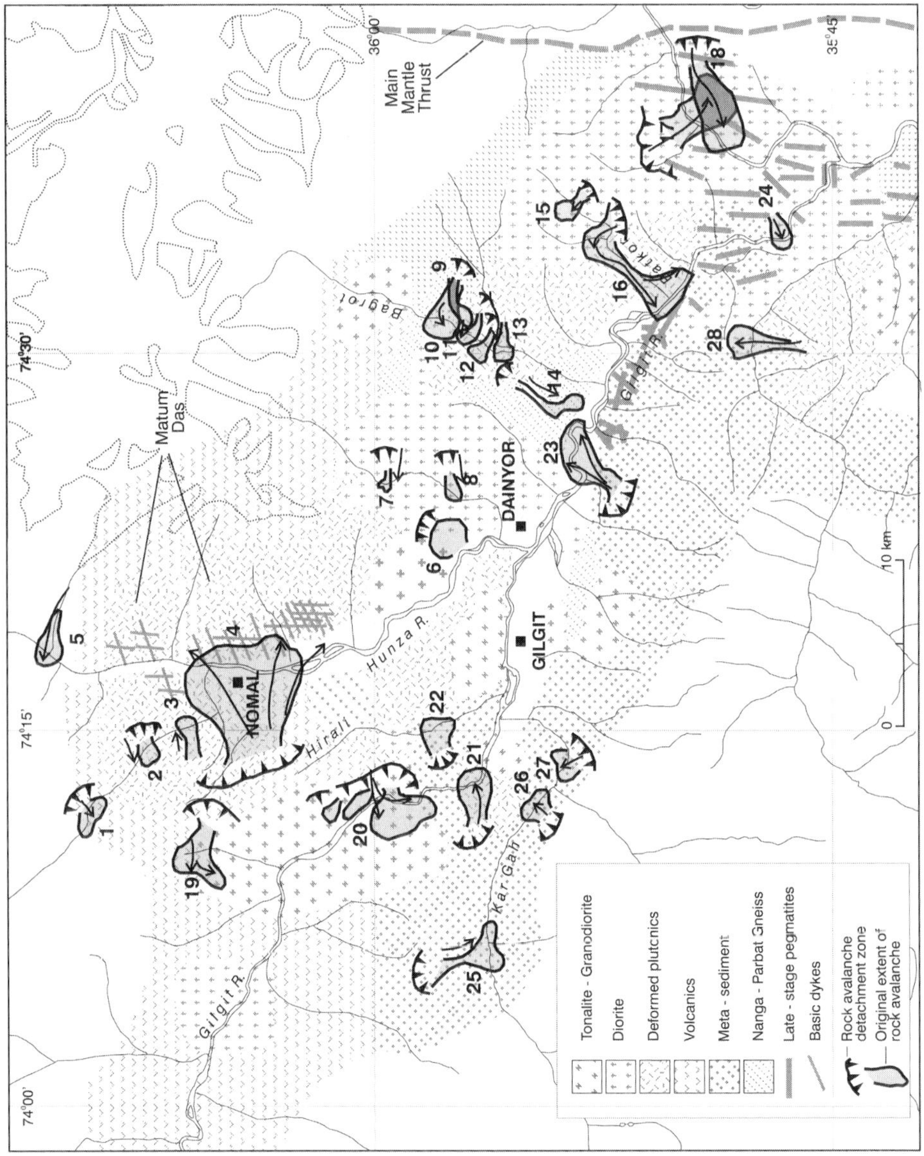

Fig. 2. The lower Gilgit–Hunza basin study area showing rockslide–rock avalanches in relation to the regional geology (after Petterson 1984).

Table 1. *Catastrophic rock-slope failures identified in the lower Gilgit–Hunza basins in order of magnitude. Estimates of original and present-day volumes, descents and run-out distances are based on field observations and satellite imagery (LANDSAT)*

Landslide (# Fig. 2)	Volumes* (km^3)		Elevations (m)		Drop (H)	Run-out† (km^2)
	Orig.	Eroded	Crest (H_{max})	Base (H_{min})		
Nomal (4)	45.0+	24.0	3450	1520	1930	11.0
D. Chauki (23)	3.5	0.5	3450	1375	2075	6.5
Shuta W (17)	2.5	1.8	3700	1390	2310	5.5
Shuta E (18)	1.7	1.5	3400	1320	2080	6.5
Batkor (16)	1.5	0.7	3090	1300	1790	9.5
Naltar Resort (1)	1.2	0.5	4450	2370	2080	4.5
L'r Henzul (21)	0.9	0.45	2950	1480	1470	7.4
Jaglot (5)	0.9	0.7	4090	1650	2440	6.3
Jut (25)	0.8	0.3	4250	2650	1600	3.5
Kargah I (27)	0.8	0.5	3900	1610	2290	4.5
Hope I (10)	0.8	0.5	3250	1910	1340	3.6
Balas (28)	0.52	0.31	4300	3250	1050	6.0
U. Henzul (20)	0.45	0.07	2920	1530	1390	5.0
Bargu (19)	0.34	0.24	3650	1610	2040	6.4
Naltar village (2)	0.3	0.1	3550	2080	1470	3.0
Bilchar II (12)	0.2	0.1	3200	1800	1400	2.4
Kargah II (26)	0.18	0.14	2900	2200	700	1.5
Hirali (22)	0.16	0.14	2470	2005	465	1.9
Bilchar I (11)	0.14	0.08	2900	1800	1100	2.3
Lower Naltar (3)	0.12	0.07	3620	1880	1560	2.7
Kothi (13)	0.1	0.04	2950	1670	1280	2.1
Manu Gah I (8)	0.1	0.5	3520	2170	1350	2.4
M G.-Barit (9)	0.06	0.035	3690	2415	1275	2.3
Jalalabad (14)	0.04	0.0	2890	1475	1415	3.8
U. Batkor (15)	0.03	0.01	4270	2630	1645	2.9
Hope II (9)	0.02	0.0	3250	2100	1150	2.4
Totals	61.55	33.4				

*The accuracy of volume estimates vary according to degree of burial, complexity of emplacement and erosion, and is extrapolated from visible morphology and vertical sections, and experience with comparable events elsewhere.
†Run-out distances, based on surviving deposits, and may be underestimated because of post-landslide removal of outermost material.

sharp bend seems to locate the detachment areas of the Naltar Resort (#1 on Fig. 2), Batkor (16), Shuta West (17), Upper Henzul (20), Hirali (22), Jut (25) and Balas (28) events. Collapses opposite or just below the entry of tributary ice apply to Nomal (4), Bircha (13), Batkor (16), Dhak Chauki (23) and Kargah I (27). In a few cases the source slope of the landslides was already set back from valley walls up- and downstream, presumably by enhanced undercutting: for example, the source slope of the Dhak Chauki event opposite the entrance of Hunza ice. It contrasts with and, in a way, balances the protected slope between Dainyor and Bagrot, where abundant ice-margin sediments accumulated.

The interfluve between the lower Hunza and Gilgit rivers is of special interest. During the LGM it was completely covered by ice. Extensive areas of glacially moulded and scoured bedrock are preserved; moraines, glacio-fluvial and kame terrace deposits. Gilgit and Hunza ice overflowed the interfluves to create a tongue of diffluent ice that moved down the small, intervening Hirali valley. There are terminal moraine complexes in the upper Hirali and a landslide from its glacially undercut western slope. The lower Hirali contains a classic misfit stream, including a 60 m-deep bedrock gorge, that records meltwater from retreating Gilgit–Hunza ice. At the valley head are the detachment zones of the Nomal and Upper Henzul landslides with extensive sackung features. They are termed landslide 'complexes' because there are several distinct landslide units and several types of mass movement.

The Nomal event involved deep-seated rotational movements and a vertical drop that exposes a headwall some 500 m high. There are sackung features immediately above the headwall (Fig. 5a, b). The Upper Henzul complex is of smaller mass but covers almost as large an area. There is one rock avalanche, and two detached blocks of equal or greater volume lodged part way down the source slope. There are also well-developed sackung features at the head of the

Table 2. *Gilgit–Hunza catastrophic rock slope failures and rock avalanches, the geological formations in their source areas and lthologies (after Petterson 1984)*

Event/basin	Geological formation	Lithologies
Nomal (Hunza R.)	Matum Das and N. Gilgit Ridge plutons	volcanics, tonalite, gneiss
Shuta W (Indus R.)	Indus Confluence pluton	diorite
Shuta E (Indus R.)	Indus Confluence pluton	diorite
Batkor (Gilgit. R.)	Indus Confluence pluton	diorite
U. Henzul (Gilgit. R.)	Shirot pluton	volcanics. gneiss, granodiorite
Dhak Chauki (Gilgit R.)	Kohistan terrane	metasedimentaries
Naltar Resort (Naltar R.)	Matum Das pluton	tonalite, volcanics
Lower Henzul (Gilgit R.)	Henzul pluton	gneiss, granite
Jut (Kargah)	Kohistan terrane	metasedimentaries
Hope I (Bagrot R.)	Bagrot pluton	gneiss, diorite
Hope II (Bargot R.)	Bagrot pluton	gneiss, diorite
Bilchar I (Bagrot R.)	Bagrot pluton	gneiss, diorite
Bilchar II (Bagrot R.)	Bagrot pluton	gneiss, diorite
Kothi (Bagrot R.)	Bagrot pluton+	gneiss, diorite, metaseds
Jaglot (Hunza R.)	Matum Das pluton	tonalite, volcanics
Gulapor (Gilgit R.)	Shirot pluton	granodiorite
Parri remnant (Gilgit R.)	Indus Confluence pluton (?)	diorite, gneiss
Jutal remnant (Hunza R.)	Dainyar pluton (?)	diorite
Bargu (Gilgit R.)	Matum Das pluton	tonalite, volcanics
Kargah I (Kargah R.)	Kohistan terrane	metasedimentaries
Kargah II	Kohistan terrane	metasedimentaries
Hirali	Northern Gilgit Ridge pluton	gneiss
Naltar village (Naltar R.)	Matum Das pluton	tonalite
Lower Naltar	Matum Das pluton	tonalite
Balas (Sai R.)	Gashu pluton	geniss, metasedimentaries
Upper Batkor	Indus Confluence pluton	diorite, gneiss
Jalalabad (Gilgit R.)	Kohistan terrane	metasedimentaries
Manu Gah I	Dainyar pluton	diorite
Manu Gah-Barit	Dainyar pluton	diorite

Fig. 3. The Batkor event. Deposits in the upper valley and detachment zone (top of the picture). The rock avalanche travelled north here before impacting the opposing valley wall and turning through 180° to travel down into the Gilgit valley (Hewitt 2001, and Fig. 10 below).

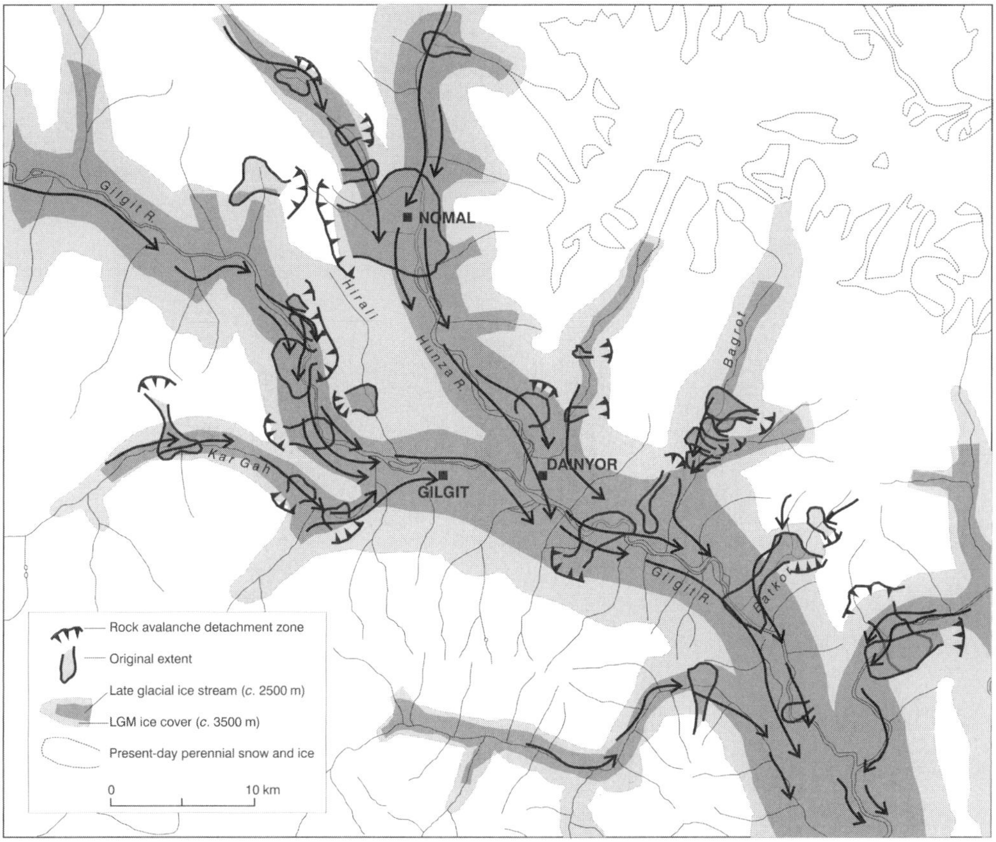

Fig. 4. Rockslide–rock avalanches in the lower Gilgit–Hunza basins related to the level of the last glacial maximum (LGM), and glacier-flow patterns. The LGM of around 3500 m (Kuhle 2006) is shown schematically based on present-day contours. An inferred late-glacial pattern of main ice streams is shown at a roughly 2500 m level.

slope (Fig. 6a, b). In both cases these are parallel or subparallel to landslide detachment zones, and cut through glacial deposits and scoured bedrock. The two landslide complexes are located at sites of marked undercutting by the glaciers. Upper Henzul is at a sharp bend in the Gilgit valley; Nomal immediately below the inflow of Naltar valley ice on the Hunza side (Fig. 4).

Slope stability may well have been influenced by post-glacial permafrost and its decay, wetter conditions or seismic events. The interfluves above the landslide detachment zones have remnants of a former periglacial activity, probably involving permafrost. It could have acted to retard or accelerate failure at sites of post-glacial instability conditioned by glacial action. Evidently these landslides raise questions about the difficulties of disentangling a paraglacial signal from other factors or, indeed, of identifying it with certainty. Deglaciation here involved thinning of ice by hundreds to more than 1200 m. At maximum extent, the main Gilgit, Hunza and Indus ice, separately or together, reached at least 150 km downstream of Gilgit town (Haserodt 1989; Kuhle 2001). Upstream, the main valley glaciers have retreated hundreds of kilometres. While a rapid collapse is conceivable, it may well have taken several millennia after the LGM before trunk glaciers finally disappeared from the study area. Detachment zone areas would have been ice-free long before lower slopes. If paraglacial instabilities developed quickly many landslides may have travelled onto surviving trunk glaciers, to be dispersed and lost from the inventory. Conversely, as indicated in Figure 4, thinner, late glacial ice streams may have played a critical role in destabilizing particular slopes.

The sackung features suggest a wide distribution of adjustment rates or lags, probably complicated by geological and climatic variables mentioned earlier. There is no well-defined, systematic change of landslides incidence in up- or downstream directions or according to the heights of detachment

Fig. 5. Nomal Complex seen from the head of the detachment zone (3200 m). (**a**) Looking southwards along the cliff face. The surface of the main mass is 400–500 m below to the left and there are glacial deposits to the right. The Hunza valley cuts through the middle ground, its unseen floor at 1500 m. (**b**) Looking northwards and down to the plateau-like surface of the main rotational landslide mass at approximately 2500 m, and towards the Naltar valley (right background), which it dammed.

zones (Fig. 7). Rather, they seem quite irregular in all these respects within the study area. A set of reliably dated events may reveal statistically significant relations to the pattern of deglaciation. For the moment, the existence of a paraglacial relaxation signature remains speculative. Meanwhile, post-landslide developments are in some ways more important for the regional geomorphology.

Landslide–interruption complexes

The impacts of the landslides themselves relate especially to cross-valley rock-avalanche barriers that have blocked the rivers. In the whole Upper Indus Basin more than 200 landslide dams have been identified to date, including barriers of exceptional size, stability and longevity (Hewitt 2006*c*). The nature of the landslide interruptions and the

Fig. 6. Sackung around the head of the detachment zone of the Upper Henzul Complex. (**a**) Short vertical cliffs in bedrock, steps suggesting blocks or slices of rock that have moved down the cliff and, further on, graben-like depressions. (**b**) Major sackung depression (large arrows) across the interfluve above the Upper Henzul rock-avalanche detachment zone (small arrow).

role along the Indus streams has been described at length elsewhere (Hewitt 1998, 2002*b*). Their impacts in the Gilgit–Hunza study area serve as a basis for considering the implications for paraglacial responses.

Most, if not all, of these rock avalanches initially impounded a lake. Multi-year lacustrine deposits are seen against the upstream flanks of the Upper Henzul, Dhak Chauki and Batkor landslides. The dam at Nomal was over 1000 m high, judging from lacustrine deposits draping its upper surface at 2490 m (Fig. 8). Such a lake would stretch 100 km up the Hunza valley under present conditions. The Naltar valley was also dammed, leading to a distinctive interruption feature. The Naltar River was diverted from the landslide lake across a bedrock

Fig. 7. View down the Bagrot valley showing four different rock avalanches (large arrows) that have blocked the valley and are in various stages of breaching and degradation. Lacustrine beds record former landslide lakes (small arrows).

Fig. 8. Multi-year lacustrine beds preserved in transverse depressions across the surface of the Nomal landslide at approximately 2500 m.

spur to cut a rock gorge that joins the Hunza River facing upstream. The original junction of the rivers is buried under the landslide. The lower Gilgit valley was also drowned by at least nine other large, post-glacial landslides that blocked the Indus River below their junction (Fig. 9).

Almost the entire Gilgit and Hunza rivers, and most of their tributaries, flow in and over valley fill that records interruption by large landslides. Epicycles of sedimentation when the dams were intact were followed by trenching and remobilizing of sediment as they were breached. With so many

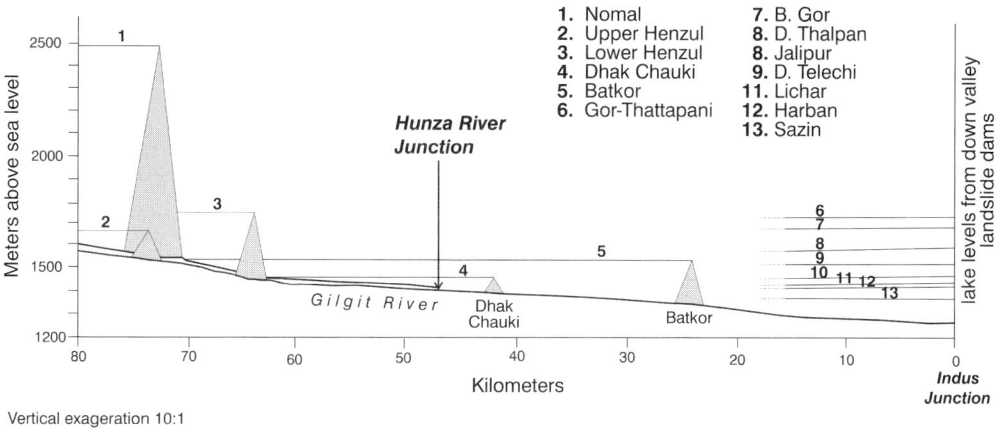

Fig. 9. Landslide dams and landslide lake levels along the lower Gilgit and Hunza streams based on remnants of cross-valley deposits (Hewitt unpublished field notes).

events, independent in scope, location and timing, the landslide interruptions produced a chronically fragmented drainage system (Fig. 7). This explains a typical feature: the alternation of steeper and narrower with gentler, more open stream sections. Whitehouse (1983) describes a similar situation in the South Central Alps of New Zealand, where 'River gradients are commonly low above the [rock avalanche] deposit with marked aggradation, while deeply incised gorges may occur below or through the deposit'.

A phenomenon described by Ferguson (1984) may relate to these conditions. In a study of the Hunza River he observed that sediment inputs from today's glacier basins, which supply 80% or more of the water, were an order of magnitude smaller than sediment transport at the junction with the Gilgit. The difference was attributed to entrainment of valley fill. If so, the supply along the lower Hunza would come mainly from materials originally aggraded behind landslide barriers. Hewitt (1982) had observed large day-to-day variations in sediment concentrations at given discharges, suggesting sediment availability and short-term events mobilizing valley fill are critical (Shroder *et al.* 1998).

Other processes add to and compound the phenomenon of drainage disruption. There are frequent and large debris flows. In every decade since the 1960s they have blocked the main Hunza and Gilgit rivers, in at least one case for several decades (Hughes & Nash 1986). Other disruptions occur due to flash floods, avalanche dams and rapidly aggrading sediment fans (Hughes 1984). Across the northern rim of the basins glaciers interrupt and occasionally dam the rivers (Hewitt 2001). However, in almost all cases the other disturbances affect rivers already responding to larger and longer-lived landslide interruptions (Hewitt 2002*b*).

Discussion

In addition to the landslides, it seems important to ask whether the landslide-interrupted fluvial systems could involve or extend the role of paraglacial adjustments. An answer is complicated, however, by prevailing views of late Quaternary events in the region.

Reinterpretations

Developments identified here with catastrophic landslides have been the main focus of landform and Quaternary studies. Early European visitors emphasized the vast quantities of valley-fill sediments along the Upper Indus valleys (e.g. Strachey 1853). The widespread occurrence of lacustrine sediments and river terraces was discussed, and the term 'alluvial fan' coined for a ubiquitous landform (Drew 1873). Coarse, boulder-capped deposits across valley floors, now regarded as rock avalanches, were treated as keys to Quaternary history (Norin 1925). Dainelli (1922) described these features as 'characteristic morphological elements within the ancient valley floor'. However, he and most others attributed them to late-glacial events moderated by tectonics (Searle 1991; Kalvoda 1992). The landslides were missed or their deposits misinterpreted as glacial (Hewitt 1999). Quite different explanations were given for the landslide–interruption features, most of the lacustrine sediments being attributed to impoundment by ice dams (Burgisser *et al.* 1982).

In the Gilgit–Hunza valleys rock-avalanche deposits have been mapped as glacial moraines or tills, and used as evidence of late-glacial ice positions (Paffen *et al.* 1956; Li Jijun *et al.* 1984; Owen 2006). The extensive lacustrine deposits up- and downstream of Gilgit town were attributed to a 'Glacial Lake Gilgit' impounded by an ice dam (Kamb & Haserodt 2004). The Dhak Chauki rock avalanche has been identified as moraines or till from a late-glacial terminus (Desio & Orombelli 1983). Owen & Derbyshire (1993) assumed it represented the 'Hunza Glacier' that dammed the valley after the Gilgit ice had receded. Instead, the lacustrine beds record post-glacial landslide dams, and more than one. There are well-defined sets against the upstream flanks of the Upper Henzul, Dhak Chauki and Batkor rock avalanches. On the Gilgit itself, only Batkor is high enough to account for the more extensive lacustrine beds such as occur around the Kargah junction. However, other barriers, downstream on the main Indus, were also high enough (Figs 9 and 10).

Sediment fans in the study area have been attributed to paraglacial 'resedimentation' (Derbyshire & Owen 1990). Mainly they record sediment-transport relations between main valleys and tributaries disrupted by landslides. It has involved either tributaries continuing to deliver large loads of coarse sediment to main valleys blocked by landslides or large volumes of sediment coming from degrading landslide–interruptions in the tributary valley. The Bagrot fan illustrates the latter, consisting mainly of sediments remobilized with the breaching of landslide barriers and dumped in the main valley (Fig. 7).

The results described here are incompatible with the late Quaternary interpretations that have prevailed to date. A different framework and sequence of events emerges with implications for a paraglacial component. Of particular concern is a view that the last glacial maximum (LGM) in the study area dates back 50 000–60 000 years BP (Shroder 1993; Owen 2006). In effect, this leaves a more than 40 000 year hiatus or retarding of post-glacial and paraglacial action: for example, since ice last overrode the detachment zone areas of the Upper Henzul and Nomal landslides. The present interpretation supports a more conventional view; a regional LGM roughly coincident with the Wisconsinian or Wurm glaciations and not earlier than 15 000–20 000 BP (Kuhle 2006). It suggests a more dynamic post-glacial environment and geomorphic processes in keeping with the exceptional rates of erosion observed today.

The paraglacial and disturbance regimes

In formerly glaciated areas of the Karakoram, as elsewhere, there are countless features or transitional changes that would not arise without glaciation. However, the original notion of paraglacial adjustment not only focused on one aspect, post-glacial sedimentation, it also inferred a unique spatial and temporal response (Ryder 1971). In extending the scope to include landslides, Dadson & Church (2005) expected a similar 'signature'.

Can this be usefully extended further to include the landslide-fragmented river systems of the Upper Indus Basin? It would go beyond responses

Fig. 10. The Batkor rock avalanche where it enters and once dammed the Gilgit River. The Batkor (short arrows) and Gilgit (long arrow) streams have since breached the dam.

directly conditioned by glaciation. The release of sediment stored in landslide impoundments, for example, is several steps removed. In Church's (2002) terms it does involve a series of 'reservoirs'. Eventually, it may be possible to decode a relaxation signature coinciding with, or similar to, that of paraglacial sedimentation. However, today's more significant features are uniquely patterned by 'reservoirs' very dissimilar in size, shape, timing and history, and by interactions among them. Is it helpful to subordinate this to a, as yet hypothetical and net, long-term signature? Does that not divert attention from distinctive landforms and adjustments that have been the focus of past investigations and of this one?

In the basins of the Karakoram, as noted, there is a spectrum of landscape disturbances, each probably with its own response patterns. Each process system – glacial, fluvial, mass wasting, aeolian and lacustrine – is being directly and indirectly conditioned by others. However, in the study area, and along most of the Indus streams, catastrophic landslides have affected all these different sources of disturbance. Sediment is stored in and released from reservoirs whose form, location and fate depend mainly on the landslide barriers. Even paraglacial sedimentation – and considerable volumes of former glacial deposits have been or are being mobilized – is part of this. It is, however, overwhelmed by epicycles of aggradation and trenching in the fragmented drainage system. Sediment delivery is reorganized by and through the landslide interruptions. Individual disturbances may be geologically transient, lasting centuries or a few thousands of years. However, the repetition and multiplicity of events has regulated morphological evolution over all of post-glacial time, at least.

Past interpretations have been preoccupied with impoundment histories too, but mainly with sudden failures of natural dams. These have been seen to explain 'megaflood' deposits and the 'Indus erratics', although attributed to glacier dams rather than landslides (Desio & Orombelli 1983; Hewitt 1982, 2006*c*; Shroder *et al.* 1998). While the only historic landslide dam of 1841 did fail catastrophically, now it appears many of the landslides created relatively long-lived dams and gradually or intermittently degraded impoundments.

Church (2002) described paraglacial sedimentation itself as one class of 'fluvial disturbances', and this seems the appropriate view here. Moreover, the range and scope of disturbances in the Upper Indus Basin can be said to comprise a *disturbance regime* (Hewitt 2006*b*). As with the paraglacial hypothesis, developments are *directly conditioned by other geomorphic events*. 'Disturbance' refers to specific, extended interference in the action of one or more Earth surface processes by another, and landforms whose location and history are dependent on the disturbance. Resulting landforms tend to be transient, polygenetic and exhibit 'morphological heterogeneity' (Benda *et al.* 2003).

The disturbances create a 'regime' if they operate on temporal and spatial scales comparable with those commonly identified with climatic regimes, geotectonic provinces or the internal dynamics of major geomorphic systems. This is the case for the Upper Indus Basin in the late Quaternary.

'Disturbance regime' may seem an oxymoron but, like some others – dynamic equilibrium, mature-from-birth and deterministic chaos – it identifies a conceptual relation to, but *departure from*, some prevailing notions. It qualifies the view that a disturbance refers only to an anomaly or extreme and singular phenomenon, or that whatever it disturbs is *the* system or state towards which everything is tending. In the past, such thinking encouraged the misinterpretation of the landslides and related features. Premature attempts to define them in terms of broad trends in the tectonic and climatic environment, mainly glaciation, were compounded by circular reasoning. Phenomena such as former lakes, sediment fans, stream terraces and paraglacial sedimentation were thought themselves to reflect co-ordinated, basin-wide responses. Rather, fluvial features do not vary systematically with stream order, slope or other drainage network properties, or record unidirectional changes. Earth surface processes in this region have created a series of transitional, fragmented adjustments, at least with respect to those landforms and sediments that have been the focus of existing interpretations (Owen & Derbyshire 1993).

Partitioned and transitional landscapes

Another observation, evident in the Gilgit study area, further qualifies this interpretation. Important as the landslides and interrupted fluvial zone are, they account for barely 30% of the area (Figs 10 and 11). The detachment, transport and emplacement areas reconstructed for 26 landslides cover about 17% (*c.* 230 km^2). Roughly half comprises material deposited in valley-fill areas. The remainder of the latter, between the landslides but affected by them, comprise about 15%. What of the remaining, more than 70%, of the landscape?

Another, roughly 15% of the study area comprises gentler interfluves, such as surround the Hirali Basin. Today these are arid or semi-arid environments. Salt weathering is indicated by many tafoni, and there is evidence of aeolian deposition and deflation everywhere. Brief, ephemeral drainage, that occurs in spring snow melt and rare summer rainstorms, has gullied superficial deposits

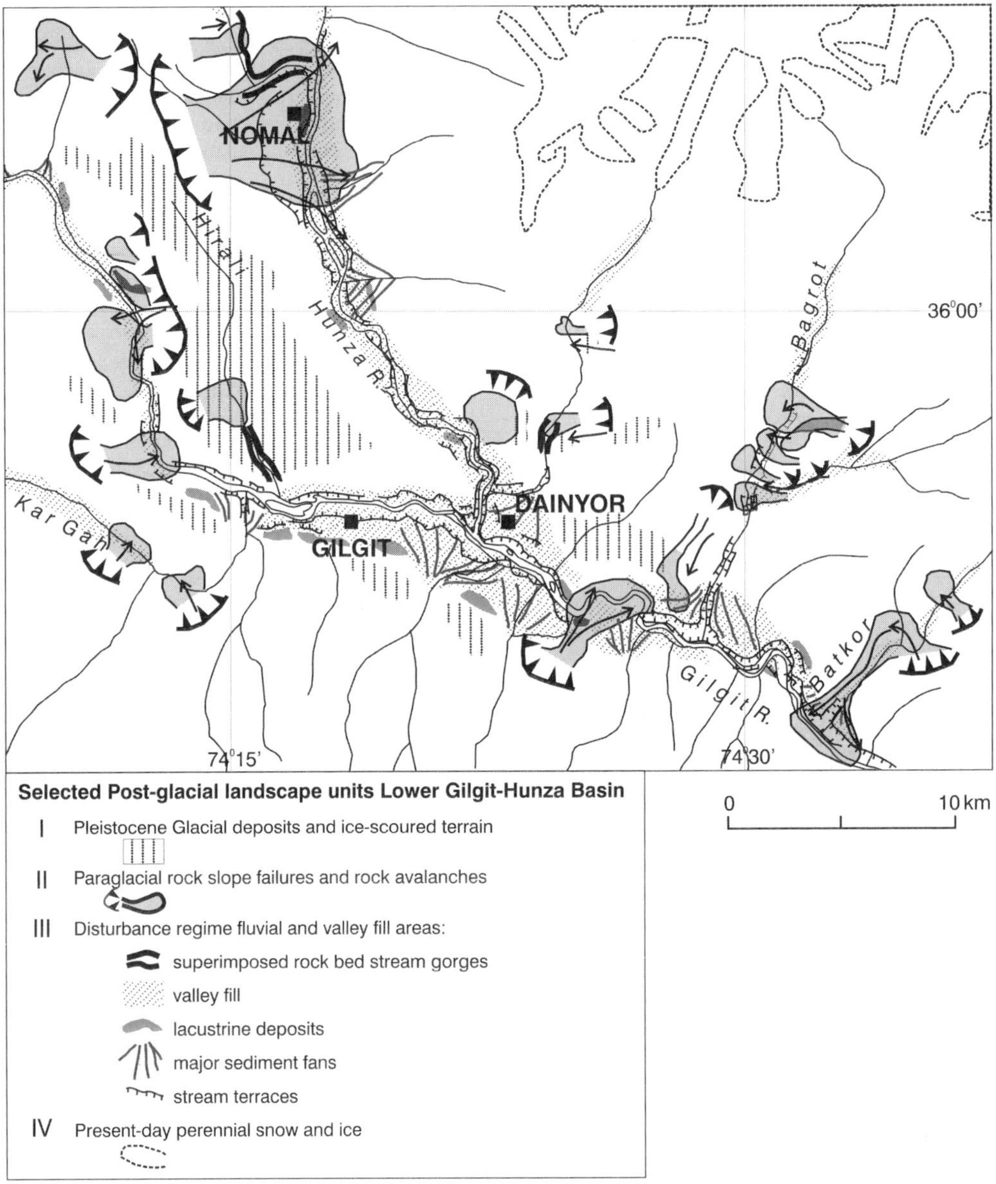

Fig. 11. Transitional and partitioned landscapes: post-glacial morphological units in the lower Gilgit and Hunza basins based on field observations, extended with the aid of satellite imagery.

on steeper slopes. Otherwise, the ice-sculptured bedrock exposures and glacial deposits show little modification since deglaciation.

About 10% of the area, mainly in the Bagrot and Manu Hag valleys, lies in present-day ice climates of the high Karakoram where snow avalanches, rock-wall processes and glaciers ensure high rates of primary erosion (Hewitt 1993). The remaining, roughly 40% consists of steep, mostly dry, rock walls. If their steepness is a legacy of Pleistocene glaciations, only a few areas retain marks of ice action. Considerable post-glacial activity is indicated by weathering, rock falls, and steep chutes with talus and debris cones below. Sackung features at the heads of many of these slopes may signal a paraglacial component, yet to be investigated.

What these divisions suggest is a landscape partitioned into series of morphological units where post-glacial transitions differ greatly in character and rate. Below about 4000 m, intense late Quaternary dessication has been the overwhelming factor. It has helped in this partitioning, cutting off much of the landscape from all but rare or intermittent drainage, but enhancing the relative importance of the main river valleys. Only the fraction of the rock walls and valley fill directly undercut by rivers are subject to rapid and efficient removal of material.

Again, the rock-avalanche emplacement areas are divided up according to the burial, segmentation and removal of material by axial drainage. Emplacement on arid valley floors makes distance from valley walls or changeable river channels critical. Tens of square kilometres of their boulder-covered surfaces are virtually unchanged since the landslides occurred. They became part of the slowest 'queues' of debris, along with debris flow, rock fall and other erosional materials deposited on arid terrace and fan surfaces out of reach of axial drainage. There is a further temporal segregation between areas affected by stream flow in most years, and those only reached by rare and exceptional floods: by winter snow avalanches, spring debris flows, or by regional or local intense rain storms.

There could hardly be a greater contrast with the humid ice climates of the high Karakoram. These are responsible for the intense cascade of moisture and sediment that descends from today's glacial zone (Hewitt 1993). They supply 90% of river flow through the study area. However, developments along these axial drainage channels have been reorganized by locally derived landslide interruptions. The consequences are very different from the widely assumed correlation between rapid rates of uplift, intensified landsliding *and* stream incision (Howard 1996). There is a more complicated relationship between these three variables. The Gilgit and Hunza rivers flow largely in valley fill, thanks to catastrophic landslides – that may be directly related to glaciation and only indirectly to tectonics – and there is little evidence of any net incision in the Holocene. The same applies throughout the main trans-Himalayan Indus, except for a short section where it crosses the rapidly uplifting Nanga Parbat–Haramosh massif (Burbank *et al.* 1996; Hewitt 2006*b*).

These observations may also seem at odds with the most common portrayals of the Karakoram, typified as a 'high-energy' environment. The enormous relief is usually emphasized, the extreme steepness of slopes and globally high rates of uplift and denudation (Miller 1984). In the study area total available relief is about 5000 m. Immediately to the east and south is, perhaps, the steepest terrain on Earth. Around Nanga Parbat (8125 m) slopes rise more than 7000 m in 20 km horizontally (Fig. 1). Yet, the landslides studied rarely involve more than 1200–1500 m of relief (Table 1). Their detachment-zone heads are generally below 4500 m, and many are below 3500 m. Some of the very largest, like the Nomal, Upper Henzul or Batkor events, descended from lesser, more rounded watersheds. This is partly an artefact of the choice of study area, focusing on landslides from at and below the LGM.

However, work in other trans-Himalaya Upper Indus valleys has revealed similar densities and scope of catastrophic rock-slope failures originating at all elevations (Hewitt 2002*a*). Some evidence suggests rockslide/rock avalanches are most frequent in existing glacier basins (Hewitt 2009). Six out of the seven known historical events in the Karakoram originated there. But the much poorer chances of observation and survival of rock avalanches deposited on glaciers mean few prehistoric events are known (McSaveney 2002).

In this respect, it seems likely that the study area is not representative of the whole region, which may be further partitioned in ways that generate different temporal and spatial distributions of catastrophic rock-slope failures. They may reflect different altitude- and exposure-related topo-climates. There is recent deglaciation to consider, not only late Pleistocene, and slopes subject to vigorous local steepening by post-glacial stream action. Distinct geological terranes or greater seismicity and active faulting add to the mix of spatially varying factors that may overlap with, or overwhelm and replace, a paraglacial signature.

Conclusions

Catastrophic rock-slope failures in the Karakoram may well involve paraglacial adjustments. A large fraction of known events originated at or below the limit of ice at the LGM. About one-third of those from the lower Gilgit–Hunza valleys occur at sites with strong modifications of valley-wall geometry by ice streams. The evidence is circumstantial rather directly measured, but as convincing as that reported from most other regions. Evidence is lacking to show or refute a paraglacial signature in the incidence of landslides. It could be resolved with a sufficient number of reliable dates for sets of landslide events, but that is a work-in-progress. It is likely that any paraglacial adjustments would be slowed or accelerated by climatic changes accompanying or following deglaciation. They include growth and degradation of mountain permafrost, warming and dessication.

Meanwhile, the findings show catastrophic landslides have influenced late Quaternary developments in other, perhaps more significant, ways. Their contribution to denudation in post-glacial time is huge; likewise, to the unroofing of plutons in the Ladakh–Kohistan batholith. Their blocking of the rivers and related disturbances are decisive for landforms of the fluvial zone, and in regulating sediment storage and transport. They also contribute to a partitioning of the landscape into areas responding to post-glacial conditions in different ways. In other parts of the basin, glaciers and rock glaciers, debris flows, rapidly aggrading sediment fans and dune fields also interrupt rivers continuously or repeatedly (Hewitt 2001). It is suggested that these phenomena create a disturbance-regime geomorphology, of which paraglacial adjustments are just one manifestation.

The disturbance regime and paraglacial hypotheses are a challenge for two prevailing views of Earth surface processes. The first is that they are essentially response systems determined by climatic or 'exogenetic', and geotectonic or 'endogenetic', controls. Over geological time and for broad regional landscapes, no doubt these controls do prevail. Yet, in the Karakoram, despite extreme manifestations of both, their influence in the late Quaternary has been buffered and reconfigured by responses among geomorphic processes. These create their own, distinctive morphogenetic developments.

The second challenge relates to the prevailing specialization, in particular geomorphic processes. Paraglacial adjustments refer to the interactions of different Earth surface processes, and these are even more important and varied in a disturbance-regime geomorphology. Distinctive sets of interactions among other geomorphic processes were shown to have put their stamp on much of the region. Catastrophic rock-slope failures emerge as major controlling influences in post-glacial time–with or without a paraglacial signature.

I thank Drs J. Clague and W. Mitchell for comments, two reviewers for many helpful suggestions and Dr M. G. Petterson for geological information about the study area. Funding for parts of the research was provided by the International Development Research Centre (IDRC) Ottawa, Canada and Wilfrid Laurier University's Office of Research. P. Schaus prepared the maps and diagrams.

References

ABDRAKHMATOV, K., EVANS, S. G., HERMANNS, R., SCARASCIA-MUGNOZZA, G. & STROM, A. L. (eds). 2004. *Security of Natural and Artificial Rockslide Dams: Extended Abstracts Volume.* NATO Advanced Research Workshop, Bishkek, Kyrgyzstan.

ABELE, G. 1974. Bergsturze in den Alpen, ihre Verbreitung, Morphologie und Folgengeerscheinungen. *Wissenschaftliche Alpenvereinshefte*, **25**, 203.

BALLANTYNE, C. K. 2002. Paraglacial geomorphology. *Quaternary Science Reviews*, **21**, 1935–2017.

BALLANTYNE, C. K. & BENN, D. I. 1996. Paraglacial slope adjustment during recent deglaciation and its implications for slope evolution in formerly glaciated environments. *In*: ANDERSON, M. G. & BROOKS, S. M. (eds) *Advances in Hillslope Processes*, Volume 2. John Wiley, New York, 1173–1195.

BENDA, L., VELDHUISEN, C. & BLACK, J. 2003. Debris flows as agents of morphological heterogeneity at low-order confluences, Olympic Mountains, Washington. *Geological Society of America Bulletin*, **115**, 1110–1121.

BLIKRA, L. H., LONGVA, O., BRAATHN, A., ANDA, E., DEHLL, J. F. & STALSBERG, K. 2006. Rock slope failures in Norwegian fjord areas: Examples, spatial distribution and temporal patterns. *In*: EVANS, S. G., SCARASCIA-MUGNOZZA, G., STROM, A. & HERMANNS, R. L. (eds) *Landslides from Massive Rock Slope Failure*. NATO Science Series. Springer, Dordrecht, 475–496.

BOVIS, M. J. 1990. Rock-slope deformation at Affliction Creek, southern Coast Mountains, British Columbia. *Canadian Journal of Earth Sciences*, **27**, 243–254.

BOVIS, M. J. & STEWART, T. W. 1998. Long-term deformation of a glacially undercut rock slope, southwest British Columbia. *In*: MOORE, D. & HUNGR, O. (eds) *Proceedings of the 8th International Congress, International Association for Engineering Geology and the Environment*. A.A. Balkema, Rotterdam, 1267–1276.

BURBANK, D. W., LELAND, J., FIELDING, E., ANDERSON, R. S., BROZVIC, N., REID, M. R. & DUNCAN, C. 1996. Bedrock incision, rock uplift and threshold hillslopes in the northwestern Himalayas, *Nature (London)*, **379**, 505–510.

BURGISSER, H. M., GANSSER, A. & PIKA, J. 1982. Late Glacial lake sediments of the Indus valley area, northwestern Himalayas. *Eclogae Geologische Helvetica*, **75**, 51–63.

BUTLER, R. W. H. 2000. Structural evolution of the western margin of the Nanga Parbat massif, Pakistan Himalaya: insights from the Raikhot-Lichar area. *In*: KHAN, M. A., TRELOAR, P. J., SEARLE, M. P. & JAN, M. Q. (eds) *Tectonics of the Nanga Parbat Syntaxis and the Western Himalaya.* Geological Society, London, Special Publications, **170**, 51–75.

CHURCH, M. 2002. Fluvial sediment transfer in cold regions, *In*: HEWITT, K., BYRNE, M.-L., ENGLISH, M. & YOUNG, G. (eds) *Landscapes of Transition: Landform Assemblages and Transformations in Cold Regions*. Kluwer Academic, Dordrecht, 93–111.

CHURCH, M. & RYDER, J. M. 1972. Paraglacial sedimentation: a consideration of fluvial processes conditioned by glaciation. *Geological Society of America Bulletin*, **83**, 3059–3072.

CRUDEN, D. M. & HU, X.-Q. 1993. Exhaustion and steady-state models for predicting landslide hazard in the Canadian Rockies. *Geomorphology*, **21**, 18–30.

DADSON, S. J. & CHURCH, M. 2005. Postglacial topographic evolution of glaciated valleys: a stochastic landscape evolution model. *Earth Surface Processes and Landforms*, **30**, 1387–1403.

DAINELLI, G. 1922. *Studi sul Glaciale: Relazione Scientifiche della Spedizione Italiana de Filippi, nell'Himalaia, Caracorum e Turchestan Cinese (1913–1914)*, Volume 3. Zanichelli, Bologna.

DERBYSHIRE, E. & OWEN, L. 1990. Quaternary alluvial fans in the Karakoram mountains. *In*: RACHOKI, A. H. & CHURCH, M. (eds) *Alluvial Fans: A Field Approach*. Wiley, New York, 27–54.

DESIO, A. & OROMBELLI, G. 1983. The 'Punjab erratics' and the maximum extent of the glaciers in the middle Indus valley (Pakistan) during the Pleistocene. *Accademia nazionale die Lincei, Classe di scienzia fisiche, matematiche e naturali*, Ser.**VII**, **XVI**, Memorie, Sezione IIa, 135–180.

DREW, F. 1873. Alluvial and lacustrine deposits and glacial records of the Upper Indus Basin. *Quarterly Journal of the Geological Society, London*, **29**, 441–471.

EISBACHER, G. H. & CLAGUE, J. J. 1985. *Destructive Mass Movements in High Mountains*. Geological Survey of Canada, Paper, **84-16**.

ERISMANN, T. H. & ABELE, G. 2001. *Dynamics of Rockslides and Rockfalls*. Springer, Berlin.

EVANS, S. G. & CLAGUE, J. J. 1994. Recent climatic change and catastrophic geomorphic processes in mountain environments. *Geomorphology*, **10**, 107–128.

EVANS, S. G. & DEGRAFF, J. V. (eds). 2002. *Catastrophic Landslides: Effects, Occurrence, and Mechanisms*. Geological Society of America. Reviews in Engineering Geology, **15**.

FERGUSON, R. I. 1984. Sediment load of the Hunza River. *In*: MILLER, K. J. (ed.) *The International Karakoram Project*, Volume 2. Cambridge University Press, Cambridge, 581–598.

FORT, M. 2005. How significant is the distinction between paraglacial and seismo-tectonically induced mass-movements in active orogens? Insights from the Pamir–Karakorum–Himalayan ranges. *In*: ETIENNE, S. (ed.) *Shifting Lands, New Insights into Periglacial Geomorphology. Second Workshop, SEDIFLUX Network, Clermont-Ferrand, France*. European Science Foundation, Strasbourg, 26.

GARDNER, J. S. 1983. Rockfall frequency and distribution in the Highwood Pass area, Canadian Rocky Mountains. *Zeitschrift fur Geomorphologie*, **27**, 311–324.

GEERTSMA, M., CLAGUE, J. J., SCHWAB, J. W. & EVANS, S. G. 2006. An overview of recent large catastrophic landslides in northern British Columbia, Canada. *Engineering Geology*, **83**, 120–143.

HARRIS, C., HAEBERLI, W., MUHLL, D. V. & KING, L. 2001. Permafrost monitoring in the high mountains of Europe: the PACE project in its global context. *Permafrost and Periglacial Processes*, **12**, 3–11.

HASERODT, K. 1989. Zur pleistozanen und postglazialen Vergletscherungzwishen Hindukutsch, Karakorum und Westhimalaya. *In*: HASERODT, K. (ed.) *Beitragen und Materialen zur Regionalen Geographie*, Volume 2. Institut fur Geographie der technischen Universitat, Berlin, 182–233.

HEIM, A. 1932. *Bergsturtz und Menschenleben*. Fretz & Wasmuth, Zurich.

HERMANNS, R. L. & STRECKER, M. R. 1999. Structural and lithological controls on large Quaternary rock avalanches (sturzstroms) in arid northwestern Argentina. *Geological Society of America Bulletin*, **111**, 934–948.

HEWITT, K. 1982. Natural dams and outburst floods of the Karakoram Himalaya. *In*: GLEN, J. (ed.) *Hydrological Aspects of Alpine and High Mountain Areas*. International Association for Scientific Hydrology (IAHS) Publication, **138**, 259–269.

HEWITT, K. 1988. Catastrophic landslide deposits in the Karakoram Himalaya. *Science*, **242**, 64–67.

HEWITT, K. 1993. Altitudinal organization of Karakoram geomorphic processes and depositional environments. *In*: SHRODER, J. F. JR. (ed.) *Himalaya to the Sea: Geology, Geomorphology and the Quaternary*. Routledge, London, 159–183.

HEWITT, K. 1998. Catastrophic Landslides and their effects on the Upper Indus streams, Karakoram Himalaya, Northern Pakistan. *Geomorphology*, **26**, 47–80.

HEWITT, K. 1999. Quaternary moraines vs catastrophic rock avalanches in the Karakoram Himalaya, Northern Pakistan. *Quaternary Research*, **51**, 220–237.

HEWITT, K. 2001. Catastrophic rock slides and the geomorphology of the Hunza and Gilgit Basins, Karakoram Himalaya. *Erdkunde*, **55**, 72–93.

HEWITT, K. 2002*a*. Styles of rock avalanche depositional complex in very rugged terrain, Karakoram Himalaya, Pakistan. *In*: EVANS, S. G. & DEGRAFF, J. V. (eds). *Catastrophic Landslides: Effects, Occurrence, and Mechanisms*. Geological Society of America, Reviews in Engineering Geology, **15**, 345–378.

HEWITT, K. 2002*b*. Postglacial landform and sediment associations in a landslide-fragmented river system: the trans-Himalayan Indus streams. *In*: HEWITT, K., BYRNE, M.-L., ENGLISH, M. & YOUNG, G. (eds) *Landscapes of Transition: Landform Assemblages and Transformations in Cold Regions*. Kluwer Academic, Dordrecht, 63–91.

HEWITT, K. 2006*a*. Rock avalanches with complex run out and emplacement, Karakoram Himalaya, Inner Asia. *In*: EVANS, S. G., SCARASCIA-MUGNOZZA, G., STROM, A. & HERMANNS, R. L. (eds) *Landslides from Massive Rock Slope Failure*. NATO Science Series. Springer, Dordrecht, 521–550.

HEWITT, K. 2006*b*. Disturbance regime landscapes: mountain drainage systems interrupted by large rockslides. *Progress in Physical Geography*, **30**(3), 365–393.

HEWITT, K. 2006*c*. Rock avalanche dams on the transHimalayan Upper Indus Streams: a survey and assessment of hazards-related characteristics. *Italian Journal of Engineering Geology and Environment*, Special Issue on Security of Natural and Artificial Rockslide Dams, 61–66.

HEWITT, K. 2009. Rock avalanches that travel onto glaciers and related developments, Karakoram Himalaya, Inner Asia. *Geomorphology*, **103**, 66–79.

HEWITT, K., BYRNE, M.-L., ENGLISH, M. & YOUNG, G. (eds). 2002. *Landscapes of Transition: Landform Assemblages and transformations in Cold Regions*. Kluwer Academic, Dordrecht.

HOWARD, A. 1996. The ephemeral mountains. *Nature*, **379**, 488–489.

HSU, K. 1975. Catastrophic debris streams (sturzstroms) generated by rock falls. *Geological Society of America Bulletin*, **86**, 129–140.

HSU, K. 1978. Albert Heim: observations of landslides and relevance to modern interpretations. *In*: VOIGHT, B. (ed.) *Rockslides and Avalanches, Volume 1. Natural Phenomena.* Elsevier, New York, 70–93.

HUGHES, R. E. 1984. Yasin valley: the analysis of geomorphology and building types. *In*: MILLER, K. J. (ed.) *The International Karakoram Project*, Volume 2. Cambridge University Press, Cambridge, 253–288.

HUGHES, R. E. & NASH, D. 1986. The Gupis debris flow and natural dam, July 1980. *Disasters*, **10**, 8–14.

ISDR. 2000. *Usoi Landslide Dam and Lake Sarez: An Assessment of Hazard and Risk in the Pamir Mountains, Tajikistan.* International Strategy for Disaster Reduction. United Nations, Geneva.

KAAB, A., CHIARLE, M., RAUP, B. & SCHEIDER, C. 2006. Editorial: Climate change impacts on mountain glaciers and permafrost. *Global and Planetary Change*, **56**, vii–ix.

KAMP, U. & HASERODT, K. 2004. Quaternary glaciations in the high mountains of northern Pakistan. *In*: EHLERS, J. & GIBBARD, L. P. (eds) *Quaternary Glaciations: Extent and Chronology, Part III, South America, Asia, Africa, Australia, Antarctica.* Elsevier, Amsterdam, 293–311.

KALVODA, J. 1992. *Geomorphological Record of the Quaternary Orogeny in the Himalaya and the Karakoram.* Elsevier, Amsterdam.

KEEFER, D. K. 1984. Landslides caused by earthquakes. *Bulletin of the Geological Society of America*, **105**, 406–421.

KORUP, O. 2002. Recent research on landslide dams – A literature review with special attention to New Zealand. *Progress in Physical Geography*, **26**, 206–235.

KORUP, O., STROM, A. & WEIDINGER, J. T. 2005. Effects of large rockslide dams on sediment budgets – evidence from the Himalaya, the Tien Shan, and the New Zealand Southern Alps. *In*: *European Geosciences Union, General Assembly, Vienna*, Abstract EGU05-A-01195; NH3.02-1FR20-005.

KUHLE, M. 2001. The maximum Ice Age (LGM) glaciation of the Central- and South Karakorum: an investigation of the heights of its glacier levels and ice thicknesses as well as lowest prehistoric ice margin position in the Hindukush, Himalaya and in East Tibet on the Minya Konka-massif. *GeoJournal*, **54**, 109–396.

KUHLE, M. 2006. The past Hunza Glacier in connection with a Pleistocene Karakoram ice stream network during the last ice age (Wurm). *In*: KREUTZMANN, H. (ed.) *Karakoram in Transition: Culture, Development, Ecology in the Hunza Valley.* Oxford University Press, Karachi, 24–48.

LEGROS, F. 2006. Landslide mobility and the role of water. *In*: EVANS, S. G., SCARASCIA-MUGNOZZA, G., STROM, A. & HERMANNS, R. L. (eds) *Landslides from Massive Rock Slope Failure*, NATO Science Series, Springer, Dordrecht, 233–242.

LEWKOWITZ, A. G. 1988. Slope processes, *In*: CLARK, M. J. (ed.) *Advances in Periglacial Geomorphology.* Wiley, New York, 325–368.

LI JIJUN, DERBYSHIRE, E. & SHUYING, X. 1984. Glacial and paraglacial sediments of the Hunza Valley, North-West Karakoram, Pakistan: a preliminary analysis. *In*: MILLER, K. J. (ed.) *The International Karakoram Project*, Volume 2. Cambridge University Press, Cambridge, 496–535.

MCSAVENEY, M. J. 2002. Recent rockfalls and rock avalanches in Mount Cook National park, New Zealand. *In*: EVANS, S. G. & DEGRAFF, J. V. (eds) *Catastrophic Landslides: Effects, Occurrence, and Mechanisms.* Geological Society of America, Reviews in Engineering Geology, **15**, 35–70.

MILLER, K. J. (ed.). 1984. *The International Karakoram Project*, Volume 2. Cambridge University Press, Cambridge.

NORIN, E. 1925. Preliminary notes on the late Quaternary glaciation of the North West Himalaya. *Geografisca Annaler*, **7**, 165–194.

OWEN, L. 1989. Neotectonics and glacial deformation in the Karakoram mountains and Nanga Parbat Himalaya. *Tectonophysics*, **163**, 227–265.

OWEN, L. 1991. Mass movement deposits in the Karakoram Mountains: their sedimentary characteristics, recognition and role in Karakoram landform evolution. *Zeitschtrift fur Geomorphologie*, N.F., **35**, 401–424.

OWEN, L. 2006. Quaternary glaciation. *In*: KREUTZMANN, H. (ed.) *Karakoram in Transition: Culture, Development and Ecology in the Hunza Valley.* Oxford University Press, Karachi, 12–23.

OWEN, L. & DERBYSHIRE, E. 1993. Quaternary and Holocene intermontane basin sedimentation in the Karakoram Mountains. *In*: SHRODER, J. F. (ed.) *Himalaya to the Sea: Geology, Geomorphology and the Quaternary.* Routledge, London, 108–131.

OWEN, L. & ENGLAND, J. 1998. Observations on rock glaciers in the Himalayas and Karakoram Mountains of Northern Pakistan and India, *Geomorphology*, **26**, 199–213.

PAFFEN, K. H., PILLEWIZER, W. & SCHNEIDER, H.-J. 1956. Forschungen im Hunza–Karakorum. *Erdkunde*, **10**, 1–33.

PETTERSON, M. G. 1984. *The Structure, Petrology and Geochemistry of the Kohistan Batholith, Giglit, Kashmir, N. Pakistan.* PhD dissertation, University of Leicester.

PETTERSON, M. G., WINDLEY, B. F. & SULLIVAN, M. 1990. A petrological, chronological, structural and geochemical review of Kohistan Batholith and its relationship to regional tectonics. *Physics and Chemistry of the Earth*, **17**, 47–70.

ROUSE, W. C. 1984. Flowslides. *In*: BRUNSDEN, D. & PRIOR, D. B. (eds) *Slope Instability.* Wiley, New York, 491–522.

RYDER, J. M. 1971. Some aspects of the morphometry of paraglacial alluvial fans in South-Central British Columbia. *Canadian Journal of Earth Sciences*, **8**, 1252–1264.

SCHEIDEGGER, E. 1998. Structural predesign of mass movements. *Geomorphology*, **6**, 82–93.

SEARLE, M. P. 1991. *Geology and Tectonics of the Karakoram Mountains.* Wiley, New York.

SELBY, M. J. 1993. *Hillslope Materials and Processes*, 2nd edn. Oxford University Press, Oxford.

SHRODER, J. F. (ed.). 1993. *Himalaya to the Sea: Geology, Geomorphology and the Quaternary.* Routledge, London.

SHRODER, J. F. 1998. Slope failure and denudation in the western Himalaya. *Geomorphology*, **26**, 81–105.

SHRODER, J. F., BISHOP, M. P. & SCHEPPY, R. 1998. Catastrophic flood flushing of sediment, western Himalaya, Pakistan. *In*: KALVODA, J. & ROSENFIELD, C. I. (eds) *Geomorphological Hazards in High Mountain Areas*. Kluwer, Dordrecht, 27–48.

SHRODER, J. F., KHAN, M. S., LAWRENCE, R. D., MADIN, I. & HIGGINS, S. M. 1989. Quaternary glacial chronology and neotectonics in the Himalaya of northern Pakistan. *In*: MALINCONICO, I. L. & LILLIE, R. J. *Tectonics and Geophysics of the Western Himalaya*. Geological Society of America, Special Paper, **232**, 275–293.

STRACHEY, H. 1853. Physical geography of Western Tibet. *Journal of the Royal Geographical Society*, **23**, 1–69.

STROM, A. L. 1996. Some morphological types of long-runout rockslides: effects of the relief on their mechanism and of rockslide deposits distribution. *In*: SENNESET, K. (ed.) *Landslides: Proceedings of the 7th International Symposium on Landslides*. Balkema, Rotterdam, 1977–1982.

VOIGHT, B. (ed.). 1978. *Rockslides and Avalanches, Volume 1. Natural Phenomena*. Elsevier, New York.

WASSMER, P., SCHIEDER, J. L., POLLET, N. & SCHMITTER-VOIRIN, C. 2004. Effects of the internal structure of a rock-avalanche dam on the drainage mechanism of its impoundment, Flims Sturzstrom and Ilanz paleo-lake, Swiss Alps. *Geomorphology*, **61**, 3–17.

WHALLEY, W. B. 1984. Rockfalls. *In*: BRUNSDEN, D. & PRIOR, D. B. (eds) *Slope Instability*. Wiley, New York, 217–256.

WHITEHOUSE, I. E. 1983. Distribution of large rock avalanche deposits in the central Southern Alps, New Zealand. *New Zealand Journal of Geology and Geophysics*, **26**, 271–179.

WIECZOREK, G. F. 2002. Catastrophic rockfalls and rockslides in the Sierra Nevada, USA. *In*: EVANS, S. G. & DEGRAFF, J. V. (eds) *Catastrophic Landslides: Effects, Occurrence, and Mechanisms*. Geological Society of America, Reviews in Engineering Geology, **15**, 165–190.

YARNOLD, J. C. & LOMBARD, J. P. 1989. A facies model for large rock avalanche deposits formed in dry climates. *In*: COLBORN, I. P., ABBOTT, P. L. & MINCH, J. (eds) *Conglomerates in Basin Analysis: A Symposium Dedicated to A. O. Woodford*. Society of Economic Paleontologists and Mineralogists, Pacific Section, **62**, 9–31.

ZEITLER, P. K. 1985. Cooling history of the NW Himalaya, Pakistan. *Tectonics*, **4**, 127–151.

Climate sensitivity: implications for the response of geomorphological systems to future climate change

STEPHAN HARRISON

School of Geography, Archaeology and Earth Resources, University of Exeter, Cornwall Campus, Penryn, Cornwall TR10 9EZ, UK
(e-mail: stephan.harrison@exeter.ac.uk)

Abstract: Climate sensitivity is defined as the equilibrium temperature response of the climate system to a doubling of atmospheric CO_2 levels from pre-industrial levels. Despite three decades of debate in the climatological literature on the estimation and significance of climate sensitivity, very little appears in the geomorphological literature on the implications of this for geomorphological systems. This paper reviews the concept of climate sensitivity and applies its findings to an assessment of future landscape change in cold regions. It is concluded that paraglacial processes will become the dominant mechanism of sediment transfer in currently glaciated catchments and that this period of sediment mobilization will be the last episode of major sediment movement for geological time periods.

There is a scientific consensus (Oreskes 2004) that the mean surface temperature of the Earth has increased in recent decades, and that the warming amounts to around 0.8 °C since the beginning of the twentieth century (IPCC 2007). From this, NASA Goddard Institute of Space Studies (GISS) estimates that 2005 was the warmest year to date since reliable instrumental measurements become available, although the World Meteorological Office (WMO) and Climatic Research Unit (CRU) place it just behind 1998. Detection and attribution studies show that there is high probability (at least 90%) that this warming is largely the result of anthropogenic emissions of greenhouse gases (GHG; mainly CO_2) in the troposphere, and that the amount and rate of warming lie outside of the range of natural variation and are unprecedented within the context of the Holocene (the last 11 000 years). Continued future warming is expected to have important consequences for a range of Earth systems (including the atmosphere, cryosphere, oceans, hydrological systems and the biosphere), and there are compelling reasons to expect increases in the magnitude and frequency of some natural hazards such as floods (Huntington 2006), droughts (Mason & Goddard 2001) and landslides (e.g. Fischer *et al.* 2006), and increases in the intensity of tropical cyclones (Emanuel 2005). There are also concerns about the stability of several of the large ice sheets on Earth (e.g. Overpeck *et al.* 2006) as break-up of these has the ability to impact upon global sea levels and regulate ocean currents.

Given that climate is a major driver of many Earth systems, the rate, nature and extent of future climate changes are important variables for understanding the future behaviour of geomorphological processes and the likely evolution of geomorphological systems. As a result, it is surprising that little appears in the geomorphological literature assessing climate sensitivity as a factor in landform and landscape evolution. It is, therefore, the purpose of this paper to synthesize current understanding of climate sensitivity (defined later), and discuss briefly the implications of this for assessing the response of mountain geomorphological systems to future temperature rises.

Metrics for climate change

Metrics are of interest to geomorphologists and other Earth scientists modelling the response of Earth systems to changes in climate as they provide a baseline measurement from which to calculate change. The most widely used metric is the global mean surface temperature, which is estimated by the HADCRU (Hadley/Climatic Research Unit) data set (which excludes the polar regions) or by the NASA GISS assessment (which extrapolates to the poles). The warming of mean surface temperature is related theoretically and empirically to the idea of radiative forcing at the top of the atmosphere. Radiative forcing is measured in $W\ m^{-2}$ and can be defined as 'a measure of the influence a factor has in altering the balance of incoming and outgoing energy in the Earth–atmosphere system, and is an index of the importance of the factor as a potential climate change mechanism' (IPCC 2001).

Despite this metric having been used by the IPCC Assessment Reports and adopted by

From: KNIGHT, J. & HARRISON, S. (eds) *Periglacial and Paraglacial Processes and Environments.*
The Geological Society, London, Special Publications, **320**, 257–265.
DOI: 10.1144/SP320.16 0305-8719/09/$15.00

policy-makers, there have been a number of challenges to its use. The most credible challenge comes from the suggestion that a global heat budget measures the warming of the Earth in a more complete way than the global mean surface temperature (e.g. Pielke 2003). In addition, there are important uncertainties and limitations associated with the concept (National Academies of Sciences 2005), especially as radiative forcing fails to account adequately with regional forcings, such as by aerosols and land-use changes, that may have significant climatic effects at the local scale.

Sensitivity

One of the most widely used metrics for climate change is climate sensitivity (S), which is defined as the equilibrium global mean temperature response to a doubling of atmospheric CO_2 levels over pre-industrial levels (Dessler & Parsons 2006); in other words, what is the temperature response associated with atmospheric CO_2 levels of 550 ppm?

When the climate is perturbed from its steady state the increased outgoing radiative flux is proportional to the globally averaged surface temperature change. Gregory *et al.* (2002) use global climate models (GCMs) to show that the increase in globally averaged outgoing radiative flux when the climate is perturbed from a steady state is proportional to the globally averaged surface temperature change, ΔT. With climate change, the imbalance between radiative forcing, Q, and the radiative response, $\lambda \Delta T$, is mainly absorbed by the heat capacity of the oceans (Levitus *et al.* 2001). With t as time and F as the heat flux into the ocean, this is shown as:

$$F(t) = Q(t) - \lambda \Delta T(t). \quad (1)$$

They show that with an unperturbed steady-state climate, $Q = F = 0$ and $\Delta T = 0$. With an increased radiative forcing (Q), heat flux into the oceans (F) rises and ΔT becomes positive. With a constant radiative forcing (Q) the climate reaches a new equilibrium and F eventually reaches zero as $\Delta T = Q/\lambda$. The equilibrium climate sensitivity is defined as $\Delta T_{2x} = Q_{2x/\lambda}$, where Q_{2x} is the forcing response to a doubled CO_2 concentration (Gregory *et al.* 2002).

It is not possible to assess climate sensitivity directly owing to the lag between increased CO_2 levels and global temperatures (Gregory *et al.* 2002; Annan & Hargreaves 2006). As a result, sensitivity forms one of the most important areas of uncertainty in predictions of future climate change. There have been a number of attempts to assess the extent to which global temperatures are sensitive to changes in CO_2. However, the most recent predictions of climate sensitivity are based on subjective estimations from 1979, which put climate sensitivity at between 1.5 and 4.5 °C with a mean estimate of 3 °C. This estimate has remained essentially unchanged since 1979 (Annan & Hargreaves 2006). The lower temperature limit of S is likely to be reasonably robust, but there have been a number of estimates of the higher limit of S that show temperature rises far above 4.5 °C, and such increases would have severe implications for natural and human systems.

The standard estimate of sensitivity of about 3 °C (Charney sensitivity; Charney 1979) makes the unrealistic assumption that over time the atmospheric composition, land surface, ice cover and other important variables remain constant. More recent estimates using a more realistic model (Hansen *et al.* 2009) allow the forcings and feedbacks to vary in response to changes in temperature, and this suggests that long-term climate sensitivity may be nearer 6 °C.

There are two distinct ways in which climate sensitivity has been estimated (Schneider von Deimling *et al.* 2006). The bottom-up approach uses empirical, quantitative understandings of the physical processes forcing climate, such as changes in radiation balance, feedbacks and lapse rates to drive climate models. Owing to the uncertainties in the strength of the feedbacks (especially those concerned with the behaviour of clouds) it has not been possible to reduce the uncertainty in climate sensitivity beyond that which was derived in the late 1970s (Charney 1979). The top-down approach employs understandings of how past temperatures have changed in response to changing CO_2 forcings. However, it is not always possible to isolate the temperature response to changes in CO_2 as variations in other forcings often occurred at the same time. In addition, during past glaciations the climate was colder than now, and the feedbacks partly responsible for this were also probably different (Schneider von Deimling *et al.* 2006).

Despite these problems, there are now attempts to employ observational data to provide a more objective quantification of S and these, used in association with Bayesian approaches, have considerably reduced the uncertainty in estimating S (e.g. Hasselmann 1998; Berliner *et al.* 2000; Annan & Hargreaves 2006). Such an approach allows the integration of climate information from a number of lines of evidence and the incorporation of independent prior information into the analysis (Barnett *et al.* 2005). The latter includes information on climate models, and their responses to forcing and uncertainty in estimates of external climate forcing. Although much of this prior information may be subjective, the information that is used for analysis is explicitly declared. In addition, as Bayesian inferences are probabilistic this means that they are able to be used for policy analyses that require estimates of risks and benefits (Barnett *et al.* 2005). The use of subjective prior assumptions has been criticized, however (Frame *et al.* 2005). They demonstrate that any probabilistic estimate of climate sensitivity, and therefore of the risk that GHG stabilization at any given level may result in a 'dangerous' equilibrium warming, is critically dependent on subjective prior assumptions of the investigators, not simply on constraints provided by actual climate observations.

However, despite this, Annan & Hargreaves (2006) used a Bayesian approach to estimate the temperature response to changing forcings over a number of timeslices, and to therefore assess climate sensitivity. These time periods include the global temperature change over the past century, the temperature response to major volcanic eruptions, the climate of the Last Glacial Maximum (LGM; around 21 000 years BP) and the reduction in temperature during the Maunder Minimum period (1645–1715 AD) of the Little Ice Age (sixteenth–nineteenth centuries).

Global temperature over the last century

A number of studies have shown that the recent net warming does not give us a clear view of climate sensitivity (Forest *et al.* 2002). This is because equilibrium sensitivity may operate on longer timescales than one century, and the forcings (especially from sulphate aerosols) are not well constrained (Annan & Hargreaves 2006). As a result of variations in solar output, volcanism and the increase in anthropogenic GHG accumulations in the atmosphere from the late eighteenth century, the recent global climate has not reached a steady state. Continuous fluctuations in radiative forcing also take place, and at timescales shorter than is required for the climate to reach equilibrium (Gregory *et al.* 2002).

Annan & Hargreaves (2006) also discussed the use of the Maunder Minimum as a timeframe when the climate forcing was too low to allow for sufficient time for the climate to reach equilibrium. In addition, the temperature response is well known enough to estimate climate sensitivity. Both Crowley (2000) and Rind *et al.* (2004) estimated climate cooling at this time period using model simulations, but it is suggested that the projected cooling for both of these is likely to be too low.

Volcanic eruptions

The short-term global-scale cooling associated with volcanic eruptions can be used to estimate climate sensitivity (Yokohata *et al.* 2005; Annan & Hargreaves 2006). Wigley *et al.* (2005) used an energy balance model (MAGICC) and observations on the reduction in global temperature to estimate the climate sensitivity associated with the volcanic eruptions of Agung, El Chichon and Pinatubo. Their assessment was that climate sensitivity is around 3 °C with a range of around 1.8–7.7 °C.

Last Glacial Maximum

The climate of the LGM has been used to assess climate sensitivity. Although it is known that global temperatures were significantly lower than today, accurate reconstruction is hampered by the imprecision associated with using proxy temperature data and geochronological resolution. The radiative forcing at this time was dominated in the high latitudes by changes in ice cover (affecting lapse rates and albedo) and low GHG levels, each of which reduced radiative forcing by around 3 W m^{-2}. With further reductions of 1 W m^{-2} associated with changes in dust fluxes and vegetation (Claquin *et al.* 2003; Crucifix & Hewitt 2005), the total reduction in radiative forcing estimated by Annan *et al.* (2005) was around 6–11 W m^{-2}. Despite this, climate sensitivity at this time is likely to have been around 2.7 °C. However, Annan *et al.* (2005) used an atmospheric global climate model (AGCM) coupled with a slab-ocean model to show that this result is strongly influenced by the topographic and climate parameterization built into the GCM; consequently, estimates of climate sensitivity derived from analysis of LGM climates are uncertain (see Otto-Bliesner *et al.* 2006). They suggest that this uncertainty may be in the order of 0.8 °C. Another recent estimate of LGM sensitivity comes from Schneider von Deimling *et al.* (2006). They argued that the magnitude of LGM forcings was about 8 ± 2 W m^{-2}. Allied to reduction of temperature of between 5 and 6 °C, this allows us to estimate a Charney sensitivity of around 3 °C, with uncertainty ranging from 1.5 to 6 °C.

The analysis of Annan & Hargreaves (2006) uses multiple and independent observational evidence allied to a Bayseian statistical approach to argue that climate sensitivity is very probably in the range of 1.7–4.9 °C (with 95% probability). This estimate is more tightly bound than previous estimations, and appears to rule out the very high climate sensitivities (up to 11 °C) suggested by Stainforth *et al.* (2005) and other workers.

Future understandings of climate sensitivity are hampered by a number of factors. First, it is likely that climate sensitivity is not a constant, but varies as the properties of the Earth system varies. For example, with the recession of ice masses, the albedo feedback becomes less effective and the climate sensitivity decreases accordingly. Hence, assessing climate sensitivity using the LGM as a model may overestimate the value, as the continental-scale ice sheets that existed at the time will have heightened the importance of high-latitude albedo as a feedback to global temperatures. From this, it is clear that climate sensitivity should be defined with respect to the time-period over which it is measured.

Second, defining climate sensitivity partly depends on the scientists' decision to allocate the model processes as forcings or feedbacks. For instance, there is uncertainty as to whether indirect aerosol effects should be defined as a feedback to changes in the climate system or as a forcing. This partly reflects the problems inherent in parameterization of complex phenomena and the scale at which such phenomena are treated in GCMs. As a consequence of the differing ways in which climate models treat such parameterizations, there may be uncertainty in the climate sensitivity that results.

In essence, it may be ultimately impossible to assess climate sensitivity accurately. The uncertainties associated with reconstructing past climate change, allied to our incomplete knowledge of the changing forcings over time and the problems of random noise in the climate system, all combine to make estimates of climate sensitivity inherently uncertain.

What is more easily assessed is the likely period when atmospheric concentrations of CO_2 reach double pre-industrial levels (around 550 ppm). The exact timing is highly dependent on the future evolution of global emissions and the efficiency of carbon sinks, but IPCC (2007) estimated that this figure will be reached by between 2050 and 2070. If atmospheric concentrations of CO_2 do not stabilize by this time, then the global temperature will not reach equilibrium.

Implications for geomorphology

Despite the vigorous and long-standing debate in the climate science literature on the nature and magnitude of climate sensitivity, this issue has been almost ignored by geomorphologists and others interested in assessing the response of landscapes to future climate change. This is surprising, not least because it is clear that significant landscape change may be driven by changes in temperature and that the future rise in temperature experienced by glaciated catchments is intimately related to climate sensitivity. What we can call *geomorphological sensitivity* to temperature change is likely to be most evident in glacial, paraglacial and periglacial systems in mountainous regions.

One characteristic of climate sensitivity that is not always recognized is that the estimated climate sensitivity of around 3 °C is not evenly distributed over the Earth's surface, and that the temperature changes experienced by glacial and periglacial regions is likely to be much higher. This is a consequence of two important factors. First, the outputs of GCMs used in the Coupled Model Intercomparison Project to predict future climate change show Arctic amplification of between 1.5 and 4.5 times the global mean warming, suggesting a doubled climate sensitivity (e.g. Holland & Bitz 2003; Masson-Delmotte *et al.* 2006). Polar amplification is largely driven by positive feedbacks from reduction in albedo, increased atmospheric water vapour and weakening of the near-surface temperature inversion, as melting of ice masses progresses.

The geomorphological systems in such areas are dominated by glacial and periglacial processes, and this suggests that these are likely to be preferentially affected by such temperature increases. Second, all future climate projections show elevated temperature increases over land relative to the oceans (a consequence of the larger thermal capacity of the seas).

What is increasingly clear is that high-latitude glaciated regions will experience larger temperature increases than the projected global mean surface temperature rise forecast for the end of this century. The current response of glacial systems to the relatively modest rise in global mean surface temperatures (about 0.7 °C since 1850) is a clear reduction in glacier mass balance in most glaciated mountain regions (Meier *et al.* 2003), and this trend of glacier recession is likely to accelerate strongly. However, the temperature rise metric masks a strong latitudinal and hemispheric variability in the temperature increase. With the oceans having a large thermal inertia, recent climate warming has been largely concentrated in the northern hemisphere, and this pattern may have significant implications for the future evolution of glacial systems.

Data from GISS show that the hottest years in the instrumental record from the northern hemisphere are 2007, 2005 and 2006 (G. Foster pers. comm. 2008). Detection and attribution analysis of climate change since the nineteenth century suggests that the global warming signal can only be isolated from background variability from about 1975, and the time period since this has been termed by several commentators the 'modern global warming era'. Whilst the rise in global mean surface temperature since then has averaged at 0.018 °C $year^{-1}$, land in the northern hemisphere has warmed around 70% faster, at 0.031 °C $year^{-1}$, including over 1 °C since 1975. Within the northern hemisphere also, there has been uneven recent warming, with the lowest amount of warming occurring in the lowest latitudes (equator–24°N) at 0.7 °C since 1975. The latitude band from 24–44°N has warmed nearly 1 °C in this time. The latitude bands from 44–64°N and from 64°N to the Pole have warmed at over 1 °C and nearly 2 °C, respectively, since 1975. We can see this pattern in Figure 1 and Table 1 (G. Foster pers. comm. 2008). This trend towards increasing temperature rise with latitude in the northern hemisphere reflects, in part, the operation of polar amplification on the data set and may have profound implications for the behaviour of glacial systems in these regions.

Implications of warming trends for glacial, paraglacial and periglacial systems

The majority of mountain glaciers in the world are currently experiencing strongly negative mass balances and this is regarded as a response to the widespread warming that has occurred since the 1970s (Meier *et al.* 2008). Work in contemporary glacial and Pleistocene settings has shown that periods of glacier recession are associated with profound changes in related geomorphological systems as landscapes undergo rapid modification as they are exposed to a range of active geomorphological processes (Ballantyne 2002). Such elevated geomorphological activity is termed 'paraglacial' (Church & Ryder 1972) and forms a complex response to deglaciation; valley side slopes become debuttressed creating rockfalls (e.g. Hewitt 1988; Cossart *et al.* 2008), debris-flow activity is increased (e.g. Blair 1994; Harrison & Winchester 1997) and glacial lake outburst floods develop (e.g. Clague *et al.* 1985; Hubbard *et al.* 2005). In addition, new sets of landforms are produced, including certain types of rock glaciers (Johnson 1984) and debris cones (Harrison *et al.* 2005).

Until the past two decades or so, much work on paraglacial geomorphology concentrated on landforms and sediments developed during past periods of glacier recession, especially those following Pleistocene deglaciation (Ryder 1971; Eyles & Kocsis 1988; Owen 1991; Augustinus 1995). Other work assessed the paraglacial response to post-Little Ice Age glacier recession (e.g. Ballantyne & Benn 1994). However, the recent glacier recession driven by global warming has created a renewed phase of enhanced sediment delivery to valley bottoms, and this late-twentieth-century paraglacial phase has attracted the attention of researchers investigating the nature and rates of landscape modification (e.g. Beylich *et al.* 2006; Warburton 2007).

The implications for geomorphological processes in glaciated catchments of the likely pattern and nature of climate change over the next few decades are considerable. If the pattern of warming coincides with that projected by the majority of GCM runs, then polar amplification will force high-latitude glaciers and ice masses in the northern hemisphere to warm up more than those in other regions. The shift in mass balance that this produces will be partially offset by increased precipitation associated with increased evaporation from recently ice-free seas and latent heat release from melting ice. In the mid-latitudes of California, a marked increase in precipitation has been used to explain the positive mass balances of the glaciers of Mount Shasta, despite regional warming over the past 50 years (Howat *et al.* 2007) and this may have been driven partly by the Pacific Decadal Oscillation. Despite this, however, eventually increased temperatures will overwhelm the predicted increased precipitation forcing negative mass balances, and this has been used by Hansen *et al.* (2009) to forecast the likely evolution of the Greenland ice sheet under continued global warming exacerbated by polar amplification.

Discussion

Consideration of the nature of future climate change and its impact on glaciated catchments results in several tentative inferences. First, at the global scale, if the pattern of warming corresponds to that predicted by GCMs then southern hemisphere glaciers may be relatively more resilient than northern hemisphere glaciers to warming. Second, high-latitude glaciers are likely to undergo most reduction in their mass, especially in the northern hemisphere where polar amplification is greatest. Third, paraglacial processes will become the dominant mechanism of sediment transfer in glaciated catchments. Fourth, with future high-temperatures predicted by all GCMs running on Special Report on Emissions Scenarios (SRES), GHG forcing is likely to continue to overwhelm natural forcing mechanisms on the global climate. Therefore, it is very unlikely that catchments that undergo deglaciation will experience renewed glaciation at timescales less than 10^4–10^5 years (e.g. see Berger & Loutre 2002; Crucifix & Berger 2006 for a discussion of the likely length of the present interglacial), especially as deglaciation will initiate water vapour and albedo feedbacks that contribute to regional warming. Consequently, the contemporary phase of enhanced sediment fluxes being experienced in such regions will be the last episode of major sediment movement in such catchments over geological timescales.

However, there are some important caveats to the above inferences. The patterns of future climate change projected by GCMs are likely to mask large regional shifts in climate, and it is known that any projections are highly uncertain at

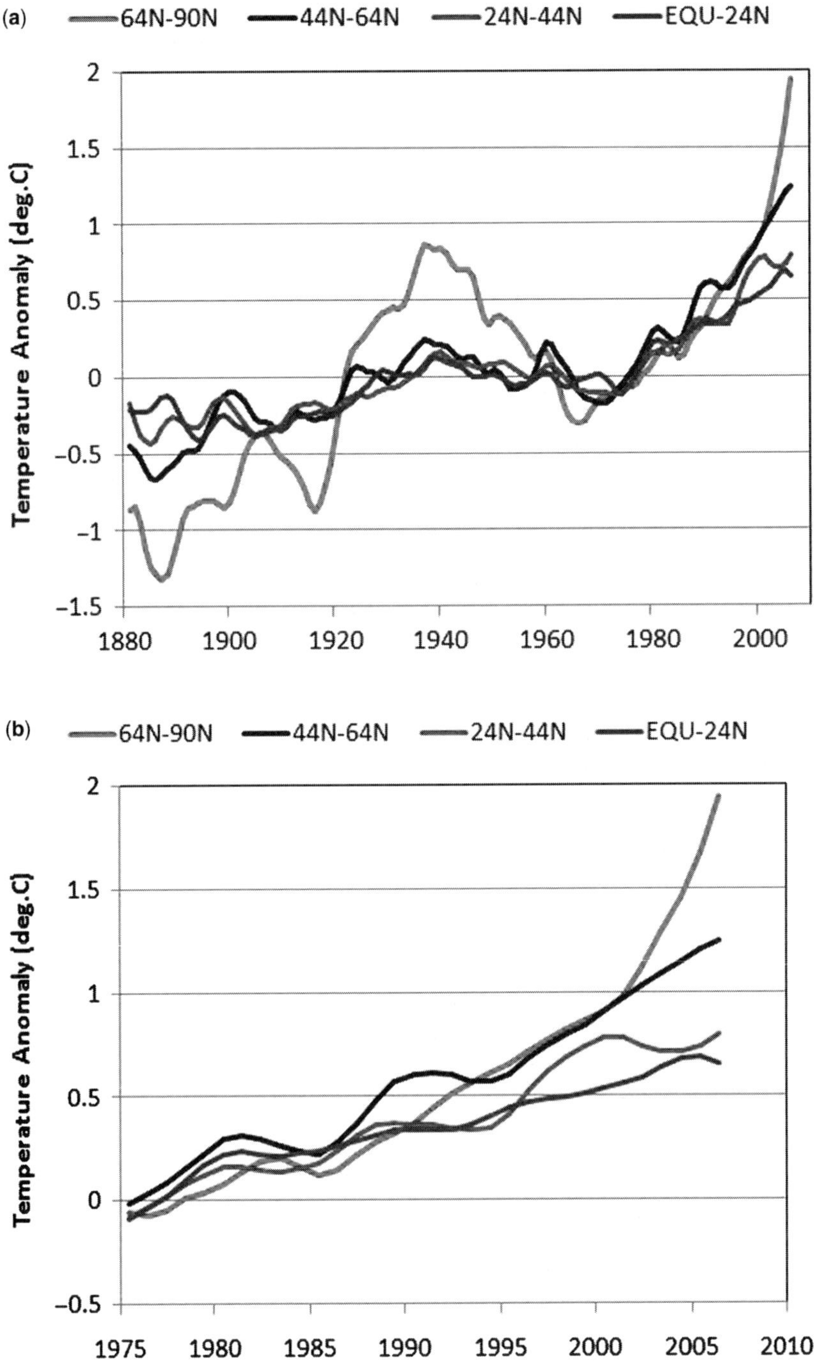

Fig. 1. (**a**, left) Modelled northern hemisphere temperature anomaly in different latitudinal zones, net rise (linear) in °C per year (G. Foster pers. comm. 2008). (**b**, right) Modelled northern hemisphere temperature anomaly in different latitudinal zones, net rise (smoothed) in °C per year (G. Foster pers. comm. 2008).

Table 1. *Rate of net warming in northern hemisphere (NH) land masses from 1975 to 2008 (°C); net warming according to the linear warming rate, and the net warming as estimated from the smoothed data (G. Foster pers. comm. 2008)*

Region	Rate 1975–2008 (°C $year^{-1}$)	Net rise (°C) (linear)	Net rise (°C) (smoothed)
NH	0.031	1.01	1.11
0–24°N	0.021	0.70	0.77
24–44°N	0.029	0.95	0.91
44–64°N	0.037	1.24	1.30
64–90°N	0.057	1.87	2.00

the regional scale (e.g. Lau *et al.* 2006; Allan & Soden 2007). Consequently, the behaviour of glaciers at such scales is equally uncertain. It has also not been possible to produce projections for precipitation amounts or timing that are as robust for those predicting temperature change.

Future trends in precipitation are more uncertain than those for temperature since precipitation is affected by large-scale elements of the climate system, the future behaviour of which is not easily modelled, and projections from regional modeling show a wide variance. For instance, projections for central and tropical South America to 2020 range from increases in precipitation of 5%, to decreases of 5%. By 2050 projections are still more uncertain, ranging from around 10% increases and decreases. Such uncertainties are a feature of projections from GCMs, and Magrin *et al.* (2007, 594) caution that 'the current GCMs do not produce projections of changes in the hydrological cycle at regional scales with confidence. In particular the uncertainty of projections of precipitation remain high'. In addition, in such dry regions as the central Andes sublimation is an important component of ablation, and this is correlated to air humidity and atmospheric moisture content. As a result, a significant component of the mass changes of such glaciers may not be related to shifts in atmospheric temperature; hence, the projection of future glacier changes under conditions of future warming is likely to be uncertain (e.g. Juen *et al.* 2007).

Similar uncertainties exist in the Arctic. As has been discussed earlier, projections of future climate change are almost wholly in agreement that warming will be greatest in the high latitudes and this polar amplification is a characteristic of most GCM runs. However, recent research using an ensemble of GCMs to predict future climate change in the Arctic has revealed wide variations in model outputs, associated with uncertainties concerning simulated cloudiness and terrestrial snow cover; both of which play a crucial role in driving glacier and permafrost behaviour. Using five GCMs and the B2 emissions scenario, Arctic temperatures are predicted to rise by around twice the global average by the late twenty-first century (from 2.8 to 4.6 °C), and this pattern is a robust result for all modeling studies (ACIA 2005). This research also suggests that mean annual temperatures will rise by around 5 °C in the Canadian Archipelago. The simulations from five models all suggest an increase in precipitation by 2071–2090 ranging from 7.5 to 18.1%. Local increases in precipitation may be as high as 35%, although there is considerable uncertainty in such projections at the local scale. As a result, for those glaciers whose mass balance has a strong precipitation component, predicting their future behaviour is more difficult (Braithwaite *et al.* 2003; Braithwaite & Raper 2007). Other uncertainties arise because the development of feedbacks, such as those where climate sensitivity affects the size of ice bodies, also operates so that the size of ice bodies affects climate sensitivity, at least at the local or regional scale.

A final, and important, point is that we must assess carefully the timescales linking climate change and different elements and regions of landscape stability. Certain elements in mountain environments may respond rapidly to shifts in climate (debris flows from exposed flanks of lateral moraines, for instance), while others respond after considerable delay (such as the failure of rock slopes to deglaciation, which may occur decades to millennia after glacier recession).

Implications

Climate sensitivity of around 3 °C will mask considerable regional variations that will have important geomorphological consequences. In high-latitude areas of the northern hemisphere, polar amplification of warming is expected to accelerate the current trend of glaciers towards negative mass balances. As a result, paraglacial processes will dominate sediment movement in such regions. We may also be able to conclude that for currently glaciated catchments, future sediment transfer rates will approach the 'geological norm' for the first time at least in the late Quaternary. With enhanced radiative forcing being driven by high concentrations of long-lived atmospheric GHGs and combined with the Earth's future orbital relationship with the Sun over the next tens of thousands of years, renewed glaciation in many climatically sensitive mountain regions will probably not occur for at least this time. The current phase of paraglaciation will therefore be the last over these timescales.

I would like to thank Dr G. Foster for the climate analyses presented in Table 1 and Figure 1. I also thank the referees, including Drs P. Hughes, N. Betts and J. Warburton, whose useful suggestions improved the manuscript considerably.

References

ACIA. 2005. *Arctic Climate Impacts Assessment*, Cambridge University Press, New York.

ALLAN, R. P. & SODEN, B. J. 2007. Large discrepancy between observed and simulated precipitation trends in the ascending and descending branches of the tropical circulation. *Geophysical Research Letters*, **34**, L18705, doi:10.1029/2007GL031460.

ANNAN, J. D. & HARGREAVES, J. C. 2006. Using multiple observationally-based constraints to estimate climate sensitivity, *Geophysical Research Letters*, **33**, L06704, doi:10.1029/2005GL025259.

ANNAN, J. D., HARGREAVES, J. C., OHGAITO, R., ABE-OUCHI, A. & EMORI, S. 2005. Efficiently constraining climate sensitivity with ensembles of paleoclimate simulations. *SOLA*, **1**, 181–184.

AUGUSTINUS, P. C. 1995. Rock mass strength and the stability of some glacial valley slopes. *Zeitschrift für Geomorphologie*, **39**, 55–68.

BALLANTYNE, C. K. 2002. Paraglacial geomorphology. *Quaternary Science Reviews*, **21**, 1935–2017.

BALLANTYNE, C. K. & BENN, D. I. 1994. Paraglacial slope adjustment and resedimentation following recent glacier retreat, Fåbergstolsdalen, Norway. *Arctic and Alpine Research*, **26**, 255–269.

BARNETT, T., ZWIERS, F. *ET AL.* 2005. Detecting and attributing external influences on the climate system: A review of recent advances. *Journal of Climate*, **18**, 1291–1314.

BERGER, A. & LOUTRE, T. 2002. An exceptionally long interglacial ahead? *Science*, **297**, 1287–1288.

BERLINER, L. M., LEVINE, R. A. & SHEA, D. J. 2000. Bayesian climate change assessment. *Journal of Climate*, **13**, 3805–3820.

BEYLICH, A. A., ETIENNE, S. *ET AL.* 2006. The European Science Foundation (ESF) Network SEDIFLUX – An introduction and overview, *Geomorphology*, **80**, 3–7.

BLAIR, R. W. 1994. Moraine and valley wall collapse due to rapid deglaciation in Mount Cook National Park, New Zealand. *Mountain Research and Development*, **14**, 347–358.

BRAITHWAITE, R. J. & RAPER, S. C. B. 2007. Glaciological conditions in seven contrasting regions estimated with the degree-day model. *Annals of Glaciology*, **46**, 297–302.

BRAITHWAITE, R. J., ZHANG, Y. & RAPER, S. C. B. 2003. Temperature sensitivity of the mass balance of mountain glaciers and ice caps as a climatological characteristic. *Zeitschrift für Gletscherkunde und Glazialgeologie*, **38**, 35–61.

CHARNEY, J. G. 1979. *Carbon Dioxide and Climate: A Scientific Assessment*. Climate Research Board, National Academy, NAS Press, Washington, DC.

CHURCH, M. & RYDER, J. M. 1972. Paraglacial sedimentation: a consideration of fluvial processes conditioned by glaciation. *Geological Society of America Bulletin*, **83**, 3072–3095.

CLAGUE, J. J., EVANS, S. G. & BLOWN, I. G. 1985. A debris flow triggered by the breaching of a moraine-dammed lake, Klattasine Creek, British Columbia. *Canadian Journal of Earth Sciences*, **22**, 1492–1502.

CLAQUIN, T., ROELANDT, C. *ET AL.* 2003. Radiative forcing of climate by ice-age atmospheric dust. *Climate Dynamics*, **20**, 193–202.

COSSART, E., BRAUCHER, R., FORT, M., BOURLÈS, D. L. & CARCAILLET, J. 2008. Slope instability in relation to glacial debuttressing in alpine areas (Upper Durance catchment, southeastern France): Evidence from field data and ^{10}Be cosmic ray exposure ages. *Geomorphology*, **95**, 3–26.

CROWLEY, T. J. 2000. Causes of climate change over the past 1000 years. *Science*, **289**, 270–277.

CRUCIFIX, M. & BERGER, A. 2006. How long will our interglacial be? *Eos*, **87**, 352–353.

CRUCIFIX, M. & HEWITT, C. D. 2005. Impact of vegetation changes on the dynamics of the atmosphere at the Last Glacial maximum. *Climate Dynamics*, **25**, 447–459.

DESSLER, A. E. & PARSONS, E. A. 2006. *The Science and Politics of Global Climate Change: A Guide to the Debate*. Cambridge University Press, Cambridge.

EMANUEL, K. A. 2005. Increasing destructiveness of tropical cyclones over the past 30 years. *Nature*, **436**, 686–688.

EYLES, N. & KOCSIS, S. 1988. Sedimentology and clast fabric of subaerial debris flow facies in a glacially-influenced alluvial fan. *Sedimentary Geology*, **59**, 15–28.

FISCHER, L., KAAB, A., HUGGEL, C. & NOETZLI, J. 2006. Geology, glacier retreat and permafrost degradation as controlling factors of slope instabilities in a high-mountain rock wall: the Monte Rosa east face. *Natural Hazards and Earth System Sciences*, **6**, 761–772.

FOREST, C. E., STONE, P. H., SOKOLOV, A. P., ALLEN, M. R. & WEBSTER, M. D. 2002. Quantifying uncertainties in climate system properties with the use of recent climate observations. *Science*, **295**, 113–117.

FRAME, D. J., BOOTH, B. B., KETTLEBOROUGH, J. A., STAINFORTH, D. A., GREGORY, J. M., COLLINS, M. & ALLEN, M. R. 2005. Constraining climate forecasts: The role of prior assumptions. *Geophysical Research Letters*, **32**, L09702, doi:10.1029/2004GL022241.

GREGORY, J. M., STOUFFER, R. J., RAPER, S. C. B., STOTT, P. A. & RAYNER, N. A. 2002. An observationally based estimate of the climate sensitivity. *Journal of Climate*, **15**, 3117–3121.

HANSEN, J., SATO, M. *ET AL.* 2009. Target atmospheric CO_2: Where should humanity aim? http://arxiv.org/abs/0804.1126.

HARRISON, S. & WINCHESTER, V. 1997. Age and nature of paraglacial debris cones along the margins of the San Rafael Glacier, Chilean Patagonia. *The Holocene*, **7**, 481–487.

HARRISON, S., WINCHESTER, V., WARREN, C. R. & PASSMORE, D. G. 2005. Quantifying rates of paraglacial sedimentation: an example from Chilean Patagonia. *Zeitschrift fur Geomorphologie*, **49**, 321–334.

HASSELMANN, K. 1998. Conventional and Bayesian approach to climate-change detection and attribution. *Quarterly Journal of the Royal Meteorological Society*, **124**, 2541–2565.

HEWITT, K. 1988. Catastrophic landslide deposits in the Karakorum Himalaya. *Science*, **242**, 64–67.

HOLLAND, M. M. & BITZ, C. M. 2003. Polar amplification of climate change in the Coupled Model Intercomparison Project. *Climate Dynamics*, **21**, 221–232.

HOWAT, I. M., TULACZYK, S., RHODES, P., ISRAEL, K. & SNYDER, M. 2007. A precipitation-dominated, mid-latitude glacier system: Mount Shasta, California. *Climate Dynamics*, **28**, 85–98.

HUBBARD, B., HEALD, A. *ET AL.* 2005. Impact of a rock avalanche on a moraine-dammed proglacial lake: Laguna Safuna Alta, Cordillera Blanca, Peru. *Earth Surface Processes and Landforms*, **30**, 1251–1264.

HUNTINGTON, T. G. 2006. Evidence for intensification of the global water cycle: Review and synthesis. *Journal of Hydrology*, **319**, 83–95.

IPCC. 2001. *Climate Change 2001: The Scientific Basis, Contribution of Working Group I to the Third Assessment Report of the Intergovernmental Panel on Climate Change.* Cambridge University Press, Cambridge.

IPPC. 2007. *Climate Change 2007: The Physical Science Basis.* Summary for Policymakers. Intergovernmental Panel on Climate Change, Geneva.

JOHNSON, P. G. 1984. Rock glacier formation by high-magnitude low frequency slope processes in the south-west Yukon. *Annals of the Association of American Geographers*, **74**, 408–419.

JUEN, I., KASER, G. & GEORGES, C. 2007. Modelling observed and future runoff from a glacierized tropical catchment (Cordillera Blanca, Perú). *Global and Planetary Change*, **59**, 37–48.

LAU, K. M., SHEN, S. S. P., KIM, K. M. & WANG, H. A. 2006. Multimodel study of the twentieth-century simulations of Sahel drought from the 1970s to 1990s. *Journal of Geophysical Research*, **111**, D07111, doi:10.1029/2005JD006281.

LEVITUS, S., WANG, J., DELWORTH, T. L., DIXON, K. W. & BROCCOLI, A. J. 2001. Anthropogenic warming of the Earth's climate system. *Science*, **292**, 267–270.

MAGRIN, G. C., GARCÍA, G. *ET AL.* 2007. Latin America. *In*: PARRY, M. L., CANZIANI, O. F., PALUTIKOF, J. P., VAN DER LINDEN, P. J. & HANSON, C. E. (eds) *Climate Change 2007: Impacts, Adaptation and Vulnerability. Contribution of Working Group II to the Fourth Assessment Report of the Intergovernmental Panel on Climate Change.* Cambridge University Press, Cambridge, 581–615.

MASON, S. J. & GODDARD, L. 2001. Probabilistic precipitation anomalies associated with ENSO. *Bulletin of the American Meteorological Society*, **82**, 619–638.

MASSON-DELMOTTE, V., KAGEYAMA, V. M. *ET AL.* 2006. Past and future polar amplification of climate change: climate model intercomparisons and ice-core constraints. *Climate Dynamics*, **26**, 513–529.

MEIER, M. F., DYURGEROV, M. B. & MCCABE, G. J. 2003. The health of glaciers – Recent changes in glacier regime. *Climate Dynamics*, **59**, 123–135.

MEIER, M. F., DYURGEROV, M. B. *ET AL.* 2008. Glaciers dominate eustatic sea-level rise in the 21st century. *Science*, **317**, 1064–1067.

NATIONAL ACADEMIES OF SCIENCES. 2005. *Radiative Forcing: Expanding the Concept and Addressing Uncertainties.* National Academies Press, Washington, DC.

ORESKES, N. 2004. Beyond the ivory tower. The scientific consensus on climate change. *Science*, **306**, 1686.

OTTO-BLIESNER, B. L., BRADY, E. C., CLAUZET, G., TOMAS, R., LEVIS, S. & KOTHAVALA, Z. 2006. Last Glacial Maximum and Holocene climate in CCSM3. *Journal of Climate*, **19**, 2526–2544.

OVERPECK, J. T., OTTO-BLIESNER, B. L., MILLER, G. H., MUHS, D. R., ALLEY, R. B. & KIEHL, J. T. 2006. Paleoclimatic evidence for future ice-sheet instability and rapid sea-level rise. *Science*, **311**, 1747–1750.

OWEN, L. A. 1991. Mass movement deposits in the Karakoram Mountains: Their sedimentary characteristics, recognition and role in Karakoram landform evolution. *Zeitschrift für Geomorphologie*, **35**, 401–424.

PIELKE, R. A. SR. 2003. Heat storage within the earth system. *Bulletin of the American Meteorological Society*, **84**, 331–335.

RIND, D., SHINDELL, D., PERLWITZ, J., LERNER, J., LONERGAN, P., LEAN, J. & MCLINDEN, C. 2004. The relative importance of solar and anthropogenic forcing of climate change between the Maunder Minimum and the present. *Journal of Climate*, **17**, 906–929.

RYDER, J. M. 1971. The stratigraphy and morphology of paraglacial alluvial fans in south-central British Columbia. *Canadian Journal of Earth Sciences*, **8**, 279–298.

SCHNEIDER VON DEIMLING, T., HELD, H., GANOPOLSKI, A. & RAHMSTORF, S. 2006. Climate sensitivity estimated from ensemble simulations of glacial climate. *Climate Dynamics*, **27**, 149–163.

STAINFORTH, D. A., AINA, T. *ET AL.* 2005. Uncertainty in predictions of the climate response to rising levels of greenhouse gases. *Nature*, **433**, 403–406.

WARBURTON, J. 2007. Sediment budgets and rates of sediment transfer across cold environments in Europe. *Geografiska Annaler*, **89A**, 95–100.

WIGLEY, T. M., AMMAN, C. M., SANTER, B. D. & RAPER, S. B. 2005. Effect of climate sensitivity on the response to volcanic forcing. *Journal of Geophysical Research*, **110**, D09107, doi:10.1029/2004JD005557.

YOKOHATA, T., EMORI, S., NOZAWA, T., TSUSHIMA, Y., OGURA, T. & KIMOTO, M. 2005. Climate response to volcanic forcing: Validation of climate sensitivity of a coupled atmosphere–ocean general circulation model. *Geophysical Research Letters*, **32**, L21710, doi:10.1029/2005GL023542.

Index

Note: Page numbers in *italics* denote figures, while those in **bold** denote tables.